RELATIVITY, ASTROPHYSICS AND COSMOLOGY

ASTROPHYSICS AND SPACE SCIENCE LIBRARY

A SERIES OF BOOKS ON THE RECENT DEVELOPMENTS

OF SPACE SCIENCE AND OF GENERAL GEOPHYSICS AND ASTROPHYSICS

PUBLISHED IN CONNECTION WITH THE JOURNAL

SPACE SCIENCE REVIEWS

VOLUME 38

RELATIVITY, ASTROPHYSICS AND COSMOLOGY

PROCEEDINGS OF THE SUMMER SCHOOL HELD, 14–26 AUGUST, 1972
AT THE BANFF CENTRE, BANFF, ALBERTA

Edited by

WERNER ISRAEL

Theoretical Physics Institute, Physics Department,
University of Alberta, Edmonton, Alberta

D. REIDEL PUBLISHING COMPANY

DORDRECHT-HOLLAND / BOSTON-U.S.A.

Library of Congress Catalog Card Number 72–97957

ISBN-13:978-94-010-2641-3 e-ISBN-13:978-94-010-2639-0
DOI: 10.1007/978-94-010-2639-0

Published by D. Reidel Publishing Company
P.O. Box 17, Dordrecht, Holland

Sold and distributed in the U.S.A., Canada, and Mexico
by D. Reidel Publishing Company, Inc.
306 Dartmouth Street, Boston,
Mass. 02116, U.S.A.

Softcover reprint of the hardcover 1st edition 1973

TABLE OF CONTENTS

PREFACE

The 1972 Banff lectures attempted a systematic exposition of the ideas underlying recent developments in general relativity and its astronomical applications at a level accessible and useful to graduate students having some previous acquaintance with the subject. To our regret, it was not possible to include any printed record of Peebles' beautiful lectures on observational cosmology or of the many stimulating seminars on special topics contributed by the participants. What remains is nevertheless a reasonably self-contained and compact introduction to Einstein's theory in its modern incarnation, and we hope it will be found useful by the many physicists, astronomers, and mathematicians who wish to update and deepen their understanding of the theory.

On behalf of the organizing committee, I should like to express appreciation to a number of people whose help was crucial to the success of the enterprise: to Jan van Kranendonk, who initiated the idea of a Banff summer school on general relativity; to him and to David Rowe and Don Betts for inspiration and moral support; to our indefatigable secretaries Olwyn Buckland and Leslie Hughes; and to Garry Nash, Richard Sigal, Tim Spanos, and Gordon Wilson who helped in a variety of ways to keep the wheels running. How much we owe to the splendid cooperative effort of the lecturers will be clear to any reader of the following pages.

WERNER ISRAEL

Left to right: Karel Kuchař, Jim and Judy Hartle, Ray Sachs, Alison and Jim Peebles, Anita and Jürgen Ehlers, Uta and Bernd Schmidt, Birgitte and Dieter Brill.

ACKNOWLEDGEMENTS

These Proceedings were held under the auspices of the Theoretical Physics Division, Canadian Association of Physicists, with financial support from NATO, the National Research Council of Canada and the University of Alberta.

ORGANIZING COMMITTEE

W. Israel, H. P. Künzle, G. Ludwig, P. Rastall, H. Schiff, R. J. Tait, and J. R. Trollope.

LIST OF PARTICIPANTS

Acquista, Charles, Stevens Institute of Technology, Hoboken, New Jersey.
Anderson, Terry L., University of Nebraska, Lincoln, Nebraska.
Aurilla, Antonio, University of Alberta, Edmonton, Alberta.
Becker, Louis, Northwestern Illinois University, Chicago, Illinois.
Brill, Dieter R., University of Maryland, College Park, Maryland.
Brouwer, Wytze, University of Alberta, Edmonton, Alberta.
Camenzind, M., University of Berne, Switzerland,
Campbell, Gerald A., University of Texas, Austin, Texas.
Capri, A. Z., University of Alberta, Edmonton, Alberta.
Carlberg, Raymond, University of Saskatchewan, Saskatoon, Saskatchewan.
Chapman, Timothy C., University of Windsor, Windsor, Ontario.
Chase, James, Nipissing College, North Bay, Ontario.
Chia, Tzu Tit, Queen's University, Kingston, Ontario.
Chow, Tai Low, Stanislaus State College, Turlock, California.
Cooke, James H., University of Texas, Arlington, Texas.
Cooperstock, F., University of Victoria, Victoria, British Columbia.
Crampin, Michael, King's College, Strand, London, England.
Das, A., Simon Fraser University, Burnaby, British Columbia.
Davies, P. W. C., University of Cambridge, Cambridge, England.
Demmie, Paul, University of Pittsburgh, Johnstown, Pennsylvania.
Duncan, Donald P., University of Texas, Austin, Texas.
Easwaran, Mary, University of Alberta, Edmonton, Alberta.
Eardley, Douglas, University of California, Berkeley, California.

Eby, Peter B., Marshall Space Flight Center, Huntsville, Alabama.
Ehlers, J., Max Planck Institut für Astrophysik, München, West Germany.
Emberson, Richard, Amherst College, Amherst, Massachusetts.
Esposito, F. Paul, University of Cincinnati, Cincinnati, Ohio.
Estabrook, Frank, California Institute of Technology, Pasadena, California.
Evans, Geoffrey, University of Toronto, Toronto, Ontario.
Evans, Ronald K., University of Victoria, Victoria, British Columbia.
Fischer, Arthur, University of California, Berkeley, California.
Fishbone, Leslie, University of Maryland, College Park, Maryland.
Garwin, Charles, University of California, Berkeley, California.
Gautreau, Ronald, Newark College of Engineering, Newark, New Jersey.
Gelman, Yoram, State College, Framingham, Massachusetts.
Goethals, Michel, Institut de Physique Theorique, Heverlee, Belgique.
Goldreich, Peter, California Institute of Technology, Pasadena, California.
Greenberg, Philip, Massachusetts Institute of Technology, Cambridge, Massachusetts.
Haggerty, Michael, University of Texas, Austin, Texas.
Hale, Daniel P., Marshall Space Flight Center, Huntsville, Alabama.
Hartle, James B., University of California, Santa Barbara, California.
Hartman, Richard, University of California, Santa Barbara, California.
Hasan, Mazhar, Northern Illinois University, Dekalb, Illinois.
Horndeski, Gregory, University of Waterloo, Waterloo, Ontario.
Hurwitz, Harlan, Columbia University, New York, New York.
Huschilt, James, University of Windsor, Windsor, Ontario.
Hussein, M. S., Universitade de Sao Paulo, Sao Paulo, Brazil.
Ihrig, Edwin, University of Toronto, Toronto, Ontario.
Inomata, Akira, State University of New York, Albany, New York.
Israel, Werner, University of Alberta, Edmonton, Alberta.
Jones, Bernard, Princeton University, Princeton, New Jersey.
Jones, Janet E., Princeton University, Princeton, New Jersey.
Junevicus, Gerald, University of Victoria, Victoria, British Columbia.
Kantowski, Ronald, University of Oklahoma, Norman, Oklahoma.
Kaufman, Sheldon, Queensborough Community College, Bayside, New York.
Keeping, E. S., University of Alberta, Edmonton, Alberta.
Kegeles, Lawrence, University of Pennsylvania, Philadelphia, Pennsylvania.
King, Andrew R., University of Cambridge, Cambridge, England.
Klarfeld, Joseph, Queens College, Flushing, New York.
Knudson, John, University of British Columbia, Vancouver, British Columbia.
Kovacs, Sandor J., California Institute of Technology, Pasadena, California.
Kuchař, Karel V., Princeton University, Princeton, New Jersey.
Kumar, Sudhir, University of Notre Dame, Notre Dame, Indiana.
Künzle, H. P., Universtiy of Alberta, Edmonton, Alberta.
Lee, David L., California Institute of Technology, Pasadena, California.
Lee, Kotik K., Syracuse University, Syracuse, New York.

Lerner, David E., University of Pittsburgh, Pittsburgh, Pennsylvania.
Leroy, Jules, Universite Libre de Bruxelles, Bruxelles, Belgium.
Lewis, John K., Instituut voor Theoretische Physica, Utrecht, Netherlands.
Lightman, Alan, California Institute of Technology, Pasadena, California.
Lind, Robert W., Syracuse University, Syracuse, New York.
Longo, Stephen A., La Salle College, Philadelphia, Pennsylvania.
Lubin, Michael, Westeyan University, Middletown, Connecticut.
Lucey, Carol A., Jamestown Community College, Jamestown, New York.
Ludwig, Garry, University of Alberta, Edmonton, Alberta.
Madore, John, Institut Henri Poincaré, Paris, France.
Manoukian, Edward, University of Alberta, Edmonton, Alberta.
May, Thomas, San Diego State College, San Diego, California.
McCarthy, Patrick, University of British Columbia, Vancouver, British Columbia,
McLenaghan, Ramond, University of Waterloo, Waterloo, Ontario.
Melvin, M. A., Temple University, Philadelphia, Pennsylvania.
Miketinac, M. J., University of Saskatchewan, Regina, Saskatchewan.
Miller, Bonnie D., University of Chicago, Chicago, Illinois.
Miller, Gary G., University of Victoria, Victoria, British Columbia.
Moncrief, Vincent, University of Maryland, College Park, Maryland.
Morgan, Thomas, University of Nebraska, Lincoln, Nebraska.
Morganstern, Ralph, Avanzados del Instituto Politecnico Nacional, Mexico.
Nagai, Ernesto, Michigan State University, East Lansing, Michigan.
Nash, G., University of Alberta, Edmonton, Alberta.
Newman, Ezra, University of Pittsburgh, Pittsburgh, Pennsylvania.
Nutku, Yavuz, University of Maryland, College Park, Maryland.
Olson, Donald, University of California, Berkeley, California.
Page, Donald N., California Institute of Technology, Pasadena, California.
Pechlaner, Edgar, Simon Fraser University, Burnaby, British Columbia.
Peebles, P. J. E., Princeton University, Princeton, New Jersey.
Porter, Richard, State University of New York, Stony Brook, New York.
Poss, Howard L., Temple University, Philadelphia, Pennsylvania.
Preston, Ralph, Laurentian University, Sudbury, Ontario.
Quale, Andreas, Oslo University, Oslo, Norway.
Ramsden, David, Newmarket, Ontario.
Rastall, Peter, University of British Columbia, Vancouver, British Columbia.
Ravindra, Ravi, Dalhousie University, Halifax, Nova Scotia.
Ray, John R., Clemson University, Clemson, South Carolina.
Robinson, David, University of London, Strand, England.
Roeder, Robert, University of Toronto, Toronto, Ontario.
Russo, Onofrio L., Newark College of Engineering, New Jersey.
Sachs, Rainer K., University of California, Berkeley, California.
Schechter, Paul, California Institute of Technology, Pasadena, California.
Schendel, Jack, Miami University, Oxford, Ohio.

Schiff, Harry, University of Alberta, Edmonton, Alberta.
Schmidt, Bernd, Universität Hamburg, Hamburg, West Germany.
Sejnowski, Terrence, Princeton University, Princeton, New Jersey.
Shamaly, A., University of Alberta, Edmonton, Alberta.
Sigal, Richard, University of Alberta, Edmonton, Alberta.
Skinner, Ray, University of Saskatchewan, Saskatoon, Saskatchewan.
Smeyers, Paul, Astronomisch Instituut, Leuven, Belgium.
Smith, Lyman A., Queens University, Kingston, Ontario.
Sommerfield, M., University of Michigan, Ann Arbor, Michigan.
Spanos, Timothy, University of Alberta, Edmonton, Alberta.
Spero, Alan, Wesleyan University, Middletown, Connecticut.
Srulovicz, Peter, Queen's University, Kingston, Ontario.
Tait, R. J., University of Alberta, Edmonton, Alberta.
Taub, A. H., University of California, Berkeley, California.
Tripathi, Vishwa, Banaras Hindu University, Varanasi, India.
Trollope, J. R., University of Alberta, Edmonton, Alberta.
Tupper, Brian, University of New Brunswick, Fredericton, New Brunswick.
Turcillo, Joseph, University of California, Burbank, California.
Van Riper, Kenneth, University of Pennsylvania, Philadelphia, Pennsylvania.
Vardalas, John, University of Alberta, Edmonton, Alberta.
Vigmato, Agostino, Viale del Parco Mellini, Roma.
Wahlquist, Hugo, California Institute of Technology, Pasadena, California.
Wald, Robert M., Princeton University, Princeton, New Jersey.
Wertz, James, Moorhead State College, Moorhead, Minnesota.
Wilson, Alan K., University of Victoria, Victoria, British Columbia.
Wilson, Gordon A., University of Alberta, Edmonton, Alberta.
Witten, Louis, University of Cincinnati, Cincinnati, Ohio.
Witting, Harvey, Stevens Institute of Technology, Hoboken, New Jersey.

SURVEY OF GENERAL RELATIVITY THEORY*

JÜRGEN EHLERS

Max-Planck-Institut für Physik und Astrophysik, München, G.F.R.

0. Introduction

The purpose of these lectures was to give a survey of Einstein's theory of gravitation. Since special astrophysical topics have been covered by other lecturers and the principal mathematical tools of general relativity were also considered separately, I tried to concentrate on the general structure of the theory and the connection between physical and mathematical concepts and ideas; examples and applications have been mentioned briefly only.

In the first two sections I wished to show how convincing the main ideas of general relativity are, and how the spacetime framework of the general theory can be synthesized out of a few assumptions which are idealizations of empirically very well established facts, following [16]. It is not true, in my opinion, that Einstein's theory is phenomenologically less well founded than, say, classical electrodynamics; the number of qualitatively different, quantitatively controllable, measureable 'effects' on which it is based is admittedly smaller, but the *precision* and *generality* of the few facts on which it is based are impressive enough. The introduction of a non-integrable, linear connection to account for the combined inertial-gravitational 'guiding field' is, in view of the universality of free fall and the inhomogeneity of real gravitational fields, as well motivated physically as the introduction of vector fields to describe electromagnetism. If this is accepted, the universal local validity of special relativity almost forces one towards a curved, pseudoriemannian spacetime.

Of course, a theory cannot be deduced from empirical facts; nevertheless an inductive (or 'constructive') axiomatics as attempted in Section 2 gives perhaps a better insight into the theory than a deductive reasoning which starts with involved basic laws.

In Part II I tried to give a reasonably complete presentation of the classical description of matter in curved spacetime as far as it now exists. The parts on continuum mechanics, and even more so those on radiation and on extended bodies, are much less systematic than those on kinetic theory; but at least the main ideas are considered, and he who wants to go into details has at least something to build on, and a guide to the literature.

The section on approximation methods is a sketch only; a good, self-contained presentation of these methods seemed beyond the possibilities of these lectures.

The elementary topics of spherically symmetrical spacetimes serve to illustrate the working of the field equation, the matter model, and the junction conditions.

* This work was supported in part by the National Science Foundation (Grant No. GP-34639X). The author did part of this work while visiting the University of Texas at Austin. He wishes to thank this institution and his colleagues at the Center For Relativity for their kind hospitality.

W. Israel (ed.), Relativity, Astrophysics and Cosmology, 1-125. *All Rights Reserved*
Copyright © 1973 *by D. Reidel Publishing Company, Dordrecht-Holland*

Originally it was planned to include a brief section on the derivation of the Kerr solution. Any presentation which would have heen interesting – and not just a mere calculation – would have taken up too much additional space, and so I have dropped this topic.

Throughout the lectures I followed the sign conventions proposed by Misner, Thorne and Wheeler, with a signature $(+++-)$ of the metric. The units and dimensions are so chosen that $c=G=\hbar=k=1$, where c is the speed of light, G Newton's constant of gravitation, $\hbar$ the quantum of angular momentum, and k Boltzmann's constant. All quantities are then pure numbers.

Basic concepts from differential geometry such as manifold, linear connection, curvature, (pseudo-) Riemannian metric and the formalism of tensor calculus are assumed to be known; in cases of doubt the reader may consult, e.g., refs. [23], [28], [50], [60] or [61]. Mostly the standard index notation is used, with the understanding that the quantities themselves are the essential objects, not their components (see, in particular, [32] for a formalization of this point of view). Occasionally I shall write U instead of U^a for a 4-vector, and use similar abbreviations.

Partial differentation is written $(\ )_{,a}$ or ∂_a, covariant differentiation is written $(\ )_{;a}$; absolute differentiation (along a curve) D/du. Also $(\ )_{;ab}=(\ )_{;a;b}$, $(\ )^{;a}=g^{ab}(\ )_{;b}$ etc. Round (square) brackets denote symmetrization (antisymmetrization); also $A^{[a}B^{b]}=$ $=\frac{1}{2}(A^aB_b-B^aA_b)$. d is the exterior derivative operator, $\wedge$ the exterior (or wedge) product of forms. The differential of a map f is f_*, the associated pull back f^*. Contractions and inner products are occasionally indicated by a dot: $u\cdot v=u_a v^a$. The tangent space of a manifold M at p is M_p, its dual M_p^*. A *path* (= one-dimensional submanifold) has no preferred parameter, a *curve* does. Vectors are identified with linear differential operators, as is now customary. A function is C^r if it is r times continuously differentiable.

No attempt was made to give exhaustive references; those given are thought to supplement the lectures well, or have been of particular help for the preparation of these lectures.

1. Quasi-Historical, Empirical Motivation of the General Theory of Relativity

I. *Physical and Mathematical Foundations of Einstein's Theory of Gravitation*

1.1. Introduction

The purpose of this section is to outline, in a quasi-historical manner, a line of reasoning which leads to general relativity via Newtonian mechanics and the special theory of relativity. Since the aim is not to describe accurately the true historical development, but rather to review the main empirical and theoretical arguments which may lead one to believe that Einstein's theory is a good model for a certain range of natural phenomena, the approach is called quasi-historical, to distinguish it from the constructive-axiomatic approach presented in Section 2. Whereas there we select, for the purpose of logical clarity, some empirical facts as 'basic' and use them, in a necessarily idealized way, to build up the theory systematically, here we try to assemble the broad, somewhat disordered mixture of facts and concepts from different areas of physics which from the background and source of any systematic approach; and for this purpose it seems best to follow roughly the historical development. Whenever it seems useful to accentuate the role of certain facts or theoretical assumptions, a modern description, even a post general-relativistic one, will be used to elucidate, e.g., the Newtonian theory of gravity.

The principal thesis of this section is: An attempt to construct a theory which contains in a consistent way the empirically supported structural ingredients of both Newtonian mechanics including gravity theory and of special relativity theory leads, with considerable force, to Einstein's general theory of a curved spacetime.

1.2. Newtonian spacetime, mechanics, and gravity theory*

Newtonian physics rests on some assumptions about the structure of spacetime which enable one to consider, in a quantitative way, the motions of bodies. These assumptions which form the basis of *nonrelativistic classical kinematics*** are the following:

Spacetime – the arena in which material processes are imagined to take place – is a 4-dimensional, real, smooth *manifold* M. On M there is defined, uniquely up to linear transformations $t \rightarrow at+b$ with $a>0$, an *absolute time* t, a smooth map from M into the real line R whose gradient vanishes nowhere. t defines the (absolute, i.e., observer-independent) *simultaneity* relation $t(a)=t(b)$ between *events* $a, b \in M$, the *time-order* ($t(a)>t(e)$ etc.) between events, and the *durations* $t(a)-t(e)$ of processes happening between e and a; t represents mathematically the readings of *standard clocks*.

The simultaneity relation defines on M a *stratification*: M is partitioned into simultaneity sections or *space sections* each consisting of a maximal class of simultaneous events. This stratification describes the *causal structure* of Newtonian spacetime: The space section S_e containing e separates the causal future of e from its causal past. The circumstance that S_e is the common boundary of the future and the past of e reflects the hypothesis implicit in Newtonian theory (rigid bodies, instantaneous action at a distance) that there are arbitrarily fast signals.

Each space section is further assumed to have the structure of a Euclidean 3-space with a (positive definite) *metric* $\mathrm{d}l^2$; the latter is assumed to vary smoothly if one passes from one space section to another one. $\mathrm{d}l^2$ is assumed to represent length measurements with *standard rods*.

A structure $(M, t, \mathrm{d}l^2)$ as characterized in the preceding three paragraphs and illustrated in Figure 1 is called a *Galilean manifold*; $(t, \mathrm{d}l^2)$ is a *Galilean metric* on M. a and b are simultaneous as are c and d. a is later than e. The half-space with $t>t(e)$ is the future, that with $t<t(e)$ the past of e. c and d have the distance $\mathrm{d}l$, e and a determine the duration $t(a)-t(e)$. P represents a particle.

This *chronogeometry* is well known to be a very successful framework for the

* The spacetime-geometrical formulation of nonrelativistic mechanics and gravity theory outlined in this subsection has been elaborated in detail; see [1], [2], [3] and the references given therein.
** The adjective 'classical' will be used throughout as the alternative to 'quantal'.

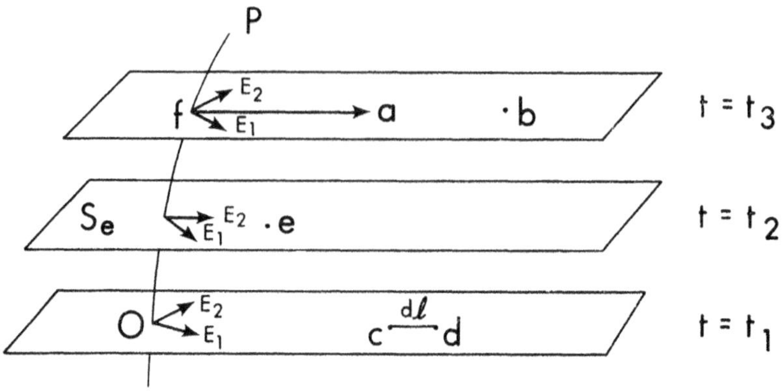

Fig. 1. A Galilean manifold.

description of a large variety of phenomena; in fact, many important branches of physics (such as the quantum theory of atoms and molecules) can presently be carried out quantitatively only on its basis.

For the remainder of this section we assume M to be a Galilean manifold.

In M one can distinguish between spacelike vectors (like **ab** in Figure 1) and future or past directed timelike vectors (i.e., those vectors A with $A(t)>0$ and $A(t)<0$, respectively; for this notation, see Schmidt's lectures, this volume, p. 289). The spacelike vectors are the familiar 'three-dimensional vectors' used in the traditional treatments of Newtonian Physics. Similarly, one can speak of timelike or spacelike paths or curves (see remarks on conventions and terminology, p. 2). Between spacelike vectors there is defined an inner product, whereas timelike vectors have only a duration associated with them. Accordingly, only simultaneous events have an unambiguous spatial distance, in consequence of the 'relativity of space', i.e., the absence of any objective criterion for two nonsimultaneous events to happen at 'the same' spacepoint.

The history, or motion, of a *particle* P is represented as a timelike path in M (see Figure 1).

A *Galilean frame of reference* (warning: this term will be used here not for inertial frames, but for the much more general preferred frames of Galilean kinematics) is determined by a timelike path P, an event o on P, and a smoothly varying orthonormal triad $(\mathbf{E}_\lambda)$, $\lambda=1, 2, 3$ of spacelike vectors along P (see Figure 1). P, $\mathbf{E}_\lambda$ may be thought of as realized by marks on an arbitarily moving, rigid *body of reference*, e.g. the Earth.

The *coordinates* x^λ, x^4 of an (arbitrary) event a with respect to a Galilean frame $(P, o, \mathbf{E}_\lambda)$ are defined as follows. Let f be the (unique) event on P which is simultaneous with a. Then $\mathbf{fa}=\mathbf{x}^\lambda(a)\mathbf{E}_\lambda$ (summation $\sum_{\lambda=1}^{3}$, implied), and $x^4(a)=t(a)$ where the absolute time function t is chosen such that $t(o)=0$.

We assume that units for time and length have been chosen once and for all, and do not consider here the effect of changes of units.

The transformations $(x^a)\rightarrow(x^{a'})$ relating coordinates of two Galilean frames are of the form

$$x' = D(t)\cdot x + d(t), \quad t' = t + \tau \tag{1.1}$$

where $x = \begin{pmatrix} x^1 \\ x^2 \\ x^3 \end{pmatrix}$ etc., $\tau \in \mathbb{R}$, and $D(t)$ is a time-dependent orthogonal matrix. These transformations form the *kinematical group* $\mathfrak{K}$.

On this basis the kinematics of particles and rigid and deformable bodies can be erected. It is important to recall that at this kinematical level there are *no preferred motions* (world lines); velocities, accelerations, angular velocities etc. are all relative quantities defined with respect to Galilean frames, but having no 'absolute', frame-independent meaning or value. In particular, a time-differentiation operator d/dt acting on vectors or tensors is defined only for each such frame, but not in an absolute way.

Collisions between particles, if considered as point-events due to interaction with zero range, can be described already at this kinematical level. With each particle there is associated a positive, constant *inertial mass m*, characterized by the property that in each collision, the sum of the linear momenta $m\dot{\mathbf{x}}$ of the particles involved is conserved:

$$\sum_{\text{initially}} m_A \dot{\mathbf{x}}_A = \sum_{\text{finally}} m_B \dot{\mathbf{x}}_B . \tag{1.2}$$

The velocities are to be taken 'immediately' before or after the collision, i.e., as limits from the past and the future, respectively, taken at the time of the collision event. (1.2) is a frame-independent, $\mathfrak{K}$-invariant relation if and only if mass is conserved,

$$\sum m_A = \sum m_B , \tag{1.3}$$

as follows from the transformation law for velocities implied by (1.1).

The transition from kinematics to *dynamics* consists essentially of singling out a particular class of motions as *standard motions* which are considered as force-free, and then to define for all motions differing from standard motions *forces* in terms of accelerations relative to these standard motions. In order that this idea works there must be one and only one standard motion (world line) through each event and time-like direction (velocity) at that event. The kinematical reason for the possibility of this definition of forces is the easily established *theorem*: In Galilean spacetime the relative acceleration between two particles passing through an event with the same direction (velocity) is a frame-independent, spacelike vector. The generalized form of *Newton's second law* is accordingly written

$$m(\ddot{\mathbf{x}} - \mathbf{a}(\mathbf{x}, \dot{\mathbf{x}})) = \mathbf{F} \tag{1.4}$$

where $\mathbf{x}(t)$ describes the motion of an arbitrary particle relative to some Galilean frame, and $\mathbf{a}(\mathbf{x}, \dot{\mathbf{x}})$ is the acceleration of a standard motion specified by $\mathbf{x}$ and $\dot{\mathbf{x}}$; $\mathbf{F}$ is the force.

From the point of view of spacetime geometry the choice of a class of standard motions amounts to giving M the structure of a *pathspace*, which just means that with any given point and direction at that point there is associated a unique path passing through that point in the specified direction.

The traditional way of introducing a class of preferred motions is to postulate the *law of inertia*: There exist *free motions* which are characterized, with respect to a special class of Galilean frames of reference, the *inertial frames*, as having zero acceleration,

$$\ddot{\mathbf{x}} = 0 . \tag{1.5}$$

It is then easy to show that the transformations between inertial frames are those special transformations (1.1) for which

$$\dot{D} = 0 , \quad \ddot{d} = 0 . \tag{1.6}$$

They form the *Galilean group* $\mathfrak{G}$, a subgroup of the kinematical group $\mathfrak{K}$,

$$\mathfrak{G} \subset \mathfrak{K} . \tag{1.7}$$

With respect to inertial frames, (1.4) simplifies to the familiar form of the second law:

$$m\ddot{\mathbf{x}} = \mathbf{F} . \tag{1.8}$$

On this basis the usual formalism of classical mechanics can be constructed.

Since the transformations (1.1), (1.6) are linear, it is meaningful to call two 4-vectors **A** and **B** *parallel* if they have the same components with respect to one (and hence with respect to all) inertial coordinate systems. This relation is, of course, independent of the choice of any path connecting the events at which **A** and **B** are defined. Hence, one may say that *the law of inertia defines on spacetime an integrable symmetric connection whose (timelike) geodesics represent the free motions*, namely the connection whose components Γ^a_{bc} all vanish in an inertial coordinate system.

What is the empirical basis of the law of inertia? To realize a free motion one should take a particle that has no electric charge and no electric and magnetic multipole moments; in the case of a macroscopic particle one should eliminate drag forces by having the particle move in a vacuum, etc. But even then the particle will accelerate towards massive bodies such as the Earth. Assuming that each particle has a *passive gravitational mass* μ besides its inertial mass m (introduced in (1.2)) such that the gravitational force acting on it is $\mu\mathbf{g}$ where $\mathbf{g}$ is the *gravitational field strength*, we obtain relatively to an inertial frame

$$m\ddot{\mathbf{x}} = \mu\mathbf{g}(\mathbf{x}, t) + \mathbf{F} , \tag{1.9}$$

where **F** is any additional, e.g. elastic, force and **g** is assumed to depend on **x** and t only. Equations (1.1) and (1.9) imply that the motion of the particle is described relatively to an arbitrary, noninertial, Galilean reference frame by

$$\ddot{\mathbf{x}}(t) = \frac{\mu}{m}\mathbf{g}(\mathbf{x}(t), t) + \frac{1}{m}\mathbf{F} - \mathbf{A}(t) - \\ - 2\boldsymbol{\omega}(t) \wedge \dot{\mathbf{x}}(t) - \dot{\boldsymbol{\omega}}(t) \wedge \mathbf{x}(t) - \boldsymbol{\omega}(t) \wedge (\boldsymbol{\omega}(t) \wedge \mathbf{x}(t)) . \tag{1.10}$$

Here $\mathbf{A}(t)$ is the acceleration, $\boldsymbol{\omega}(t)$ the angular velocity of the arbitary (e.g. laboratory) frame relatively to some inertial frame.

If it were possible to measure, by observing the motions of particles relatively to some given laboratory frame, the quantities μ/m, $\mathbf{g}$, $\mathbf{A}$, $\boldsymbol{\omega}$ (for $\mathbf{F}=0$ or an otherwise known $\mathbf{F}$), then the usual form of the law of inertia with its postulated inertial frames would have a clear empirical foundation. That would be the case if particles with different 'specific gravitational masses' μ/m would exist; for then measurement of $\mathbf{x}$, $\dot{\mathbf{x}}$, $\ddot{\mathbf{x}}$ for various freely falling ($\mathbf{F}=0$) particles would, according to (1.10), permit the unique determination of $\mathbf{g}$, $\mathbf{A}$, $\boldsymbol{\omega}$, and μ/m. If, however, the ratio μ/m were a universal constant so that one could, without loss of generality, put $\mu=m$, Equation (1.10) shows that whereas $\boldsymbol{\omega}$ as well as $\mathbf{g}-\mathbf{A}$ could still be measured, it would be impossible to find $\mathbf{g}$ and $\mathbf{A}$ separately.

Experimental evidence supports very strongly that the second possibility corresponds to reality, i.e. that *the orbits of all neutral, freely falling test particles in a given gravitational field are independent of the mass and composition of the test particles*, being determined uniquely by an initial event and initial velocity. Although the direct evidence for this statement is not very precise, extremely good indirect evidence is provided by the null experiments of Eötvösz; Dicke, Roll, and Krotkov; and Braginski. (The evidence is indirect since in these experiments the test particles are, in fact, not freely falling, but suspended from a torsion balance, and the theoretical analysis on which the famous numbers 10^{-11} or 10^{-12} are based do not only use the law (1.10) with $\mathbf{F}=0$, but the full force- and torque equilibrium conditions of Newtonian mechanics for a system in which inertial, gravitational, and elastic (i.e. electromagnetic) forces are essential. For a more detailed account of these experiments, see Brill's lectures and the references quoted there.)

Accepting this fundamental experimental result of the *universal equality of inertial and passive gravitational mass*,

$$m = \mu\,, \tag{1.11}$$

one is led with Einstein to *abandon the law of inertia in its traditional form* as stated above, simply because *there are no free particles in the sense of that law*, all particles being coupled in the same way to the gravitational-inertial field, and to replace it by (say) *Galileo's law* (nonrelativistic form of the *weak principle of equivalence*): There exist *free fall motions*, which are characterized, with respect to a special class of Galilean frames of reference, the *Newtonian frames*, by the equation of motion

$$\ddot{\mathbf{x}} = -\nabla\phi\,, \tag{1.12}$$

in which $\phi(\mathbf{x}, t)$, the *gravitational potential*, is a frame-dependent, smooth, real valued function.

Taking the free fall motions instead of the (fictitious) free motions as the standard motions of dynamics, one is led to specialize (1.4) to

$$m(\ddot{\mathbf{x}} + \nabla\phi) = \mathbf{F} \tag{1.13}$$

as the modified form of Newton's second law, in which $\mathbf{F}$, the (non gravitational) force, is again to be considered as a frame-independent, spacelike vector. Thus, in this modified form of Newton's second law *the gravitational field is tied to the acceleration*, just as in general relativity theory (see below).

Newtonian frames as defined by Galileo's law can in principle be (and are essentially also in practice) recognized as such by the absence of Coriolis and centrigufal accelerations in the equation of motion of a particle. This equation has, according to (1.13), the form (1.10) with $\mu = m$ and $\mathbf{g} = -\nabla\phi$. Such frames are therefore also called *nonrotating frames*.

It is easy to prove that the transformations relating the coordinates of two Newtonian frames to each other are those transformations (1.1) for which

$$\dot{D} = 0. \tag{1.14}$$

These Newtonian transformations, which in the modified dynamics take the place of the Galilean transformations of traditional dynamics, are intermediate in generality between the kinematical transformations and the Galilean transformations. They form the *Newtonian group* $\mathcal{N}$, and we have

$$\mathfrak{G} \subset \mathcal{N} \subset \mathfrak{K}. \tag{1.15}$$

The translational state of motion of the Newtonian frames is completely arbitrary. Under a Newtonian transformation, the gravitational potential does not transform as a scalar, but according to

$$\phi(\mathbf{x}, t) \to \phi'(\mathbf{x}', t') = \phi(\mathbf{x}, t) - \mathbf{x}' \cdot \mathbf{A}, \tag{1.16}$$

as follows from (1.10), (1.11), (1.12), and (1.14). Consequently, the field strength $-\nabla\phi$ is not a frame-independent vector, but changes under a Newtonian transformation by the amount $-\mathbf{A}$. Hence gravitational fields cannot be separated from translational-acceleration fields (Einstein's elevator argument); the splitting of the lefthand side of (1.13) into an inertial term $m\ddot{\mathbf{x}}$ and a gravitational term $m\nabla\phi$ is frame-dependent and has no objective significance.

The free fall law (1.12) can be rewritten as

$$\frac{\mathrm{d}^2 x^a}{\mathrm{d}t^2} + \Gamma^a_{bc} \frac{\mathrm{d}x^b}{\mathrm{d}t} \frac{\mathrm{d}x^c}{\mathrm{d}t} = 0 \tag{1.17}$$

if we put, in Newtonian coordinates,

$$\Gamma^\lambda_{44} = \phi_{,\lambda} \quad \Gamma^a_{bc} = 0 \text{ otherwise}. \tag{1.18}$$

In (1.17) we recognize the *equation of geodesics* of a connection Γ defined on M by Equations (1.18). This *Newtonian connection* is symmetric ($\Gamma^a_{[bc]} = 0$), and the restriction of the absolute time t to any free fall geodesic is an affine parameter, according to (1.17). Moreover, the space sections $t = $const. are totally geodesic with respect to Γ, and the restriction of Γ to a space section is identical with the Euclidean connection defined by the spatial metric $\mathrm{d}l^2$. (Equation (1.17) can be used in any curvilinear coordinate system of M, provided the components Γ^a_{bc} of the connection are properly transformed according to the rules of differential geometry. In particular, all gravitational and inertial terms in (1.10) can be identified with components of Γ.)

In general, a Newtonian connection is not integrable, i.e., the associated parallel transport of 4-vectors is path-dependent. This can be seen either by computing the *curvature tensor* of Γ by means of (1.18) which gives, in Newtonian coordinates

$$R^{\lambda}{}_{4\mu 4} = \phi_{,\lambda\mu}, \quad R^{a}{}_{bcd} = 0 \text{ otherwise}, \tag{1.19}$$

or it can be understood more directly and intuitively by examples such as the following due to Heckmann and Schücking [4]. Consider two test particles initially at rest on the surface of a heavy spherical body $\underline{B}$. Imagine that a tunnel is drilled through $\underline{B}$ along a diameter, and that one of the test particles is released so that it oscillates in free fall within the tunnel, whereas the second particle is thrown vertically upwards with an initial velocity such that it returns to the initial point precisely at the time when the first particle has completed one oscillation. Figure 2 is a spacetime diagram of this process.

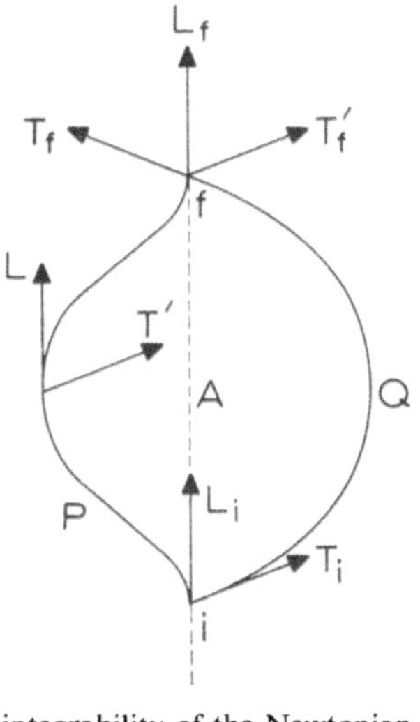

Fig. 2. Nonintegrability of the Newtonian connection.

A and B are the world lines of the two endpoints of the tunnel, C is the world line of the center of $\underline{B}$. P, Q are the orbits of the two test particles. Parallel transport of the initial tangent vector L_i of P from i to f along P leads to the final tangent vector L_f of P at f, since P is a geodesic. Similarly, parallel transport of the tangent vector T_i to Q from i to f along Q produces the tangent vector T_f. If, however, T_i is transported parallely from i to f along P, a vector $T_f' \neq T_f$ is obtained, since for reasons of symmetry the whole process will take place within the timelike plane containing A, B, C, P, Q, and T' and L, the parallely displaced vectors along P, always remain linearly independent. Parallel transport of spacelike vectors, however, is path-independent, according to (1.18) and (1.19).

Summarizing one may thus say that *the Eötvösz-Dicke experiment suggests that the gravitational field is not a* (frame-independent) *vector field, but a non-integrable, symmetric connection*, whose geodesics are the free fall trajectories, a conclusion which anticipates one of the main ingredients of Einstein's general theory.

A structure $(M, t, \mathrm{d}l^2, \Gamma(\phi))$ consisting of a Galilean manifold and a Newtonian connection (1.18) may be called a *Newtonian spacetime*. Whereas t and $\mathrm{d}l^2$ are absolute elements of such a spacetime, given once and for all, Γ is a contingent element, a dynamical variable having different values in different situations. Thus, part of the geometric structure of M is 'physicalized' as a consequence of the weak principle of equivalence.

Although we have abandoned the notion of exact inertial frames, we may, in a Newtonian spacetime, define *local inertial frames* as those Newtonian frames whose spatial origin is freely falling, i.e. for which

$$\phi_{.\lambda}(\mathbf{o}, t) = 0 \tag{1.20}$$

or, equivalently,

$$\Gamma^a_{bc}(\mathbf{o}, t) = 0\,. \tag{1.21}$$

Such frames have approximately the properties of inertial frames in a sufficiently small spatial neighbourhood of the origin $\mathbf{x}=0$; they represent Einstein's elevators. In mathematical language such coordinates are called geodesic with respect to Γ at $(\mathbf{o}, t)$.

Whereas $-\nabla\phi$ can be 'transformed away' at an event and thus does not provide a frame independent, 'intrinsic' local measure of a gravitational field, such a measure is obtained by considering the *relative accelerations* between neighbouring, freely falling test particles. Denoting the relative position vector of such a pair by $\Delta\mathbf{x}$, we obtain from (1.12)

$$\Delta\ddot{\mathbf{x}}^{\lambda} = -\,\phi_{.\lambda\mu}\Delta x^{\mu} + \mathcal{O}(\Delta\mathbf{x}^2)\,. \tag{1.22}$$

The *tidal field* $-\phi_{.\lambda\mu}$ which appears in this equation is, according to the transformation law (1.16) or also according to (1.19), a frame independent, spacelike tensor field. In view of (1.19) one recognizes in (1.22) the *equation of geodesic deviation* to be considered in more detail in subsection 2.10. The tidal field determines and is determined by the curvature tensor (1.19) of the Newtonian connection and can thus be identified with it, another result which will reappear in Einstein's theory.

So far we have not considered how a gravitational field is generated. In the usual formulation of the Newtonian theory of gravity one postulates *Poisson's equation*

$$\nabla^2\phi = 4\pi G\varrho \tag{1.23}$$

relating ϕ to the density ϱ of matter which acts as the source of ϕ. Equation (1.23) expresses locally the *inverse-square character* of the gravitational force. To indicate the accuracy with which it is established by comparing theory and observation of the motions of the planets of the solar system we recall that in 1894 Hall found that in order to explain the 'anomalous' perihelion advance of Mercury the exponent $n=-2$ had to be replaced by $n=-2-\varepsilon$ where $\varepsilon=1.6\times10^{-7}$.

It is useful to reformulate the contents of (1.23) as follows:

$$\text{trace (tidal field)} = -\ \phi_{,\lambda\lambda} = -\ 4\pi G\varrho\,. \tag{1.24}$$

Since $-\phi_{,\lambda\mu}$ transforms Δx^{μ} into $\Delta \ddot{x}^{\mu}$ – Equation (1.22) – a simple argument shows that the volume ΔV of a small cloud of test particles which are released without initial relative velocities and then start falling freely in a gravitational field will change such that initially

$$\Delta \dot{V} = 0\,, \quad \Delta \ddot{V} = -\ \phi_{,\lambda\lambda}\Delta V\,. \tag{1.25}$$

Thus, under the conditions stated, the contents of Newton's $1/r^2$-law or Equation (1.23) can be restated as the relation

$$\Delta \ddot{V} = -\ 4\pi G\varrho\Delta V\,. \tag{1.26}$$

In particular, outside of sources we have

$$\Delta \ddot{V} = 0\,, \tag{1.27}$$

and it is easy to show that *this law concerning test body motion characterizes the r^{-2} force law* within the class of central forces. Hence, any theory of gravitation must reproduce this law, at least in the weak-field, slow motion limit, in order to agree in that limit with the empirically well-supported Newtonian theory. This remark is very useful in motivating Einstein's gravitational field equation and to relate that equation to the classical law (1.23).

In Equation (1.23) the source density for the ϕ-field or, what is equivalent in view of Equations (1.18) and (1.19), for the spacetime curvature associated with the Newtonian connection, is $G\varrho$. One could introduce instead of $G\varrho$ a density $\bar{\varrho}$ of *active* (i.e., field producing) *gravitational mass*, measure the latter by means of a Cavendish-experiment, and then define G as the ratio active gravitational mass/inertial mass of some body. It is then meaningful and important to find out experimentally whether this G is a universal constant, as was tacitly assumed in (1.26). This seems indeed to be the case, at least to an accuracy of 5×10^{-5}, as shown by the *Kreutzer experiment*. (See Brill's lectures, this volume, p. 132 and the reference given there.) We therefore assume henceforth that (1.23) holds, with ϱ = inertial mass, and we choose the unit of mass such that

$$G = 1\,. \tag{1.28}$$

The structure $(M, t, \mathrm{d}l^2, \Gamma(\phi))$ together with (1.17), (1.18), (1.23), supplemented by laws governing the behaviour of matter such as conservation of mass, linear momentum etc., constitutes the basis of the modified form of nonrelativistic, classical mechanics. The advantages of this modified form of the theory as compared to the standard version are that it incorporates the weak principle of equivalence, it permits the treatment of unbounded and, in particular, spatially homogeneous self-gravitating systems as considered in cosmology, and it paves the way to Einstein's theory of gravity. In connection with the second point it may be remarked that there is no need, for the formulationof the basic laws of the theory, to choose a particular boundary condition restricting the behaviour of ϕ at spatial infinity. The standard requirement $\phi \to 0$ at infinity, which would, because of (1.16), reduce the Newtonian group $\mathcal{N}$ to the Galilean group $\mathfrak{G}$ and thus would reintroduce strict inertial frames, is incompatible with homogeneous or asymptotically homogeneous cosmological matter distributions and has no physical justification.

It may finally be pointed out that not only the whole body of non-relativistic classical mechanics and thermodynamics, but also nonrelativistic quantum mechanics can be incorporated into the framework of Newtonian spacetime [5]; this fact considerably diminishes the allegedly fundamental importance of the Galilean group for low-energy phenomena, except if gravity is to be neglected.

1.3. Minkowskian spacetime, electrodynamics, and high-energy phenomena

Whereas nonrelativistic mechanics including gravity as outlined above describes quite well *slow motion and hence weak gravity phenomena* at all scales, it is empirically *untenable for fast motions and high-energy phenomena*, and especially it is unable to account for rest mass zero particles or fields, as is well known.

To connect mechanics and gravity theory with observation, at least the *elements of optics* have to be added to the theory, since otherwise the theoretically computed orbits of, say, the planets cannot be related to the observations of planetary positions which make use of light travelling from the planets to us. The naive optics in which light rays are represented by straight lines in the space sections of Newtonian spacetime ('$c = \infty$'), though sufficient for a rough account of planetary position measurements, was disproved already by Rømer's discovery (in 1676) of the annual variations of the apparent period of one of Jupiter's moons. Interpreted as a Doppler effect, this discovery established the finiteness of the velocity of light, and this together with the subsequent discoveries of the aberration of starlight (Bradley, 1728) and of planetary aberration (retardation correction of planetary positions) suggested the hypothesis that light propagates with the speed $c \approx 3 \times 10^{10}$ cm s^{-1} along straight lines with respect to some Newtonian frame of reference which coincides, at least approximately, with the center-of-mass frame of the solar system. Thus, light propagation seemed to single out a preferred, absolute ether frame with respect to which light rays (world lines of photons) are described by

$$\ddot{\mathbf{x}} = 0, \quad \dot{\mathbf{x}}^2 = c^2. \tag{1.29}$$

From the point of view of spacetime geometry these equations may be interpreted as defining on M, in addition to the structures t, $\mathrm{d}l^2$, $\Gamma(\phi)$ provided by kinematics and gravity, a *conformal structure* $\mathfrak{C}$, i.e., a field of *null cones* given by

$$\mathrm{d}\mathbf{x}^2 - c^2\,\mathrm{d}t^2 = 0. \tag{1.30}$$

This additional structure reduces the Newtonian group $\mathcal{N}$ to that subgroup which, in the notation of (1.1), is given by

$$\dot{D} = 0, \quad d = 0, \tag{1.31}$$

and which is nothing but the direct product of the Euclidean group (of translations and rotations of Euclidean 3-space) with that of time translations. One thus arrives at the spacetime of the 'rigid ether' theory which has an *absolute space* as well as an absolute time, i.e. in which M is in a natural way a direct product of space E^3 and time $T(=E^1)$.

The ether theory just mentioned as well as variants of it had to be abandoned in view of the negative results of several *ether drift experiments* (see, e.g. [6] and Brill's lectures, this volume, p. 127), combined with other facts such as aberration, Fizeau-dragging etc., and since there are also good experimental data against emission theories [6], one is forced to modify at least some of the structures t, dl^2, Γ, $\mathfrak{C}$ assigned to spacetime. The fundamental, successful step taken by Einstein in 1905 was to *give up the assumption that there exists an absolute time t*, but to *keep the conformal structure* $\mathfrak{C}$ which is directly and precisely supported not only by the facts of light propagation, but also by many other electromagnetic phenomena, in accordance with the Maxwell-Lorentz theory.

If t is thrown away, then also the spatial metric as introduced at the beginning of Subsection 1.2 loses its meaning (see Figure 1), i.e. *the whole concept of a Galilean metric* (t, dl^2) *and with it the old kinematics has to be given up.*

The question then arises: Which structure besides the conformal structure $\mathfrak{C}$, if any, should be introduced to replace the Galilean metric?

If one disregards temporarily gravity in order to separate the difficulties, and returns to the old concept of free motion, it is plausible to postulate:

There exist inertial spacetime coordinate systems $(x^a)=(x^\lambda, t)$ such that (a) light propagation is characterized by (1.30) and (b) free motions satisfy (1.5), and such motions exist at least for all initial data

$$(\underset{0}{x^a}, \underset{0}{\dot{x}^a}) \quad \text{with} \quad \dot{x}^\lambda \dot{x}^\lambda < V^2, \quad \text{where} \quad 0 < V \leqslant c.$$

This postulate (as well as numerous equivalent ones) leads to the *spacetime structure of special relativity*, first formulated explicitly by Minkowski in 1908:

Spacetime is a smooth 4-manifold M which has a *Lorentzian, flat metric* η so that in inertial coordinates

$$\eta = (\eta_{ab}) = \text{diag.}(+1, +1, +1, -1). \tag{1.32}$$

The *null cones* of η describe *light propagation* (in vacuo), the *timelike straight lines* describe *free motions*, and

$$\int d\tau = \int \sqrt{-\eta_{ab}\, dx^a\, dx^b} = \int \sqrt{1 - v^2}\, dt, \tag{1.33}$$

taken along a timelike path representing a particle, gives the *proper time* as measured by a standard clock associated with that particle.

I shall not review here in detail the arguments leading from the above postulates to Minkowski's geometry, but refer to textbooks, in particular Weyl [7] and Fock

[8]. One has to show that requirement (a) determines the coordinates exactly up to transformations of the conformal group O(4,2) and (b) determines them up to projective transformations. Thus both experiments together determine the coordinates precisely up to dilatations ($x^a \to \lambda x^a$) and Poincare transformations, and these transformations in turn are characterized as those leaving the Minkowskian metric (1.32) invariant up to a constant scale factor. There are many other ways to obtain Minkowski's metric from physically plausible postulates; for references, see Breitenbacher [9].

In (1.32) we have used a unit convention to be kept throughout these lectures, viz.

$$c = 1\,. \tag{1.34}$$

The Minkowskian spacetime geometry is illustrated in Figure 3.

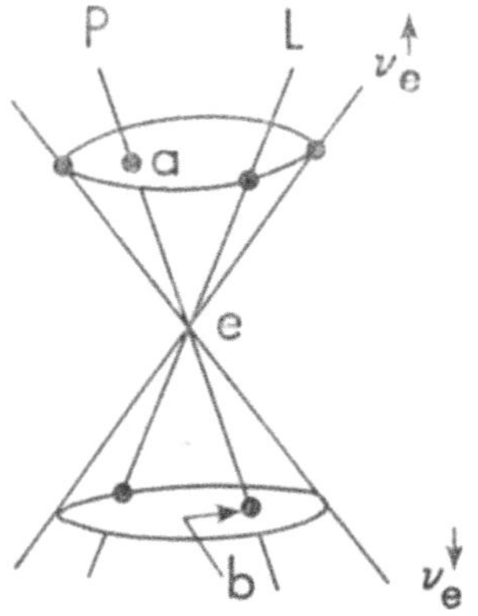

Fig. 3. Minkowskian geometry. ν_e is the null cone ($\equiv$ light cone) of e, P is a free particle ($\equiv$ timelike straight line), L is a light ray ($\equiv$ null straight line).

Without going into details we recall some important assertions of special relativity theory:

Inertial coordinates are determined precisely up to linear transformations of the form

$$x^{a'} = L^{a'}{}_b x^b + d^{a'}\,, \tag{1.34}$$

where $d^{a'}$ and $L^{a'}{}_b$ are constants and the matrix $L = (L^{a'}{}_b)$ satisfies the pseudo-orthonality relation

$$L^T \cdot \eta \cdot L = \eta \tag{1.35}$$

(L^T is the transpose of L.) These transformations form the *Poincaré* group $\mathscr{P}$, a ten-dimensional Lie group. *Inertial frames of reference* $(P, o, \mathbf{E}_a)$ can be physically defined in close analogy to Galilean frames (p. 4); P is a free particle, o an event on P, and $(\mathbf{E}_a)$, $a = 1, 2, 3, 4$, a basis of the space of 4-vectors, defined and parallel along P, orthogonal with respect to η, with $\mathbf{E}_4$ tangent to P. Inertial coordinates x^a are associated with an inertial frame in an obvious way.

Simultaneity is an *inertial-frame dependent* relation; decompositions of spacetime into space sections depend on an observer, i.e. a timelike line, or state of motion.

The *causal structure* of Minkowskian spacetime is determined by its null cones. A process at e (see Figure 3) can influence only events in or on the future null cone $v_e^\uparrow$ (like a), and can be influenced only by events in or on the past null cone $v_e^\downarrow$ (like b).

The causal structure is totally unlike that of a Galilean manifold, as a comparison of Figures 1 and 3 reveals, and this dissimilarity in the causal structures constitutes the essential difference between nonrelativistic and (special or general) relativistic spacetimes. The difference of the corresponding preferred coordinate transformations should be considered as a secondary, merely technical consequence of this basic difference of intrinsic structure.

Since the transformations (1.34) are linear, Minkowski spacetime is endowed with an *integrable affine connection*: this corresponds to disregarding gravity.

Maxwell's equations

$$F_{[ab,c]} = 0, \quad F^{ab}{}_{,b} = J^a \tag{1.36}$$

are invariant with respect to $\mathscr{P}$. In fact, $\mathscr{P}$ (together with dilatations $x^a \to \lambda x^a$) can be characterized as the group of those transformations which preserve (1.36) and which map $\mathbb{R}^4$ onto itself globally and continuously [7].

Apart from optics and electrodynamics of moving bodies which led to the discovery of special relativity there are numerous additional experimental confirmations of it; we mention the *time dilation* effect, now established with 0.1% accuracy by measurements of lifetimes of muons in storage rings [6], and the multitude of data concerning the kinematics of *high energy collisions*. (One may, in fact, consider the collision laws as an alternative to optical-electromagnetic effects in deciding between the (homogeneous) Galilean and Lorentz groups. For a simple analysis based on very weak assumptions, see [10]. Less direct, but at least equally impressive successes of the special theory of relativity resulted from its (unfortunately still not totally satisfactory) combination with quantum theory; examples are *Dirac's electron theory*, the *spin-statistics theorem*, the *PCT-theorem*, the *Poincaré-group classification of elementary particles*, and the *Lamb shift*.

There can be no doubt, therefore, that the spacetime geometry based on (1.32) has a high truth-value. More precisely, one can say that the *Minkowskian inner product*

$$A \cdot B = \eta_{ab} A^a B^b \tag{1.37}$$

between 4-vectors or, equivalently, the *homogeneous Lorentz group* ($d^{a'} = 0$ in (1.34)) implied by (1.32) describes an extremely well established local structural property of spacetime, its *metric* (which replaces the Galilean metric (t, $\mathrm{d}l^2$)). On the other hand, the facts mentioned above, particularly those referring to atomic or subatomic phenomena, do not give much support to the assumption of global inertial frames or to the inhomogeneous transformations (1.34). Stated more explicitely, the experiments referred to above do not permit a precise determination of the state of motion of those real frames of reference which should be identified with the inertial frames of the theory; they do not, e.g., decide whether a frame attached to the Earth's surface or a frame attached to a freely falling particle is a 'better' inertial frame. (This point has been emphasized by Thorne *et al.* [6]).

In other words, *the existence of a metric η at each event* ('momentum space geometry') *is experimentally well supported, whereas the relative orientations of these metrics at different events are not well determined* by the facts mentioned so far. To remove this uncertainty we have to take into account different facts, to be discussed in the next section.

Before proceeding with the analysis of spacetime geometry, we recall one further important result of special relativity needed later. With any system of particles, any fluid or solid, and with any kind of field there is associated a second rank tensor T^{ab}, the *stress energy momentum tensor* or *matter tensor*, whose components in any inertial frame have the following physical meaning:

$$(T^{ab}) = \left(\begin{array}{c|c} \text{pressure} & \text{momentum} \\ \text{tensor} & \text{density} \\ \hline \text{energy flux} & \text{energy} \\ \text{density} & \text{density} \end{array}\right) \tag{1.38}$$

T^{ab} can be described as the current density associated with the 4-momentum distribution of the system. It can also be characterized as follows: The energy density ϱ_V of any physical system as measured by an observer with 4-velocity $V^a (= \mathrm{d}x^a/\mathrm{d}\tau$, $\mathrm{d}\tau =$ proper time; $V_a V^a = -1)$ depends on V^a according to

$$\varrho_V = T_{ab} V^a V^b . \tag{1.39}$$

The *conservation of energy and momentum* is expressed with respect to inertial coordinates by

$$T^{ab}{}_{,b} = 0 . \tag{1.40}$$

1.4. Pseudoriemannian spacetime, Einstein's description of gravity.

Whereas the special theory of relativity accounts very well for local, laboratory-scale electromagnetic phenomena and atomic and subatomic collision processes etc., it disregards gravity, and, as indicated above, its inertial frames are experimentally ill defined. In view of our discussion of gravity at the Newtonian level these two weaknesses should be regarded as related to each other.

We concluded in Subsection 1.2 that in the presence of true, inhomogeneous gravitational fields there are no exact inertial frames, but only approximate local inertial frames attached to freely falling particles, and that these local inertial frames are accelerated relatively to each other. This suggests that the inclusion of gravity into relativity theory requires a change of the spacetime geometry such that the system of timelike straight lines is deformed into a system of relatively curved lines representing relatively accelerated, freely falling particles.

Since special relativity was based essentially on electromagnetism, one might ask whether there are electromagnetic phenomena which permit an experimental determination of the state of motion of the electromagnetically preferred frames, and whether these frames coincide with the local gravitational-inertial frames. We argue (compare [11], [6]) that the *terrestrial redshift measurements* provide an affirmative answer.

In these experiments, γ-ray photons were emitted from nuclei situated at the bottom A of a tower, which then moved to the top B of the tower, where similar nuclei were ready to absorb them. Frequency shifts between A and B were measured by testing the resonance between emission and absorption, using compensating Doppler shifts by moving the source at A up and down (see Figure 4)

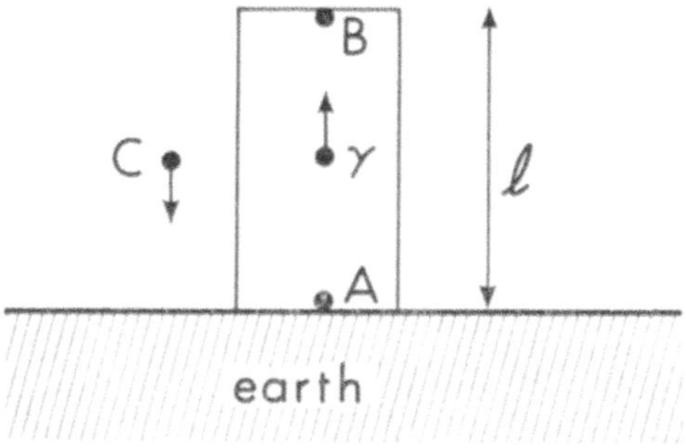

Fig. 4. Terrestrial redshift experiment.

Assuming that special relativity is valid to a sufficient approximation in the spacetime region of such an experiment we ask: Is an inertial frame in which Maxwell's equations are valid attached to (a) the Earth, or (b) to a freely falling particle C (see Figure 4), or (c) to something else?

To find the answer, consider the spacetime diagrams of the experiment based on assumption (a) – Figure 5a – or on assumption (b) – Figure 5b.

Fig. 5. Spacetime diagrams of the terrestrial redshift measurement.

In case (a), A and B are parallel, straight timelike lines, C is a hyperbola (uniformly accelerated motion in Minkowski spacetime), and the 4 velocities of emitter and absorber at emission e and absorption a are parallel. In case (b), A and B are hyperbolae with common null asymptotes (belonging to a rigid congruence of lines which are orbits under a timelike rotation), C is a timelike straight line, and the 4 velocities of emitter and absorber at e and a are not parallel.

It is a simple and useful exercise in special relativity to infer from these diagrams that in case (a) there should be no frequency shift between A and B, whereas in case (b) there should be a redshift

$$\frac{\Delta\lambda}{\lambda} = \frac{gl}{c^2} + \text{much smaller terms}. \tag{1.41}$$

Experiment clearly decides in favour of (b); Equation (1.41) is verified with an accuracy of 1% [6]. We may consider this experiment as an *electromagnetic measurement of* g, the acceleration of a local inertial frame relatively to an Earth-attached frame; and we may state the result as showing that *local inertial frames are identical with the frames in which Maxwell's equations are valid locally.*

The preceding analysis assumes that the nuclei at A and B may be considered as ideal clocks measuring $\int d\tau$, in spite of their acceleration g (in case(b)). That assumption is justified because of the smallness of the inertial forces due to g in comparison to those nuclear forces responsible for the level structure of the nuclei, and can also be inferred rather directly from other experiments (see [6]).

Equation (1.41) can also be obtained from an energy argument ([11], [12]) which is, however, less compelling since it depends on additional, somewhat doubtful assumptions.

The above conclusion regarding preferred frames extends the result that there are no exact, global inertial frames, but only approximate, local inertial frames, from the domain of mechanics to electrodynamics, and it suggests the further generalisation known as the *strong principle of equivalence* (compare [12], pp. 4–5): There exist in the neighbourhood of each event e preferred coordinates, called inertial at e. For each class of physical phenomena except gravity (mechanics, electrodynamics, atomic physics ...) a set of basic, local laws can be selected which, if expressed in terms of inertial coordinates and evaluated at e, take on some standard form, including standard values of constants, independent of the spacetime location and the gravitational tidal field. (The local inertial coordinates are approximately realisable by freely falling, nonrotating, small laboratories, as will be discussed in detail in Subsection 2.10.)

If this principle, or only that part of it which refers to mechanical and electrodynamical phenomena, is accepted, special relativity with its Minkowskian spacetime is recognized to be a *local approximation* valid only as long as gravitational field gradients (tidal forces) are negligible. At the same time this local approximation is elevated to a *universally valid* one, and the strong principle of equivalence implies that nongravitational, local physical laws, if known in the special-relativistic, gravity-free approximation, can be carried over to the general case without change, if properly formulated. Herein lies the heuristic value of the principle. (Moreover, the principle may be considered as a characterization (definition) of gravity: It is that structure of spacetime [field] for which the strong principle of equivalence is meaningful and valid–provided there is some such structure.)

The conclusion that special relativity theory is not exactly valid if inhomogeneous gravitational fields are taken into account is further supported by the observation that electromagnetic waves are deflected by the gravitational field of the Sun (for references, see [6] and Brill's lectures, this volume, p. 136). This fact shows, quite independently of the amount of the deflection, that *a Minkowskian null cone structure cannot account for the propagation of electromagnetic waves in gravitational fields; gravity is associated with a deformation of the field of null cones.* This statement complements our earlier conclusion about the deformation of the system of timelike straight lines by gravitational fields.

In view of the spectacular successes of the special theory of relativity (the Poincaré group) in microscopic physics one may still ask whether it is possible to maintain a flat, Minkowskian spacetime metric η, if one weakens the physical interpretation of this metric. Since it is not compatible with the redshift measurements to assume that η determines nuclear proper times (this follows from a slight modification of the reasoning presented in connection with Figure 5a, see Schild [11], pp. 30–31); and it is not compatible with the light deflection measurements to assume that the null cones associated with η describe the propagation of light; and since the interpretation of the timelike geodesis of η is beset with the difficulties which have been discussed extensively in Subsection 1.2, there does not seem to be any direct interpretation of η which is compatible with experiments. Therefore, either one introduces η as a purely abstract, mathematical quantity without any direct physical interpretation, or one insists that the spacetime metric should have an interpretation similar to the one which motivated the introduction of η in the first place, and accordingly abandons η in favour of a more general, curved metric.

So far we followed the second plan due to Einstein which seems both more convincing and simpler than the first plan and which will also be adopted in the sequel. As to the first, so called *Poincaré-invariant or field theoretic approach*, I should like to mention only that it ultimately also leads to the conclusion that the physically relevant, measurable metric which determines the propagation of fields is (pseudo-) Riemannian, and that the flat metric is not only not measurable, but not even uniquely related to measurable quantities, and is therefore best discarded; similarly, the Poincaré group loses its meaning as a symmetry group. (For this approach, see [13], the references given there, and [14], [15].)

The preceding line of thought suggests that a theory of spacetime and gravitation should be such that

(a) it agrees locally with special relativity;

(b) it admits a class of preferred, relatively accelerated world lines representing free fall;

(c) it admits the definition of a tidal field tensor related to the source density of gravity by a law similar to Equation (1.23) or (1.26);

(d) it is capable of describing light deflection.

To satisfy (a) one will assume the existence of a semi-Riemannian metric g_{ab}. In order to satisfy (b) and (c) our previous analysis of the Newtonian theory suggests that we should postulate a nonintegrable, symmetric linear connection representing the gravitational-inertial field. However, in contrast to the analogous step at the nonrelativistic level, we now cannot simply leave the Minkowski metric η_{ab} unchanged and just change the connection, for in order that the strong principle of equivalence, which may be considered as a precise and strong version of requirement (a), can be meaningfully formulated in the new theory, there should exist, in some neighbourhood of each event e, coordinates x^a such that, at e,

$$g_{ab} = \eta_{ab}, \quad g_{ab,c} = 0, \quad \Gamma^a_{bc} = 0; \tag{1.42}$$

these x^a then are called *locally inertial* at e. (1.42) implies, however, by a well known theorem of differential geometry, that Γ^a_{bc} must be the (unique) *Levi-Civita connection* associated with g_{ab}. But then $g_{ab}=\eta_{ab}$ (globally) contradicts (b); the metric *has* to be a *curved* one. This reasoning shows that the need for using a non-Minkowskian metric in the theory of gravitation (which is still not accepted by some physicists) arises precisely from the requirement that special relativity be valid to a good approximation locally everywhere even in the presence of gravitation, provided one accepts that gravity is to be represented by a nonintegrable connection, for which there are (as we tried to show) very good, and not at all mysterious reasons already at the nonrelativistic level.

We therefore adopt the following *first main assumption of general relativity theory*: Spacetime is a smooth 4-manifold with a semi-Riemannian metric g_{ab} of signature $(+ + + -)$. Null geodesics represent light rays, timelike geodesics represent the world lines of freely falling particles, and the arc length along timelike lines (arbitrarily moving particles) measures the time shown by a standard clock.

This assumption satisfies indeed the requirements (a), (b),; it incorporates the empirically supported structures of the Newtonian gravity theory (nonintegrable connection, curvature) and of special relativity (light cones, Minkowskian inner product of 4-vectors) and discards the ill-founded, too special structures (absolute time, integrable connection) of those earlier theories; moreover, local inertial coordinates exist then everywhere, so that the strong principle of equivalence is applicable to connect the theory with familiar local physics. These claims will be explained in more detail in later parts of these lectures.

Applying the strong principle of equivalence in particular to the law (1.40) of special relativity, one is led to assume that in a gravitational field, the stress energy momentum tensor T^{ab} satisfies

$$T^{ab}{}_{;b} = 0, \tag{1.43}$$

for this is the covariant form of that equation which, with respect to inertial coordinates, reduces to (1.40). A less formal reason for the adoption of this *response equation for matter* will be given in Section 3.6.

In order to obtain a relation between the metric (≡gravitational potential) field and its sources which replaces Poisson's equation we recall that, at least in the limit of weak, quasistationary fields we must preserve the relation (1.26) between density ϱ and the volume ΔV of a small cloud of freely falling test particles. Now, according to our first main assumption and the *equation of geodesic deviation* (to be proven in Subsection 2.10) we have

$$\frac{D^2 V^a}{d\tau^2} = R^a{}_{bcd} U^b U^c V^d \tag{1.44}$$

for the relative acceleration $D^2 V^a/d\tau^2$ between two freely falling test particles with relative position vector V^a (see Figure 6); U^a is the 4-velocity of one of the two particles.

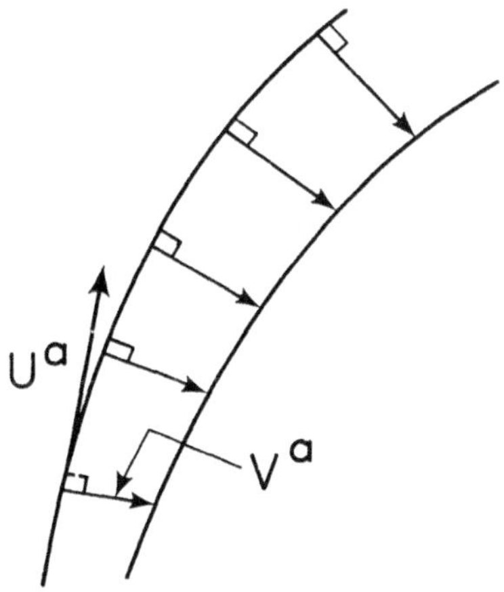

Fig. 6. Geodesic deviation.

Hence, by analogy with Newtonian theory, the tensor $R^a{}_{bcd} U^b U^c$ which transforms V^d into $D^2 V^a/d\tau^2$ is to be identified as the *tidal field* relative to an observer with 4-velocity U^a. Its trace, $-R_{ab}U^aU^b$, measures the change $\Delta \ddot{V}$ of volume of a cloud of test particles in strict analogy to the Newtonian case considered in Subsection 1.2. On the other hand we infer from Equation (1.39) and the equivalence of mass and energy that the density of the field-producing matter measured by the observer considered is given by $T_{ab}U^aU^b$. A second quantity which for a weak field is practically equal to the density is $-T = -T^a{}_a = \varrho_U - 3P_U$ (see Equation (1.38)). (In the solar system, $p \lesssim 10^{-6}\varrho$.) Hence, with an arbitrary constant λ, the quantity $(\lambda T_{ab} + (1-\lambda) g_{ab}T)U^aU^b$ corresponds very nearly to the Newtonian density; and this is the most general quantity with this property that is available in relativity. We are thus led to put

$$-R_{ab}U^aU^b \approx -4\pi(\lambda T_{ab} + (1-\lambda)\, g_{ab}T)\, U^aU^b$$

under the conditions stated. The simplest way to satisfy this approximate equation is to postulate that, for some constant λ and for all observers U^a, the above relation holds exactly, i.e. to put

$$R_{ab} = 4\pi(\lambda T_{ab} + (1 - \lambda)\, g_{ab} T). \tag{1.45}$$

Consequently, taking traces,

$$R = 4\pi(4 - 3\lambda)\, T.$$

If the two preceding equations are combined with the contracted Bianchi identity

$$R^{ab}{}_{;b} = \tfrac{1}{2} R^{,a}$$

and the response law (1.43), one obtains

$$4\pi(1 - \lambda)\, T_{,a} = 4\pi(4 - 3\lambda)\, T_{,a},$$

i.e., $(\lambda - 2) T_{,a} = 0$. Since in general $T_{,a} \neq 0$, we get $\lambda = 2$, and if this value is put into (1.45) and the Einstein tensor $G^{ab} = R^{ab} - \frac{1}{2} g^{ab} R$ is used, there results

$$G^{ab} = 8\pi T^{ab}, \tag{1.46}$$

Einstein's (and Hilbert's) fundamental gravitational field equation. We therefore (tentatively) adopt the *Second main assumption of general relativity theory*: The matter tensor T^{ab} (of all the matter and fields present) is related to the metric of spacetime through the field Equation (1.46).

With this assumption, also the requirements (c) and (d) laid down above are satisfied, and (1.43) is implied by (1.46).

No compelling justification of the assumption which led to (1.45) was given, and although we have tried to make the adoption of a pseudoriemannian spacetime plausible, other possibilities are conceivable. The ultimate justification of the two main assumptions which, together with the strong principle of equivalence, form the core of Einstein's theory of gravity, lies in the empirical verification of the consequences derivable from them. A further theoretical analysis is, however, also possible, and one such analysis will be presented in the next section.

2. A Constructive-Axiomatic Approach to General Relativity Theory

2.1. Introduction

Since Einstein's theory is presently considered the most adequate model of spacetime and gravitation of which the Newtonian theory and the special theory of relativity are only approximations valid under particular circumstances, it is reasonable to abandon the (more or less) historical approach and to attempt constructing it *ab initio*. Such an approach will be described in this section. Its main purpose is to eludicate from a different point of view and in a systematic manner, which facts and mathematical idealizations of facts are sufficient to support the edifice of general relativity theory. In particular, the following approach enables one to isolate certain relatively independent substructures which are contained in Einstein's theory, the differential-topological, conformal, projective, affine and metric structures, which are related to different observable phenomena or to different relations inferred from such phenomena. These substructures are present also in theories different from Einstein's and might also be considered as building blocks for further developments needed when Einstein's theory has to be modified.

The present section is based largely on recent work done in collaboration between F. A. E. Pirani, A. Schild and myself (ref. [16], quoted in the remainder as EPS); it is closely related in spirit and partly inspired by papers or remarks due to Castagnino [17], Geroch [18], Kundt and Hoffmann [19], Kronheimer and Penrose [20], Marzke and Wheeler [21], Reichenbach [22], Synge [23], Trautman [24], Weyl [7], Woodhouse [25].

2.2. Light rays and freely falling test particles as indicators of the geometry of spacetime

In the following approach, light rays and freely falling, spherical, nonrotating neutral test particles are considered as the tools for finding the geometric structure of spacetime. They play the role of the test charges in the elementary, inductive treatment of electrodynamics.

Examples of test particles are a stone falling near the surface of the Earth, an artificial satellite orbiting the Sun, or an atom moving quasi-classically between collisions in a dilute gas. Light rays may be wave packets of radar signals, γ-quanta in a redshift experiment, or something similar. It is assumed that their motion can be described adequately by a worldline.

As primitive concepts we use *event*, *light ray*, *particle* and *freely falling particle*. (One could do without the latter distinction, but it is useful to isolate the assumptions about free fall from those where arbitrarily moving particles – rockets, e.g. – are sufficient.) Accordingly, we assume that there are four sets

$$M = \{p, q, \dots\}, \quad \mathscr{L} = \{L, N, \dots\},$$
$$\mathscr{P} = \{P, Q, \dots\}, \quad \mathscr{P}_f \subset \mathscr{P};$$

the set M of events, called spacetime, and the sets $\mathscr{L}$, $\mathscr{P}$, $\mathscr{P}_f$ of light rays, particles and freely falling particles, respectively. Here and in the sequel 'particle' stands for 'worldline of particle'. The members of $\mathscr{L}$ and $\mathscr{P}$ are subsets of M.

To begin with, M is a *mere set without any structure*. It is by physically motivated assumptions about $\mathscr{L}$, $\mathscr{P}$, and $\mathscr{P}_f$ that a structure will be given to M. These assumptions refer to (idealized) experiments each involving light rays and particles in an otherwise empty region of spacetime. Hence the geometrical structure is obtained only for such empty regions, and it is an additional, hypothetical extrapolation to assign such a structure to all regions of spacetime, even those occupied by bulk matter of high density. Having pointed out the problematic nature of this extrapolation we shall nevertheless accept it without further examination, as is usual, for want of any available analysis.

Granted one accepts 'light ray' and 'particle' as primitive physical concepts one may ask how one can recognize 'freely falling' particles as such. A precise criterion does not seem to be known. However, most (macroscopic) particles in (nearly) vacuous regions behave according to the weak principle of equivalence, i.e. have the same orbits under the same initial conditions, and only a few particles deviate from the majority motion. To begin with, one may therefore 'operationally' define free fall as the *majority motion*, and then successively sharpen this definition in the course of the development of the theory, when criteria for forces etc. become available. (This corresponds roughly to the historical development.) What matters is that *some* universally available objects can be exhibited in reality which do behave according to the axioms about free fall (Subsection 2.4), and that the corresponding motions prove to be useful as standard motions of dynamics.

Arguments in favour of light rays and freely falling particles rather than standard clocks as primitive concepts have been given in EPS.

All axioms in Subsections 2.3–2.7 are *local*; global restrictions will be introduced only in Subsection 2.8.

2.3. Differential topology and conformal structure of M

A particle (idealized observer) may be thought of as carrying a smoothly running, but otherwise arbitrary (non-metric) clock which serves to parametrize its worldline. We thus postulate the axiom:

P. Each particle has the structure of an oriented, smooth (C^3) manifold diffeomorphic to $\mathbb{R}$.

The orientation corresponds to the *chronological order* of events contained in the history of the particle.*

The next aim is to introduce *coordinates* on M. One may think of particles as representing *observers*. Events may be observed by means of light signals reaching observers from those events, and they may be parametrized in terms of the clock readings associated with the reception events. This procedure works and is unambiguous due to the empirically well established independence of the propagation of light from the state of motion of the source [6, 9] as well as wavelength [9, 26], intensity and polarisation ('universality of light propagation').

Assume that there is a pair of particles P, P' and a subset U of M such that each event e in U can be connected to P as well as to P' by precisely two light rays (see Figure 7).

With help of (time) parameters t, t' on P, P', respectively, one can then associate numbers u, v, u', v' with each $e \in U$ (e.g., such that $u<v$, $u'<v'$). Assume that the mapping $x: U \to \mathbb{R}^4$ thus obtained is one-to-one and that the range $x(U)$ is open in $\mathbb{R}^4$. Then we shall call this mapping a *radar coordinate system* with domain U, based on P, P'. (In Minkowski space such coordinates can indeed be introduced, if P, P' are

* It is possible to analyse the empirical motivation of this axiom in more detail. See Reichenbach [22], § 6.

two non-coplanar free particles; they are analytically related to inertial coordinates in the domain complementary to $P \cup P'$.) The choice of t, t' is irrelevant; a change $t \to \bar{t}$, e.g., only induces a trivial, smooth coordinate transformation according to axiom *P*.

In spacetimes with compact space sections (such as the Einstein universe of 1917) radar coordinate systems need not exist, since there might always be more than two light rays connecting an event *e* with a particle *P*, some travelling round the space

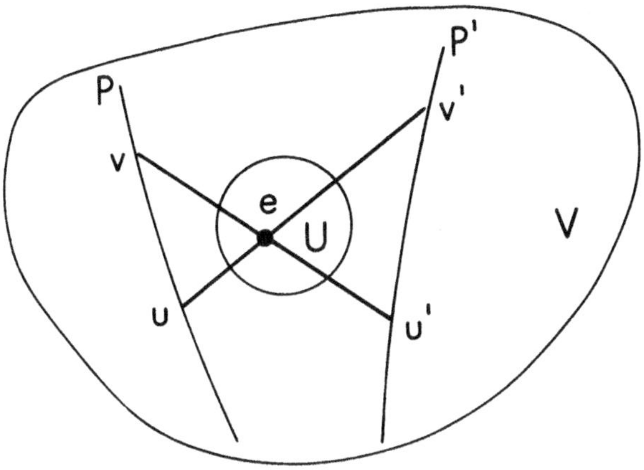

Fig. 7. Radar coordinates.

several times. In order not to exclude such spacetimes, we shall speak of a radar coordinate system x relative to V, based on P, P', with domain U, if V is a subset containing U and if x is a radar coordinate system for the sub-spacetime formed by V and the restrictions of all light rays and particles of M to V. With this precaution, we require axiom

L_1. There exists a family $F = \{(x_\alpha, U_\alpha, P_\alpha, P'_\alpha; V_\alpha)\}$ of quintuples $(x_\alpha, U_\alpha, P_\alpha, P'_\alpha; V_\alpha)$ with $U_\alpha \subset V_\alpha \subset M$, $P_\alpha, P'_\alpha \in \mathscr{P}$ such that

(a) x_α is a radar coordinate system with domain U_α, based on P_α, P'_α, relative to V_α,

(b) $M = \cup_\alpha U_\alpha$, i.e., $\{U_\alpha\}$ covers M,

(c) for each pair (α, β), x_α, x_β are smoothly (C^3) related. Any radar coordinate system $(x, U, P, P'; V)$ for M is smoothly related to the x_α's.

This axiom expresses that M can be given the structure of a *differentiable* (C^3) *manifold of dimension 4* on the basis of the sets $\mathscr{L}$, $\mathscr{P}$, and that an *atlas* can be constructed whose *local charts* are radar coordinate systems as illustrated in Figure 7. (For the teminology, see Schmidt's lectures.) This differentiable structure and the associated *topology*, in which the sets U_α, U are open, will be used henceforth. (Of course we do not have to use radar coordinates; any coordinate system smoothly connected with a radar coordinate system is also permissible.)

From the physical point of view this axiom is essentially a statement about light propagation, for it says (see Figure 7) that each event can 'see' and 'be seen' from nearby particles in a unique way. The four dimensionality of M appears as a property of the 'haystack' of paths contained in $\mathscr{L}$ and $\mathscr{P}$.

If P, P' are visualized as neighbouring radar emitters and receivers, u and v determine essentially a time and a distance of e, and u', v' furnish the additional information concerning the direction of e relative to the base P, P'. No reference to rigid bodies (theodolites) or 'metric' clocks is necessary for localising events by means of coordinates.

Having thus assigned a *manifold structure* to M we have at our disposal all the mathematical machinery for manifolds (see Schmidt's lectures, this volume, p. 289). In particular, let M_e denote the *tangent space* to M at e, and let L_e be the set of *light directions* at e. (A direction d is a one-dimensional subspace of M_e. Each nonzero vector $A \in M_e$ determines a direction d; $A, B \in M_e$ determine the same direction precisely if $B = \lambda A$ with $\lambda \neq 0$).

The next two axioms are based on the same empirical facts as L_1. They specify more precisely the structure of $\mathscr{L}$ near a given event e or a particle through e.

L_2. Any event e has a neighbourhood V such that an event $p \in V$ can be connected, within V, to a particle P by at most two light rays. Moreover, given such a neighbourhood and a particle P through e, there is another neighbourhood $U \subset V$ such that any event $p \in U$, $p \notin P$, can be connected, within V, to the particle P by precisely two light rays, which intersect P in distinct events e_1, e_2. If t is a coordinate on $P \cap V$ with $t(e) = 0$, then

$$g : p \to -t(e_1)\, t(e_2) \tag{2.1}$$

is a smooth (C^3) function on U. (See Figure 8)

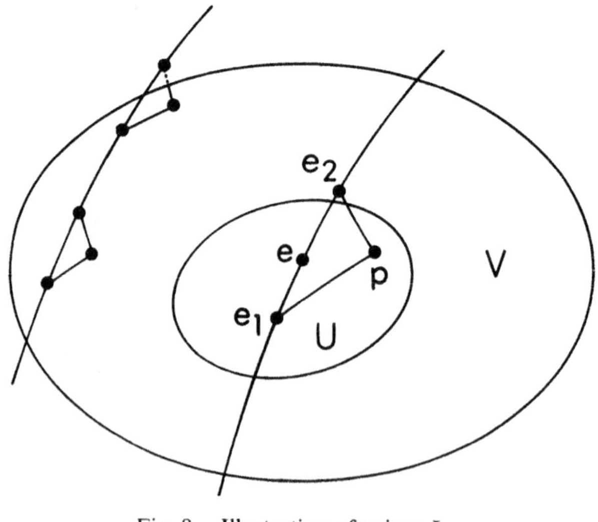

Fig. 8. Illustration of axiom L_2.

L_3. At any event e, the set L_e of light directions separates $D_e - L_e$ (the complement of L_e in D_e) into two connected components. The nonzero vectors tangent to L_e also fall into two connected components.

The last axiom is intended to express the separation of non-light-like directions into timelike ones and spacelike ones and the possibility of distinguishing between future-pointing and past-pointing vectors.

Finally we require a regularity condition:

PL. Particles and light rays are smooth paths.

(The smoothness assumptions contained in axioms L_1 and PL tie together the differential structures, or 'times' of the various particles postulated in axiom P. Assumptions of this kind are mathematically convenient, so far seemingly even indispensible idealizations inherent in all spacetime theories (kinematics) used, although it must be admitted that they are physically meaningless. It would be nice if it could be shown that they are 'harmless' and 'merely convenient' exaggerations added to a more realistic, less artificial structure.)

These five axioms imply the following

THEOREM 2.1. On M there exists a smooth (C^1), pseudoriemannian metric g_{ab} of signature $+++-$, unique up to a smooth, positive scalar factor, such that at each event e the vectors k^a tangent to light rays are characterized by

$$g_{ab}k^a k^b = 0 . \qquad (2.2)$$

For the details of the proof we refer to EPS, pp. 72–74. The idea is to show first that the function g defined in (2.1), whose vanishing tells that p can be connected to e by a light ray, has a vanishing gradient at e, $g_{,a}(e)=0$. (Otherwise, $g=0$ would define a smooth hypersurface near e, which contradicts axiom L_3.) Hence, $g_{ab}=g_{,ab}(e)$ is a tensor at e. Since $g(x^a(s))=0$ on any light ray $x^a(s)$ passing through e, two differentiations with respect to s at e give (2.2) for the tangent vector of the light ray. That Equation (2.2), in fact, characterizes the set of all light vectors and g_{ab} has the correct signature then follows from L_3. The smoothness and uniqueness up to a factor of g_{ab} follow from PL and Equation (2.2).

Theorem 2.1 says that *light propagation determines a conformal structure on M*, i.e. an equivalence class $\{\lambda g_{ab} \mid \lambda > 0\}$ of locally defined metrics or, geometrically speaking, a *field of null cones*. On its basis one can distinguish between timelike, spacelike and null vectors and paths, one can define orthogonality of vectors, and one can define the the *relative speed*

$$v = (1 - \gamma^{-2})^{1/2}, \quad \gamma = \frac{-g_{ab}u^a v^b}{\{g_{ab}u^a u^b g_{cd}v^c v^d\}^{1/2}}$$

between two motions with timelike, future-pointing tangents $u, v \in M_e$, as in special relativity. Quite generally, it is clear that all relations referring to a single tangent space of M can be formulated as in special relativity on the basis of a conformal structure;

in *this* sense the local validity of special relativity is already guaranteed by a conformal structure.

For a further development of conformal geometry, in particular, a geometrical construction of the Weyl conformal curvature tensor, I refer to Pirani and Schild [27]. We need two definitions and one easily established lemma from conformal geometry: A hypersurface is called a *null hypersurface* if it is tangent to the null cone at each of its points. A null path which is contained in some null hypersurface is called a *null geodesic*. Given a null vector k^a at a point e, there exists precisely one null geodesic through e with direction k^a.

Quantities which characterize the conformal structure of M uniquely relative to a coordinate system (x^a) are obtained by putting $\mathfrak{g}_{ab} \propto g_{ab}$, $\det(\mathfrak{g}_{ab}) = -1$. They form the components of a tensor density of weight $-\frac{1}{2}$. (see [28], and [29], Chap. 16.) Defining $\mathfrak{g}^{ab}$ by

$$\mathfrak{g}^{ab}\,\mathfrak{g}_{bc} = \delta^a_c \tag{2.3}$$

and setting

$$K^a_{bc} = \tfrac{1}{2}\mathfrak{g}^{ad}\left(\mathfrak{g}_{db,c} + \mathfrak{g}_{dc,b} - \mathfrak{g}_{bc,d}\right) \tag{2.4}$$

we can write the equation for null geodesics in the form

$$\frac{\mathrm{d}x^a}{\mathrm{d}s} = k^a, \quad \mathfrak{g}_{ab}k^a k^b = 0, \quad \frac{\mathrm{d}k^a}{\mathrm{d}s} + K^a_{bc}k^b k^c = \lambda k^a. \tag{2.5}$$

The $\mathfrak{g}_{ab}$ can be measured, once a coordinate system (e.g., a radar coordinate system) has been defined, at least in principle, by sending nine light rays with linearly independent 'squared tangents' $k^a k^b$ through e and computing the $\mathfrak{g}_{ab}$ from

$$\mathfrak{g}_{ab}k^a k^b = 0, \quad \det(\mathfrak{g}_{ab}) = -1. \tag{2.6}$$

Using more than nine light rays one can, again in principle, check the axioms stated above.

So far, we have not shown that light rays are null geodesics; the pointwise construction of $\mathfrak{g}_{ab}$ only ensures that they are null paths. We can, however, make a statement about the set $v_e = \{p \mid g(p) = 0, p \in U\}$ of events near e connectible to e by light rays. Since g is a C^3 function near e and, at e, $g = 0$, $g_{,a} = 0$, and $(g_{,ab})$ is a nondegenerate matrix, one can apply to g the Morse-Kuiper theorem [30] according to which there exists a coordinate system (y^a), C^2-related to radar coordinates (x^a) in a neighbourhood $U' \subset U$ of e, and C^3-related to (x^a) in $U' - \{e\}$, such that in U', $g(x^a) = g_{ab}(0)y^a y^b$ and $y^a(e) = 0$. Because of the signature of g_{ab} we can even arrange that $g(x^a) = \eta_{ab}y^a y^b$, where (η_{ab}) is the standard Minkowski matrix. That is, *in suitable coordinates* (y^a) *the set* v_e *'looks like' a light cone in flat spacetime*. In particular, v_e is a smooth hypersurface except at its vertex e. Thence we shall call v_e the *light cone* of e.

Maxwell's equations can be formulated in the conformal spacetime, provided an electromagnetic field is introduced as a skew-symmetric covariant tensor field F_{ab}

(a 2-form), and the current density $\mathfrak{J}^a$ is regarded as a vector density of weight 1. They are

$$F_{[ab,c]} = 0\,, \quad (\mathfrak{g}^{ac}\mathfrak{g}^{bd}F_{cd})_{,b} = \mathfrak{J}^a\,. \tag{2.7}$$

One could use these equations in conjunction with the WKB-approximation to deduce that light rays are null geodesics, but we shall postpone this to Section 6 and treat this problem within the complete, metric framework of Einstein's theory.

Similarly to Maxwell's Equation (2.7) all Poincaré-invariant wave equations belonging to vanishing rest mass can be extended to a conformal spacetime (see [31] and the references given there).

Conformal spacetimes may be considered as a special class of *causal spaces*, i.e., sets with an order structure interpretable as chronological and causal orderings (see [20]).

These remarks as well as the important investigations on asymptotic properties of spacetimes by Penrose [32] indicate the basic role of conformal structure.

2.4. Projective structure

We now return to the question – discussed already at the Newtonian level – of how to introduce a class of preferred motions. Any 'species' of test particles (e.g., electrically charged particles with a definite specific charge e/m) defines a *path structure* (p. 5) on M. The problem here is to characterize the particular path structure defined by the class of neutral, spherically symmetrical, nonrotating test-particles, i.e. to give a mathematical formulation of the weak principle of equivalence within the setting of a general spacetime. The reason for choosing this class of motions as standard motions has already been discussed in Subsection 1.2, it is the *universality of free fall.*

A minimum requirement is

F_1. Given an event e and *a* timelike direction d at e, there exists one and only one freely falling particle through e with direction d.

'Timelike' refers to the conformal structure introduced in the previous subsection, which we shall constantly use henceforth. We do not have to decide whether there are also non-spacelike free falls, all we do require at this stage is that such motions exist for subluminal initial speeds.

Beyond F_1 one wishes to express that in a sufficiently small neighbourhood of any event e the set of all free fall worldlines is asymptotically indistinguishable from the set of free motions in a Minkowski space, i.e., from the set of straight lines of an ordinary projective 4-space, so that no intrinsic gravitational field strength at an event is definable; this then expresses the weak principle of equivalence. Since a worldline not passing through e does not intersect some neighbourhood of e, it appears that such an asymptotic requirement should involve only the free fall trajectories through e. Since in ordinary projective geometry straight lines through e are given, in suitable (inhomogeneous) coordinates, by $x^a(u) = x^a(e) + u\dot{x}^a(e)$, one is led to postulate the following 'infinitesimal version of the law of inertia':

F_2. At each event e, there is a (permissible) coordinatesystem $(\bar{x}^a)$ such that every freely falling particle through e has a parameter representation $\bar{x}^a(\bar{u})$ with

$$\left.\frac{\mathrm{d}^2\bar{x}^a}{\mathrm{d}\bar{u}^2}\right|_e = 0. \tag{2.8}$$

Such a coordinate system is called projective at e. (( ) $|_e$ means 'at e'.)

It is desirable to find equivalent, coordinate free versions of this axiom, perhaps in terms of the theory of sprays, as suggested by Pirani [33].

Using arbitrary coordinates and parameters, one obtains from (2.8)

$$\left.\begin{array}{l}\ddot{x}^a + \prod^a_{bc}\dot{x}^b\dot{x}^c = \lambda\dot{x}^a \\ \left(\dot{x}^a = \dfrac{\mathrm{d}x^a}{\mathrm{d}u}\ \text{etc.}\right)\end{array}\right\}. \tag{2.9}$$

Here the $\prod^a_{bc}$ are functions of the coordinates, uniquely determined relative to a coordinate-system by means of F_1 and F_2, provided one imposes the conditions

$$\prod^a_{[bc]} = \prod^a_{ba} = 0. \tag{2.10}$$

λ depends on the parametrisation. The $\prod^a_{bc}$ at e can be determined (in principle) once a coordinate system has been chosen, by measuring the coordinate velocities and accelerations $\dot{x}^a$, $\ddot{x}^a$ of several freely falling particles passing (nearly) through e, and computing the $\prod$'s from the Equations (2.10) and

$$\dot{x}^{[a}(\ddot{x}^{b]} + \prod^{b]}_{cd}\dot{x}^c\dot{x}^d) = 0.$$

By taking more than the minimal number of particles the validity of the assumptions F_1, F_2 could be checked.

The Equation (2.9) has the same form as the standard equation for geodesics of a symmetric linear connection, expressed with respect to an arbitrary, not necessarily affine, parameter. Here, however, the quantities $\prod^a_{bc}$ subject to (2.10) do not form the components of a linear connection; they transform differently, due to (2.10). Correspondingly the paths do not admit, on the basis of the assumptions introduced thus far, an intrinsically preferred parametrisation. (One can set $\lambda=0$ in (2.9) and thereby introduce, relatively to (x^a), a special parameter, but this condition is not preserved under coordinate transformations.) Rather, there exists locally a whole class of symmetric linear connections $\{\Gamma^a_{bc}\}$ each of whose geodesics coincide, as paths, with the free fall trajectories, but with different assignments of affine parameters; namely precisely those connections for which

$$\Gamma^a_{bc} = \prod^a_{bc} + 2\delta^a_{(b}q_{c)} \tag{2.11}$$

with some functions q_c. The difference tensor of any two such connections has the form

$$(\Gamma' - \Gamma)^a_{bc} = 2\delta^a_{(b}p_{c)} \tag{2.12}$$

with some covariant vector field p_c. In other words, axioms F_1 and F_2 tell us that $\mathscr{P}_f$, the set of free fall worldlines, determines and is determined (locally) by a class of projectively equivalent – Equation (2.12) – symmetric linear connections. Such a structure is called a *projective structure*, and the paths satisfying (2.9) are called its *geodesics* (or autoparallels):

THEOREM 2.2. The set $\mathscr{P}_f$ of freely falling particles defines on M a unique projective structure such that every such particle is a geodesic.

This statement may be regarded as a mathematical formulation of the weak principle of equivalence valid both in the (modified form of) Newtonian and the relativistic theory of spacetime and gravity.

At this stage we cannot yet characterize geometrically the full set $\mathscr{P}_f$; certainly not every geodesic of the projective structure represents a particle if we exclude (as we do here) tachyons. For this purpose we have to link more closely the conformal and projective structures which so far have been introduced (almost) independently.

2.5. COMPATIBILITY OF CONFORMAL AND PROJECTIVE STRUCTURES. WEYL'S GEOMETRY

Experience, particularly high energy experiments, suggests that the conformal and projective structures are closely related to each other. A massive particle ($m>0$), though always slower than a photon, can be made to chase a photon arbitrarily closely. This suggests the following axiom of compatibility. *C*. Each event e has a neighbourhood U such that an event $p \in U$, $p \neq e$, lies on a freely falling particle P through e if and only if p is contained in the (timelike) interior of the light cone v_e (see Figure 9).

On the basis of this axiom and the previous results one can prove:

THEOREM 2.3. The set $\mathscr{L}$ of all light rays is identical with the set of all conformal null geodesics.

This theorem implies that the light cone v_e is identical with the null cone of e determined by the conformal structure.

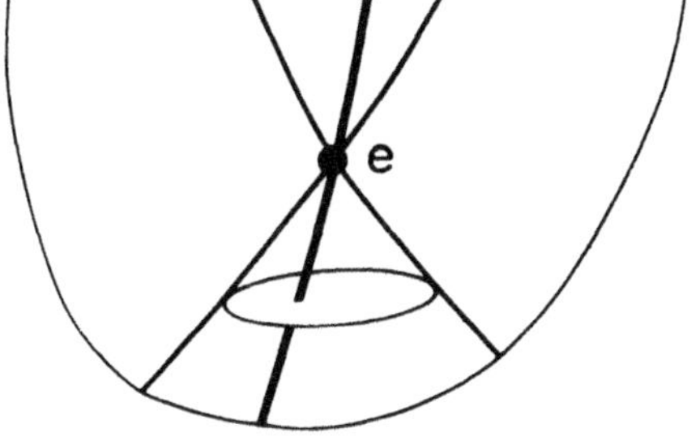

Fig. 9. Illustration of axiom C.

THEOREM 2.4. A projective geodesic which is timelike, spacelike or null, respectively, at one of its events, has this same 'causal character' everywhere.

That is, one may classify the projective geodesics into timelike, spacelike, and null geodesics.

THEOREM 2.5. The projective and conformal null geodesics are identical.

One may, consequently, speak about null geodesics without risk of confusion.

THEOREM 2.6. The set $\mathscr{P}_f$ of all freely falling particles is identical to the set of all timelike projective geodesics.

The proofs of these propositions will not be repeated here, they are contained in EPS.

Up to this point no preferred affine parameters which could serve as proper times have been introduced along particles, and no process of parallel transport of vectors has been defined. It is remarkable that such structures need not be postulated by means of additional axioms, but are already implied by the postulates laid down so far:

THEOREM 2.7. The class of projectively equivalent symmetric linear connections defining the projective structure of M contains precisely one member Γ^a_{bc} whose parallel transport preserves nullity (defined by means of the conformal structure) of vectors.

For reasons to be explained below we call this connection the *Weyl connection*. It can be computed as follows. Theorems 2.4 and 2.5 can be expressed formally in terms of the relation

$$\prod\nolimits^a_{bc} - K^a_{bc} = 5\mathfrak{g}^{ad}\mathfrak{g}_{bc}q_d - 2\delta^a_{(b}q_{c)} \tag{2.13}$$

between the projective coefficients $\prod^a_{bc}$ and the conformal quantities $\mathfrak{g}_{ab}$, K^a_{bc}; the functions q_a are thereby uniquely determined. (The q_a form the components of a 'geometric object' (see [28]), but not of a covariant vector.) Then

$$\Gamma^a_{bc} = \prod\nolimits^a_{bc} - 8\delta^a_{(b}q_{c)} \tag{2.14}$$

gives the components of the Weyl connection. A geometrical construction of the corresponding parallel transport and affine parameters will be given in the next subsection. The formal proof is given in EPS.

Summarising we can say that *light propagation and free fall define on spacetime a conformal structure and a symmetric linear connection such that light rays are* (conformal as well as affine) *null geodesics, freely falling particles are timelike* (affine) *geodesics, and parallel transport maps null vectors into null vectors* (and hence also preserves orthogonality).

It is easy to show that the geometry thus characterized is nothing but Weyl's geometry, introduced by Weyl [7] (in a slightly different, less geometrical way) in connection with the first unified field theory of gravity and electromagnetism. We can therefore also say that *light propagation and free fall assign a Weyl geometry to spacetime.*

Finally, we state the following result from EPS.

THEOREM 2.8. The differential, conformal and affine structures of M (i.e., the whole Weyl geometry) are uniquely determined by the sets $\mathscr{L}$, $\mathscr{P}_f$.

This theorem (which follows easily from a result of Hawking's) says that if two Weyl geometries have the same underlying *sets* M, $\mathscr{L}$, $\mathscr{P}_f$ then they are identical. i.e. they cannot have, e.g. inequivalent differential structures. This fact ultimately justifies considering M, $\mathscr{L}$, $\mathscr{P}_f$ as the primitive elements of spacetime.

2.6. GEOMETRICAL CONSTRUCTIONS, CLOCKS

The purpose of this subsection is to sketch how some of the properties of spacetime can be studied by means of geometrical constructions, rather than by computation; these constructions thus complement the considerations of the preceding three subsections. Since we shall work mainly with timelike and null geodesics and their intersections, the following figures represent thought experiments with freely falling particles and light rays; so we are really concerned with kinematics, not geometry in the ordinary sense.

Figure 10 recalls how one can test, by means of light rays, whether an 'infinitesimal' line element connecting events a and b is *orthogonal* to a particle P; this is nothing but a test for Einstein-simultaneity of a and b. Only a conformal structure is required for performing this test.

Figure 11 shows how a 'plane strip' containing a freely falling particle P can be constructed. One just has to draw a zig-zag-line consisting of geodesic segments intersecting P again and again. The whole zig-zag must be close to P. (Strictly speaking, a limiting procedure would be required, but in order to be intuitive and simple I shall be sloppy about that; this remark applies mutatis mutandis to all of these constructions.) Having the zig-zag one can 'fill in' further geodesics and thus obtain the strip, well defined near P. This construction can be used to *parallely displace* a tangential 2-plane element α along P to obtain, say, β; it is purely projective. (For more about projective differential geometry in this sense, see Pirani [33] and a forthcoming paper by Ehlers and Schild [34].)

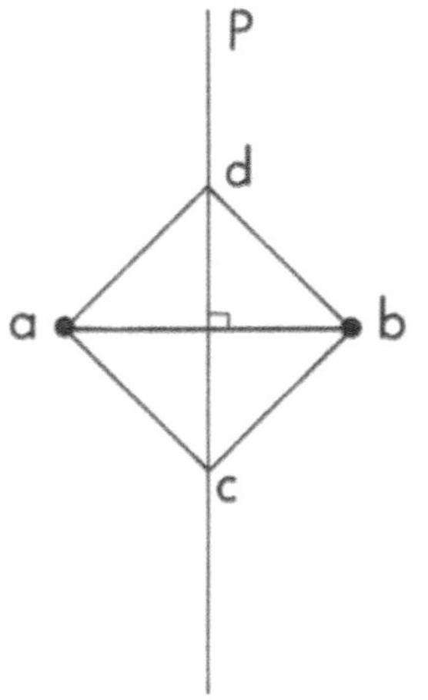

Fig. 10.

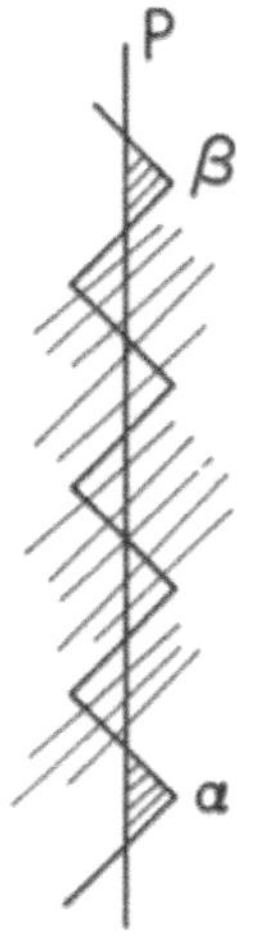

Fig. 11.

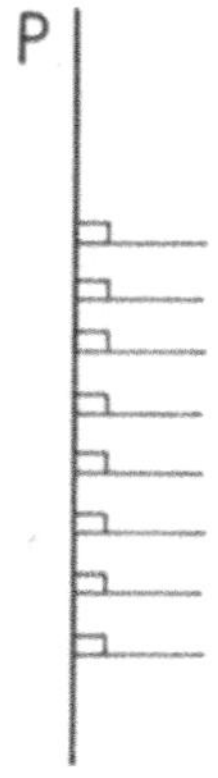

Fig. 12.

Figure 12 combines the two previous ones. It shows how one can construct a 'comb' with geodesic 'teeth' sticking out of a geodesic P, intersecting P orthogonally and contained, near P, in a strip. (Auxiliary lines from the previous constructions are not shown.)

Using a comb and constructing a path contained in the strip which intersects the teeth orthogonally, we obtain a path (particle) Q which is *parallel* to the geodesic P. (Figure 13) In general, Q will not be geodesic.

By means of a geodesic P, a parallel Q and a ladder of null geodesics (a bouncing photon), we obtain a *light clock*; its period T is defined as indicated in Figure 14.

The preceding constructions show how one can obtain a gravitational clock, i.e. how one can define geometrically a sequence of 'equidistant' points on a geodesic, by means of a projective and a conformal structure. It is now easy to construct a parallel displacement, too, by displacing the components tangential and orthogonal to P separately in the obvious way, using the preceding operations; see Figure 15.

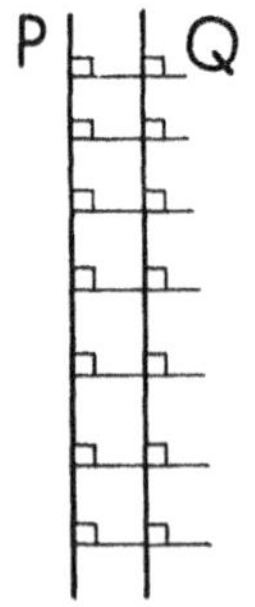

Fig. 13.

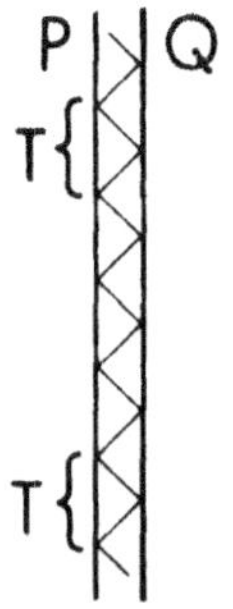

Fig. 14.

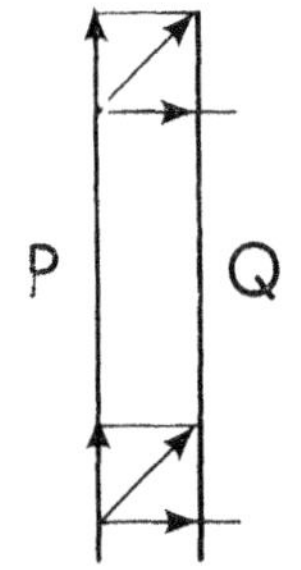

Fig. 15 ($Q \parallel P$).

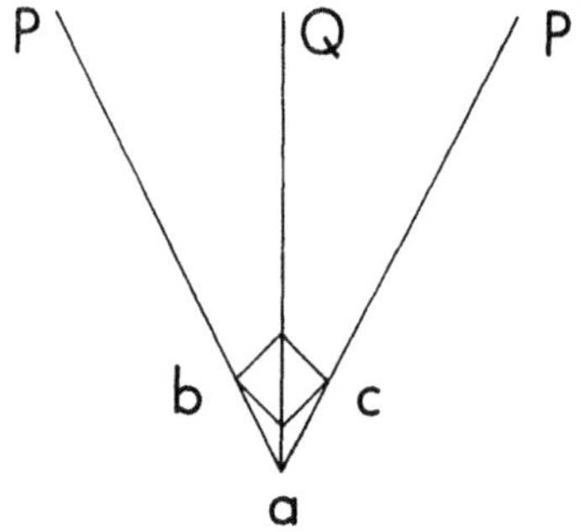

Fig. 16.

Figure 16 shows how one can test whether two segments *ab*, *ac* on intersecting geodesics P, P' are *congruent*; Q is an auxiliary geodesic through *a*.

The idea of using a light clock in setting up the relativistic chronometry is an old one due to Einstein. Einstein, however, used rigid bodies to construct such a clock, and one may wish to avoid that since (i) from the point of view of the present theory of the constitution of matter the concept of a (practically) rigid solid is not a simple, fundamental one, but requires for its justification from first principles many body quantum theory, which presupposes already much more than the elements of spacetime geometry; (ii) in a generic, general-relativistic spacetime there are no rigid congruences which could represent rigid bodies. That it is *possible* to proceed without rigid bodies and also without postulating standard clocks, in special as well as in general relativity, has already been shown by Weyl [7]. Ideal experiments to obtain gravitational standard clocks have been devised only much later (see Marzke and Wheeler [21], Kundt and Hoffmann [19]), but these still presupposed a Riemannian spacetime geometry. That one can set up such ideal clocks even on the basis of more primitive assumptions, and then use the behaviour of such clocks to analyse the conditions under which their functioning *can* be described by means of a pseudoriemannian geometry has been recognized only quite recently (Castagnino [17], EPS), and this type of analysis can probably be improved further, along the lines indicated here.

2.7. Riemannian spacetime; metric

The Weyl connection characterized in theorem 2.7 and computed in Equations (2.14), (2.13) from $\mathfrak{g}_{ab}$ and $\prod^a_{bc}$ gives rise to a *curvature tensor*

$$R^a{}_{bcd} = 2\left(\Gamma^a_{b[d,c]} + \Gamma^a_{e[c}\Gamma^e_{d]b}\right). \tag{2.15}$$

Since parallel transport in a Weyl spaces preserves nullity of vectors, the linear transformation induced in the tangent space M_e of some event e by parallel displacement around a loop at e consists of the product of a dilatation and a (homogeneous) Lorentz transformation. If this result is applied to an infinitesimal loop contained in a plane spanned by A^c and B^c, the well-known connection between curvature and such displacements (holonomy) tells us that $R^a{}_{bcd}\, A^c B^d$ must be an infinitesimal transformation of the indicated kind, whence $R^a{}_{bcd}$ decomposes according to

$$R^a{}_{bcd} = \hat{R}^a{}_{bcd} + \tfrac{1}{2}\delta^a_b F_{cd}, \tag{2.16}$$

where

$$\mathfrak{g}_{e(a}\hat{R}^e{}_{b)cd} = 0 \tag{2.17}$$

(i.e., $\hat{R}^a{}_{bcd}\, A^c B^d$ is an infinitesimal Lorentz transformation) and

$$F_{(ab)} = 0. \tag{2.18}$$

($\frac{1}{2}\delta^a_b F_{cd} A^c B^d$ is an infinitesimal dilatation.) $\hat{R}^a{}_{bcd}$ is Weyl's 'directional curvature' (Richtungskrümmung), and F_{ab} is his 'length-curvature' (Streckenkrümmung).

It is evident from the remarks preceding (2.16) that a pseudoriemannian metric g_{ab} which

(i) is compatible with the conformal structure $\mathfrak{g}_{ab}$ of the given Weyl space,

$$g_{ab} = W^2 \mathfrak{g}_{ab}, \quad W > 0, \tag{2.19}$$

and

(ii) whose Levi-Civita connection is identical to the Weyl connection,

$$\Gamma^a_{bc} = \tfrac{1}{2} g^{ad}\left(g_{db,c} + g_{dc,b} - g_{bc,d}\right), \tag{2.20}$$

exists (in a simply connected domain of M) if and only if $F_{ab} = 0$, and is then unique up to a constant scale factor. (Inserting (2.19) in (2.20), using the definition (2.4) of K^a_{bc} and comparing the result with (2.13), (2.14) shows that W is related to our earlier q_a by $(\log W)_{,a} = -5q_a$.) This last equation is *not* implied by the axioms stated in Subsections 2.3, 2.4, 2.5 which give no more than a Weyl geometry. On the basis of the stated axioms the transport of a time unit along a polygon of timelike geodesic segments (freely falling observers), which could be performed according to Figures 14 and 16, would in general be path-dependent – as indicated in Figure 17.

In the real world atoms (or nuclei or elementary particles) define units of time, and the well established identity of these submicroscopic systems and the constancy of the ratios of their masses, lifetimes and transition frequencies shows that these (concordant) *atomic units are integrably transported.* Hence if we make the hypothesis that the *gravitational standard time as defined by the Weyl geometry* (affine parameters of Γ^a_{bc}) *is identical to the atomic time*, then experience demands that we put

$$F_{ab} = 0\,. \tag{2.21}$$

Adding this as an additional *Riemannian axiom* we thus finally obtain *the standard spacetime structure of general relativity* as formulated already on p. 20.

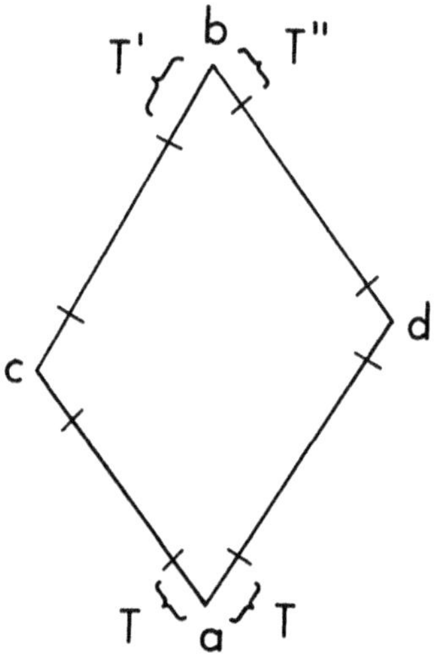

Fig. 17. Time transport in a Weyl space; $T' \neq T''$. Two congruent time intervals at a, parallely transported via c or d, respectively, to b, produce two incongruent intervals T', T'' at b. ('Second clock effect')

From the point of view of the axiomatics presented here the metric g_{ab} is composed of the more primitive elements $\mathfrak{g}_{ab}$ (conformal structure), Π^a_{bc} (projective structure), and W (see Equation (2.19)). W is a scalar density of weight $\frac{1}{4}$. Its nature is perhaps best exhibited by noting, from (2.19), that the *Riemannian volume element* of g_{ab} is given by

$$\eta = \sqrt{-g}\, d^4x = W^4 d^4x\,. \tag{2.22}$$

W could thus be called the *scale density* of spacetime, and the third ingredient of g_{ab} is recognised as equivalent to a *volume element* invariant with respect to Γ, the Weyl connection. (Incidentally, the Newtonian connection of Subsection 1.2 also admits such a volume element, simply because of $R^a{}_{acd}=0$). The Riemannian axiom could therefore also be formulated as the existence of such a Γ-invariant volume element (4-form).

It must be admitted that the motivation of (2.21) is not very satisfactory; it is an extraneous element of the theory. It seems that a deeper understanding of it could come only from a better analysis of the description of matter in a Weyl spacetime. It is quite possible that such an analysis (maybe in terms of possible wave equations for massive particles) would show that a 'reasonable' description of matter is possible only if the Weyl space is, in fact, Riemannian, and that W is somehow defined by the matter variables (action density?). However that may be, in the absence of a good argument it seems better to adopt (2.21) than to keep a quantity F_{ab} for which there is no clear physical reason; so henceforth we proceed on the assumption (2.21) and use standard Riemannian geometry. (Weyl used F_{ab} to describe the electromagnetic field, but that theory meets with serious objections and has generally been discarded.)

As a supplement to the constructions of Subsections 2.6 I add the following, well-known figure which defines the square $\mathbf{ab}^2$ of a 'small' displacement $\mathbf{ab}$, as given by the metric g_{ab}, in a geometrical way (Figure 18).

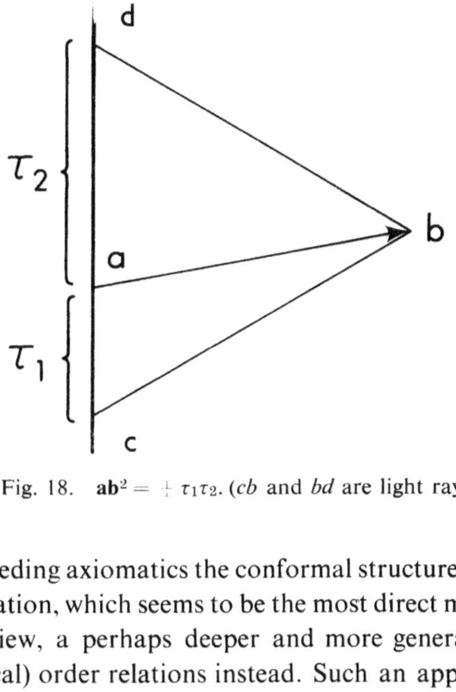

Fig. 18. $\mathbf{ab}^2 = \pm\, \tau_1\tau_2$. (*cb* and *bd* are light rays.)

Whereas in the preceding axiomatics the conformal structure of space time has been based on light propagation, which seems to be the most direct method from a phenomenological point of view, a perhaps deeper and more general approach would use causal (or chronological) order relations instead. Such an approach has indeed been elaborated by Woodhouse [25]. It is certainly attractive from a physical point of view to consider causal order as more fundamental than topology, and to base the topology on the causal order. This approach may even open the way toward spacetime models not based on the highly fictitious concept of a differentiable manifold. However, these ideas cannot be pursued here, and I refer to the literature (Robb [35], Alexandrov [36], Reichenbach [22] and, *in particular* Kronheimer and Penrose [20].)

A possibility to enrich the spacetime structure by adding a *torsion tensor* to the symmetric connection can also only be mentioned here (see Trautman [37] and the references given there.).

2.8. Orientations and Spinor Structure

From now on, the term *spacetime* will always mean a connected pseudoriemannian manifold (M, g_{ab}) often symbolised by M only.

The following remarks on orientation follow closely Geroch [38].

Consider in a spacetime M a loop L at $a \in M$ (such as *acbda* in Figure 17). If a (nonzero!) timelike vector A at a is carried around L in a continuous manner such that it always stays timelike, then on return to a the resulting vector A' determines a sign, $\text{sign}(-A \cdot A') = \varepsilon_t$. This sign is easily seen to depend (for a given metric of M, of course) on the loop L only, and it does not change if L is continuously deformed ($L \to \varepsilon_t(L)$ is, in fact, a representation of the homotopy group of M.) M is called *time orientable* if $\varepsilon_t = +1$ for all loops L. This property means that one can make a distinction between 'future' and 'past' light (half) cones continuously on M. There are then precisely two ways for making this distinction; each one is called a *time orientation* for M.

Similarly, one can take a spacelike, simple trivector $F^{abc} = A^{[a}B^{b}C^{c]}$ at a, carry it continuously around a loop L at a, preserving its spacelike, simple character. On return, one obtains (say) G^{abc}. G^{abc} can be deformed continuously at a such that, finally, $G^{abc} = \lambda F^{abc}$: let $\varepsilon_s = \text{sign}\,\lambda$. ε_s depends on (the homotopy class of) L only. If $\varepsilon_s = +1$ for all loops L, M is called *space orientable*. This means that one can distinguish on M, in a continuous manner, between 'right' and 'left' handed spacelike triads (A, B, C). Again two such assignments are then possible, and each of them is called a *space orientation* for M.

If M is both time and space orientable, we shall call it *totally orientable*; it then admits four total orientations.

It is evident that these orientation properties depend on the conformal structure of M only.

If M is totally orientable, then the underlying manifold is orientable in the topological sense.

If a theory of 'charged' particles is given on M, one can similarly introduce *charge orientability* (for electric, baryonic, ... charge).

Locally observed violations of T, P, or C symmetry (or combinations thereof) imply, if combined with the strong principle of equivalence, time, space or charge orientability of M, as can easily be verified from the definitions of these operations by imagining two sequences of overlapping coordinate domains in which the requisite experiments are performed, and which connect, via two different routes, two arbitrarily selected events. If the experimental results concerning weak and superweak interactions are accepted, it follows that our actual spacetime *is* totally orientable. At any rate we shall assume this in the sequel, choose one particular orientation, and then always work with oriented coordinate systems (x^a) and time and space oriented orthogonal tetrads (E_a) of vectors, with E_4 future directed and (E_λ) righthanded.

Closely related to the question of orientability is the question as to whether M admits globally the definition of spinor fields. It has been shown by Geroch [39] that such is the case, for noncompact M, if and only if M admits a global, continuous orthonorm tetrad. Thus existence of such a *spinor structure* implies total orientability. The existence of fermions in nature suggests that the real spacetime (as far as we permit ourselves to call such a mental construct 'real') does admit a spinor structure. One can, in fact, construct the whole pseudoriemannian structure of M by imposing a spinor structure on M as the basic geometric object; this powerful and elegant approach has been expounded by Penrose [32]. Unfortunately I can only mention this here, for lack of available 'spacetime'.

2.9. Einstein's field equation

The arguments used at the end of Subsection 1.4 to motivate Einstein's (or rather Einstein's and Hilbert's) field equation

$$G^{ab} = 8\pi T^{ab} \tag{2.23}$$

are objectionable on at least three grounds:

(a) The approximate equation preceding (1.45) was empirically well founded, by correspondence with the Newtonian theory, only for weak, quasistationary fields; its extrapolation to strong fields was hypothetical.

(b) The 4-velocity V^a in the equation just referred to was treated as arbitrary, whereas according to the Newtonian correspondence it should have been restricted to values corresponding to observers having a small velocity relative to the gravitating bodies generating the field.

(c) The extrapolation of the test-particle argument to the interiors of field sources is quite fictitious.

One can, of course, dismiss such objections as irrelevant in view of the success of the deductions based on Equation (2.23). It is more instructive and enlightening, however, to try to find answers to these objections or to *look for as many independent characterizations of Equation (2.23) as possible*, to see in which sense these equations are unique and what distinguishes the theory based on (2.23) from other possible theories. In this spirit, I should like to review some important theorems concerning (2.23).

In view of Poisson's Equation (1.23) and Maxwell's equations (written in terms of potentials), and since much more is known about second-order partial differential equations than about higher order equations, it seems reasonable to ask which gravitational field equations are possible if one assumes that they be of the form

$$V^{ab}(g_{cd}, g_{cd,e}, g_{cd,ef}) = T^{ab}, \tag{2.24}$$

where the left hand side represents a tensor valued local functional defined on the set of all Lorentzian metrics (in four dimensions) such that, with respect to any local coordinate system (x^a), the components of V^{ab} at x^a are functions of $g_{cd}, g_{cd,e}, g_{cd,ef}$, also evaluated at x^a. Let us assume in addition that the 'response equation'

$$T^{ab}{}_{;b} = 0 \tag{2.25}$$

of matter is a consequence of the field Equation (2.24) in the sense that the left hand side of (2.24) has vanishing covariant divergence (with respect to the second index) identically in $g_{ab}(x^c)$.

It has been shown by Lovelock [40] that *the assumptions stated above imply (2.24) to be of the form*

$$G^{ab} + \Lambda g^{ab} = 8\pi T^{ab} \tag{2.26}$$

(except for an additional constant multiplying G^{ab} which we have adjusted so that, for $\Lambda = 0$, the correct Newtonian limit results). Λ is a constant (with the dimension of an inverse area) called the cosmological constant.

Lovelock's theorem, the proof of which is rather lengthy and will therefore not be given here, shows that Einstein's equation, modified by the cosmological term, follows from much weaker assumptions than those made in Subsection 1.4.

In particular, it should be noticed that no symmetry assumption was made about V^{ab}, *the symmetry* $V^{[ab]} = 0$ *follows* from the stated assumptions. This means that as long as g_{ab} is considered as the complete gravitational potential (not to be supplemented by, e.g., a scalar field or a torsion tensor) and T^{ab} is regarded as the complete source satisfying (2.25), then the total matter tensor has to be symmetrical. Hence, if there were reasons for thinking that in particular situations the 'correct' total matter tensor is non symmetric – as some people claim (see, e.g., ref. [41]) – then one would have to change one of these assumptions.

A second point to be emphasized is that no assumption has been made as to whether the left-hand side of (2.24) is linear in any of its arguments. The result (2.26) shows that V^{ab} is linear in the second derivatives $g_{ab,cd}$, but nonlinear in $g_{ab,c}$ and g_{ab}. That is, the *characteristic nonlinearity, but quasilinearity of the gravitational field equation* (the self-interaction of the gravitational field) *follows* also from the stated assumptions. This nonlinearity is of fundamental significance for the following reason: Since (2.25) is implied by (2.26) (by construction), *the field equation* largely *determines the motion of the sources* – there is no room for additional 'equations of motion' (we shall come back to this point in Sections 6 and 7). Because of this circumstance *the nonlinearity of the field equation is necessary in order that the field can mediate an interaction between its sources.* (A linearly generated field like that of Maxwell can mediate interactions *only* if the field equations do *not* determine the motion of the sources, as follows immediately by use of the superposition principle.)

To substantiate the remark made above that (2.26) 'essentially' determines the motion of the sources – i.e., the spacetime configuration of the supports of the matter tensors representing, e.g., stars of planets – it is not necessary to refer to approximation methods such as that of Einstein, Infeld and Hoffmann. It is sufficient to point out that for a specified kind of source, e.g., *dust* given by $T^{ab} = \varrho u^a u^b$, the Equation (2.26) admits a well posed *Cauchy initial value problem* with 'histories' uniquely determined by initial data (see [42] or [43]), and the restriction indicated by the word 'essentially' refers to the fact that, depending on the type of matter considered, one has to add to (2.26) 'equations of state' or something similar to make the problem determinate, but *not*, in general, equations of motion like, say, the Navier-Stokes equation which would invalidate the above claim.

It is also clear that the objections raised at the beginning of this subsection do not apply to the approach contained in Lovelock's theorem which, incidentally, holds just for the dimension $n=4$, in contrast to an old, often quoted, but (for $n=4$) distinctly weaker theorem due to Weyl [7].

In view of the importance of variational principles and Euler-Lagrange-equations in mathematical physics the following theorem is of interest.

THEOREM 2.9. Einstein's field Equation (2.26) is equivalent to *Hilbert's action principle* (Hilbert, 1915)

$$\delta \int \left(\frac{1}{16\pi} R - \frac{1}{8\pi} \Lambda + L \right) \sqrt{-g}\, \mathrm{d}^4 x = 0, \tag{2.27}$$

provided a scalar L depending on the metric (and possibly its derivatives) and some matter variables and their derivatives exists such that

$$\frac{\delta(\sqrt{-g}L)}{\delta g_{ab}} = \tfrac{1}{2} T^{ab} \sqrt{-g}\,. \tag{2.28}$$

The Equation (2.26) is the only field equation of the form (2.24) which is derivable from an action principle in which the gravitational part of the Lagrangian density is a scalar differential invariant (of any finite differential order) of the metric.

The symbol δ in (2.2) denotes the *variational* (also called Euler-Lagrange) *derivative* of the calculus of variations.

The proof of the first part of this theorem consists of a straightforward calculation and need not be repeated here. I only wish to remark that the verification of the relevant equation

$$\frac{\delta(\sqrt{-g}\, R)}{\sqrt{-g}\, \delta g_{ab}} = -\, G^{ab} \tag{2.29}$$

is very simple if one uses the fact that *the difference of two linear connections*, and hence also $\delta \Gamma^a_{bc}$, *is a tensor* so that an equation involving it needs to be verified only in a locally inertial (see Equation (1.42)) coordinate system.

The second part of the theorem contains still *another characterization* of Einstein's Equation (2.26) in which *(2.25) is not assumed, but derived*. To prove it, one shows first that the variational derivative $\delta(\sqrt{-g}L)/\sqrt{-g}\delta g_{ab}$ associated with *any* scalar differential invariant of g_{ab} satisfies the relation (2.25). This proof is contained in many textbooks (see, e.g., [44] or [45]) and will therefore not be reproduced here. Therefore, field equations of the form (2.24) derivable from an action principle automatically satisfy the second assumption of Lovelock's theorem, hence the second part of theorem 2.9 follows from that theorem. (This result does not mean that $1/16\pi R$ is the only

Lagrange density, but that every other one gives the same field equation, and that is the important point.)

The foregoing considerations may create the impression that there are no 'reasonable' alternatives to Equation (2.26), as long as one considers g_{ab} as the only basic gravitational field variable and T^{ab} as the only source quantity. That is not quite true, however. For even if one accepts (2.25) as valid *for all physical sources*, it does not follow – at least I am not aware of a demonstration – that the functional V^{ab} in (2.24) satisfies $V^{ab}{}_{;b}=0$ *identically in* g_{ab}. Moreover, all the characterizations of (2.26) given here are based on the point of view that the field equation is to pick out of the set of *all* pseudoriemannian metrics g_{ab} those which are physically realisable. The first main assumption of general relativity theory as stated on p. 218 and motivated anew in this section is, however, also compatible with the proposal to first select some *subclass* from among the set of all metrics, and then to impose field equations which appear natural within that subclass. This is in fact done in some theories with stratified spacetimes (see [46]) for a discussion and for references), and there is no reason to exclude such theories a priori. However, it seems fair to say that within the (g_{ab}, T^{ab})-framework *the Einstein theory* is the simplest in the sense that it *maintains locally the highest degree of spacetime symmetry* (no universally preferred space sections or observers), and therefore it has the best chance not to contradict any ether drift experiments or the space-isotropy experiments of Hughes, Drever *et al.* (consult Brill's lectures, this volume, p. 127).

If one admits more field or source variables besides g_{ab} and T^{ab}, there arise of course new possibilities for field equations. Attempts to use all the now available experimental information to delineate the empirically permitted possibilities have been made in particular by Thorne and Will [46].

Concluding this section I wish to mention a few items which I do not treat explicitly for lack of time and energy, and make some additional comments for which there is no better place.

Besides the derivations of (2.26) based on correspondence with Newtonian theory of which I have given one variant, and the one based on Lovelock's theorem (which I consider as the most general and most convincing argument given so far, and which is an improvement of a reasoning due to Einstein and mathematically justified by Weyl), and the (strengthened version of) Hilbert's action principle, there is the derivation by Thirring, Feynman, Wyss, Deser (see [13]) which starts with a Poincaré-invariant mass 0, spin 2 field. The main importance of the latter is that it shows that there is no truly Poincaré-invariant, flat-space theory of gravity, but that in a certain formal way one may relate Einstein's theory to the ideas and formalism of those theories, so that for certain problems the techniques of flat space field theory can be exploited even for general relativity theory, just as essentially nonrelativistic techniques can sometimes be used in relativity.

Equation (2.26) shows that, from the point of view of the original Equation (2.23), $-(\Lambda/8\pi)\,g^{ab}$ appears formally as an additional matter tensor with energy density $\Lambda/8\pi$ and pressure $-\Lambda/8\pi$. Thus as long as $\Lambda/8\pi$ is clearly smaller than the average material energy density of a system the Λ-term can be ignored. Since this term disturbs the correspondence with Newtonian theory it has to be negligibly small in the solar system. In fact, present cosmological data limit Λ to the range [47]

$$|\Lambda| \lesssim 10^{-54}\,\mathrm{cm}^{-2}. \tag{2.30}$$

The ten Equations (2.26) can be split into four *constraint equations* imposing conditions on initial data and six *evolution equations* governing (in conjunction with corresponding equations for further matter or field variable) the dynamics of a system. This important fact shows up particularly clearly if one uses, instead of the g_{ab}, the so-called *Arnowitt-Deser-Misner variables* which bring the action principle (2.27) into a canonical form. For these matters I refer to Kuchar's lectures, this volume, p. 261).

Still other variables which make the *gravitational Lagrangian polynomial* are the *Peres variables* [48] $\mathfrak{g}_{ab}$, W which we used for geometrical reasons in Subsections 2.3 and 2.7.

A reformulation of (2.26) which elucidates particularly clearly its *hyperbolic, wave-equation character* and which is also useful for approximation methods is the *Landau-Lifshitz form* of (2.26) which will be mentioned in Subsection 7.2.

In Subsection 7.1 we shall mention still another equivalent of (2.26), the *Gilman-Sciama-Waylen integro-differential equation* which brings out the way in which the metric at an event can be considered as a superposition of contributions due to sources and (possibly) incoming fields (see Ellis and Sciama [49]).

2.10. Frames of Reference and (Ideal) Measurements in Curved Spacetime

In nonrelativistic spacetime we introduced Galilean coordinates (x^a) which have a simple operational significance in terms of time and distance measurements based on an arbitrarily moving reference body. In Minkowskian spacetime we considered a much more restricted class of coordinate systems only, the inertial coordinate systems, which have a similar chronometric interpretation, but require a freely moving, non-rotating reference body. In a general curved spacetime we introduced radar coordinates based on arbitrarily moving pairs of observers equipped with arbitrary, possibly 'bad' clocks; these coordinates do not have such a straightforward chronometric significance as those of the other theories.

As long as coordinates are used *only* as labels to distinguish events and to indicate their neighbourhood relations – and this is indeed their primary purpose – all these coordinates, and many others, are satisfactory and equally useful. In the Newtonian case, however, and only in that case, do the coordinates referred to above also have a close relation to the actual way in which astronomical events are located relatively to, say, an observatory on Earth which may be idealized as a rigid body in a noninertial rotational motion. In this subsection I wish to show how similar, quasi-Galilean coordinates can be introduced locally in a general, curved or flat, relativistic spacetime. In this way the local correspondence between the different spacetime theories will be elucidated, and the elements of a theory of measurement in a curved spacetime can be obtained.

We first need some mathematical tools. Let M be a m-manifold with a symmetric linear connection Γ. Consider a point $p \in M$. A vector $r \in M_p$ determines uniquely a geodesic $c: u \to c^a(u)$ starting at p (with coordinates $c^a(0)$) in the direction given by r, $\dot{c}^a(0) = r^a$. c will be defined for some range $0 \leqslant u < u_0$ of its affine parameter u. Replacing, if necessary, r by a positive multiple of it, one can arrange that $u_0 > 1$, so that the point $c(1)$ is well defined. If r varies in a suitable neighbourhood U_p of 0 in M_p, $c(1)$ will vary in a neighbourhood U of p in M. It is geometrically plausible and analytically easily established that the so defined *exponential map* $\exp_p: U_p \to U$ is diffeomorphic, i.e. smooth and invertible, at 0. Hence, choosing U_p small enough, we may take the components r^a of the *position vector* r (of $c(1)$ with respect to the 'origin' p) relative to some basis (e_a) of M_p as coordinates of $c(1) = q$. Coordinates of this kind are called *normal*.

It is convenient to use occasionally the suggestive notation $r = \mathbf{pq}$ for the position vector of q with respect to p. Thus by definition

$$\exp_p(\mathbf{pq}) = q\,. \tag{2.31}$$

Let x^a be normal coordinates with origin p. Then the curves given by $x^a(u) = ur^a$ are the geodesics through p, whence it follows that $\Gamma^a_{bc}(ur^d) r^b r^c = 0$ for $0 \leqslant u \leqslant 1$ and all (small) r^a's. Differentiation with respect to u at $u=0$ gives $\Gamma^a_{bc,u}(c)\, r^b r^c r^d = 0$. These results imply:

$$\text{at } p: \quad \Gamma^a_{bc} = 0\,, \quad \Gamma^a_{(bc,d)} = 0\,. \tag{2.32}$$

If these equations are combined with the identity

$$\Gamma^a_{bc,d} = \tfrac{2}{3}(\Gamma^a_{c[b,d]} + \Gamma^a_{b[c,d]}) + \Gamma^a_{(bc,d)} \tag{2.33}$$

and with the formula (2.15) for the curvature tensor, one gets

$$\text{at } p: \quad \Gamma^a_{bc,d} = -\tfrac{2}{3} R^a{}_{(bc)d}\,. \tag{2.34}$$

Near p we therefore have the expansion

$$\Gamma^a_{bc}(x^e) = -\tfrac{2}{3} R^a{}_{(bc)d} x^d + 0(x^a x^a)\,, \tag{2.35}$$

provided Γ is of class C^2 at p. The curvature tensor components in (2.35) are to be taken at p.

These results prove: Given an M with a symmetric Γ, then (a) there always exist, for each point p, locally geodesic coordinates, i.e. coordinates such that, at p, $\Gamma^a_{bc} = 0$; (b) there even exist coordinates such that all geodesics through some point p 'look like' straight lines in $\mathbb{R}^m$; and (2.35) illustrates how R^a_{bcd} measures the deviation from flatness.

The preceding results can be applied in particular to manifolds M with a semi-Riemannian metric g_{ab}, in which a preferred connection Γ, the Levi-Civita connection (2.20), is automatically defined. In this case (2.32) implies

$$\text{at } p: \quad g_{ab,c} = 0\,, \tag{2.36}$$

and (2.34) gives

$$\text{at } p: \quad g_{ab,cd} = \tfrac{2}{3} R_{a(cd)b} \tag{2.37}$$

so that

$$g_{ab}(x^e) = g_{ab}(0) + \tfrac{1}{3} R_{acdb} x^c x^d + 0((x^a x^a)^{3/2}). \tag{2.38}$$

Choosing (e_a) orthonormal at p we obtain, according to (2.36), coordinates which are inertial at p, and (2.38) then shows how the deviation of g_{ab} from η_{ab} (or the analogous matrix for a different signature) is governed by the curvature tensor at p.

The procedure which led to (2.35) and (2.38) can be extended to higher order terms and leads, in the case of analytic Γ^a_{bc} or g_{ab}, to convergent power series. It provides formulae for expressing the quantities $\Gamma^a_{bc,d_1 \ldots d_r}(p)$ (or $g_{ab,d_1 \ldots d_{r+1}}(p)$) linearly in terms of the quantities $R^a_{bcd;e_1 \ldots e_{r-1}}(p)$ or vice versa, and thus demonstrates the fundamental role of the tensor field $R^a{}_{bcd}(x^e)$ for an affinely connected or a semi-Riemannian manifold.

The physical application of the result (2.38) in which we are presently interested is this. Suppose a physical process which takes place in a small spacetime domain has been analyzed in the framework of special relativity theory, and one wishes to describe an analogous process occurring in a gravitational field without having a complete general-relativistic theory of the process in question, or without wishing to go through a complete general-relativistic analysis – a common situation indeed. Then one may proceed as follows. One chooses an event p in the general-relativistic spacetime model, situated where the process is assumed to happen, and identifies it tentatively with an event p' of the special relativistic model. One identifies, moreover, the flat spacetime of the latter model with the tangent space M_p of the curved spacetime M at p, and one uses the exponential map $\exp_p$ to transfer all events of interest which form part of the special relativistic model of the process in question, from M_p to M. The chronometrical relations between the events thus obtained will then, *because of (2.38)*, differ arbitrarily little from those of the original, flat space theory, provided the gravitational tidal field $R^a{}_{bcd}$, evaluated in a suitable orthonormal frame at p adapted to the process, is sufficiently small; and one can estimate the deviations if one knows the curvature. It is therefore plausible to assume that, except for errors of the order of the deviations just referred to, the model of the process obtained by applying the exponential map is a correct one. This procedure may be used to reduce the physical interpretation of general relativity, at least for most concrete physical purposes if not in a theoretically completely satisfactory way, to that of special relativity. The physical interpretation of the gravitational field comes about indirectly, through the relationships, deduced from the general theory, between locally interpreted quantities belonging to space-time regions *not* contained in a single, practically flat domain; these relations will in general differ from those which the special theory would predict. If the local physical process, which might be a measurement, involves only small relative speeds and low energies, the special relativistic description may even be replaced by a nonrelativistic one, of course. This *interpretation of general relativity by local correspondence with special relativity or Newtonian theory* is used in analysing the observations by which general relativity is tested.

Since, as we have shown, local inertial coordinates exist for each event p in any spacetime, the strong principle of equivalence formulated on p. 18 can be used to generalize tentatively non-gravitational laws from special to general relativity. Formally, this generalization amounts to replacing partial differentiations with respect to inertial coordinates by covariant differentiations with respect to arbitrary coordinates. This *principle of minimal gravitational coupling* requires more than that the general-relativistic laws should go over into those of special relativity if the curvature tends to zero, since it says that *no terms containing explicitly the tidal field* $R^a{}_{bcd}$ should be introduced into those laws. The principle is, however, unambiguous only if the laws in question are tensor (or spinor) differential equations of first order, since second and higher covariant derivatives do not commute but satisfy *Ricci's identity*

$$u^a{}_{;[bc]} = -\tfrac{1}{2}u^d R^a{}_{dbc} \tag{2.39}$$

(which is equivalent to (2.15) and may be taken as the definition of $R^a{}_{bcd}$).

Examples of unambiguous applications of the principle are: The transition from the law of free motion in special relativity to the geodesic law of free fall in general relativity; the transition from the energy-momentum conservation law (1.40) to the corresponding law (1.43); the transition from Maxwell's Equations (1.36) to

$$F_{[ab,c]} = 0, \quad F^{ab}{}_{;b} = J^a. \tag{2.40}$$

(in the first equation, it does not matter whether one writes ',' or ';' since $\Gamma^a_{[bc]}=0$.) If, however, one writes the ordinary Maxwell equations in terms of potentials,

$$\begin{aligned} &F_{ab} = 2A_{[b,a]}, \\ &A^a_{,a} = 0, \quad \Box A^a = A^{a,b}{}_b = -J^a \end{aligned} \tag{2.41}$$

one is led to the generalization

$$A^a{}_{;a} = 0, \quad A^{a;b}{}_b = -J^a \tag{2.42}$$

which is *not* equivalent to what one gets by inserting the first line of (2.41) and the Lorentz gauge condition in (2.40), which is because of (2.39)

$$A^a{}_{;a} = 0, \quad A^{a;b}{}_b - R^a{}_b J^b = -J^a. \tag{2.43}$$

In this case it is easy to decide in favour of (2.40), (2.43) and to reject the possibility (2.42), since only in the former case is the conservation of charge, $J^a{}_{;a}=0$, maintained. This choice is further supported by the fact that the familiar action principle of electrodynamics leads unambiguously to (2.40). This example shows that the rule ', → ;' does not always hold. Nevertheless, the strong principle of equivalence as formulated on p. 18 (somewhat differently than in the literature) can be maintained, since it does not specify which particular laws obey the rule ', → ;'; and in spite of this indeterminateness it is a very useful guide for finding general-relativistic laws. (The situation considered here is similar to the one encountered where one wants to find quantum laws from classical ones by correspondence. There one has the problem of ordering factors, here that of ordering covariant differentiations. In both cases, additional arguments are needed to select the correct results.)

Let us now consider an arbitrarily moving observer, idealized as a timelike line c (see Figure 19). We choose an event o on c as origin and measure proper time τ along c from o. The geodesics passing through $p \in c$ orthogonal to c form, near p, a spacelike hypersurface S_p. Along c, we also take an orthonormal triad (e_α) of vectors

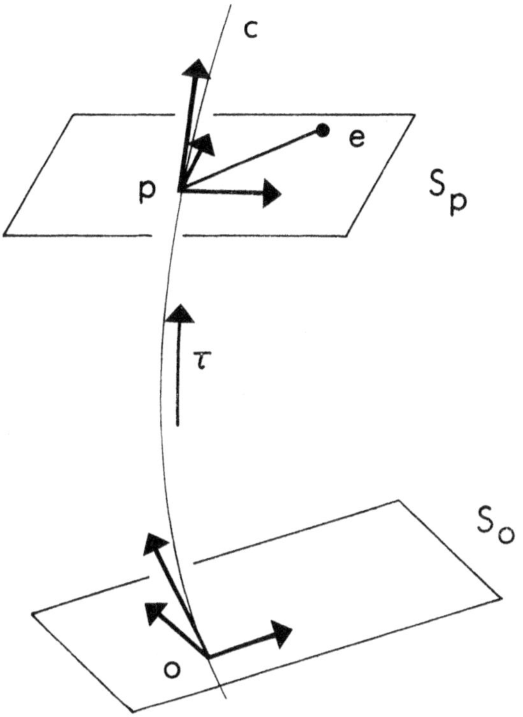

Fig. 19.

orthogonal to c, smoothly varying with τ. Any event $e \in S_p$ can be localized by means of its normal coordinates x^α with respect to $(e_\alpha(p))$. We call $(o, (e_\alpha))$ a *quasi-Galilean frame of reference* on c and call (x^α, x^4), $x^4 = \tau$, the *quasi-Galilean coordinates* with respect to that frame. Obviously they are the closest analogs of the Galilean coordinates of Subsection 1.1. which are available in a relativistic spacetime. Let U^a be the 4-velocity of c.

Since we have not restricted the variation of (e_α) along c, the (e_α) must, in general, be interpreted as rotating axes. A system of axes is called *nonrotating* if its vectors are *Fermi-transported* along c (compare, e.g. [23], [50]), i.e., if their changes $De_\alpha/d\tau$ have no spatial components:

$$\left(\frac{De_\alpha}{d\tau}\right)^a = \lambda U^a. \tag{2.44}$$

Geometrically this means that e_α at $p(\tau+\mathrm{d}\tau)$ is obtained from e_α at $p(\tau)$ by parallel displacement and subsequent orthogonal projection onto the tangent hyperplane of S at $p(\tau+\mathrm{d}\tau)$. An arbitrary frame is related to a Fermi-propagated one by a rotation matrix depending on proper time; hence one can define a pseudovector $\boldsymbol{\omega}$ measuring the angular velocity of the arbitrary frame relative to a Fermi-frame. This vector $\boldsymbol{\omega}$ can be taken to be a world vector orthogonal to c; we shall call it the *angular velocity* of the (arbitrary) frame.

From the preceding definitions one readily derives the following formulae for a quasi-Galilean coordinate system:

On c:

$$\begin{aligned} & g_{ab} = \eta_{ab}, \quad g_{ab,4} = 0, \\ & \Gamma^4_{44} = \Gamma^4_{\beta\gamma} = \Gamma^\alpha_{\beta\gamma} = 0, \\ & \Gamma^\alpha_{44} = \Gamma^4_{4\alpha} = \dot{U}^\alpha, \quad \Gamma^\alpha_{4\beta} = -\varepsilon_{\alpha\beta\gamma}\omega_\gamma. \end{aligned} \tag{2.45}$$

Here, $\dot{U}^a = DU^a/\mathrm{d}\tau$ is the 4-acceleration of c, ω^a is the angular velocity of (e_α), and $\varepsilon_{\alpha\beta\gamma}$ is Levi-Civita's permutation symbol ($\varepsilon_{[\alpha\beta\gamma]} = \varepsilon_{\alpha\beta\gamma}$, $\varepsilon_{123} = 1$). Hence *a quasi-Galilean coordinate system is inertial at all events of* c (at the spatial origin) *if and only if* c *is geodesic and the axes* e_α *are Fermi-propagated.*

We are now prepared to set up the equations of motion of a freely falling particle in a quasi-Galilean reference frame. We just have to rewrite the geodesic equation

$$\frac{\mathrm{d}^2x^a}{\mathrm{d}v^2} + \Gamma^a_{bc}\frac{\mathrm{d}x^b}{\mathrm{d}v}\frac{\mathrm{d}x^c}{\mathrm{d}v} = 0 \tag{2.46}$$

as an equation for $\mathrm{d}^2x^\alpha/\mathrm{d}\tau^2$, where τ is the proper time *along* c, not along the test particle. The $(a=4)$-equation gives the parameter-transformation $v \to \tau$. One gets

$$\frac{\mathrm{d}^2x^\alpha}{\mathrm{d}\tau^2} = \left(-\Gamma^\alpha_{bc} + \frac{\mathrm{d}x^\alpha}{\mathrm{d}\tau}\Gamma^4_{bc}\right)\frac{\mathrm{d}x^b}{\mathrm{d}\tau}\frac{\mathrm{d}x^c}{\mathrm{d}\tau}. \tag{2.47}$$

Assuming the test particle to move close to the origin and with a small speed, we evaluate (2.47) approximately, by linearising in x^α and $\mathrm{d}x^\alpha/\mathrm{d}\tau$, obtaining

$$\frac{\mathrm{d}^2x^\alpha}{\mathrm{d}\tau^2} \approx -\Gamma^\alpha_{44} - 2\Gamma^\alpha_{4\beta}\frac{\mathrm{d}x^\beta}{\mathrm{d}\tau} + \Gamma^4_{44}\frac{\mathrm{d}x^\alpha}{\mathrm{d}\tau}. \tag{2.48}$$

The Γ's, of course, are to be taken at the position of the test particle. Expanding the Γ's about the spatial origin and using (2.45) we get

$$\frac{\mathrm{d}^2x^\alpha}{\mathrm{d}\tau^2} \approx -\dot{U}^\alpha - \Gamma^\alpha_{44,\beta}x^\beta + 2\varepsilon_{\alpha\beta\gamma}\frac{\mathrm{d}x^\beta}{\mathrm{d}\tau}\omega^\gamma. \tag{2.49}$$

From the Equation (2.15) for the curvature tensor we infer

$$\begin{aligned} R^\alpha{}_{4\beta4} &= \Gamma^\alpha_{44,\beta} - \Gamma^\alpha_{4\beta,4} + \Gamma^\alpha_{\gamma\beta}\Gamma^\gamma_{44} - \Gamma^\alpha_{\gamma4}\Gamma^\gamma_{\beta4} \\ &= \Gamma^\alpha_{44,\beta} + \varepsilon_{\alpha\beta\gamma}\dot{\omega}^\gamma - \varepsilon_{\alpha\gamma\delta}\varepsilon_{\gamma\beta\varepsilon}\omega^\gamma\omega^\varepsilon. \end{aligned}$$

Inserting this into (2.49) and using vector products for 3-vectors to shorten the notation, we obtain ($v^\alpha = \mathrm{d}x^\alpha/\mathrm{d}\tau$)

$$\frac{\mathrm{d}^2 x^\alpha}{\mathrm{d}\tau^2} \approx - \dot{U}^\alpha + 2(\mathbf{v} \wedge \boldsymbol{\omega})^\alpha + (\mathbf{x} \wedge \dot{\boldsymbol{\omega}})^\alpha - (\boldsymbol{\omega} \wedge (\boldsymbol{\omega} \wedge \mathbf{x})^\alpha - \\ - (R^\alpha{}_{4\beta 4} + \dot{U}^\alpha \dot{U}_\beta)\, x^\beta . \tag{2.50}$$

(For a different derivation and further results, see [51].) This equation differs from its nonrelativistic ancestor (1.10) by the last two terms only. The very last term is a special relativistic correction; it represents a relativistic 'inertial force per unit mass' due to the fact that the instantaneous 3-space of an accelerated observer is not parallely propagated from one instant to the next, but 'wiggles', in consequence of the conventional definition of simultaneity. The next-to-last term shows the influence of the tidal field on the motion of test particles; it is the only general-relativistic term in that equation. For an observer on earth, $-\dot{U}^\alpha$ is what is usually called the acceleration of gravity, or gravitational field strength; in comparison with it the last two terms are very small, the inertial part being even much smaller than the tidal part.

Equation (2.50) shows that *the definition of* $\boldsymbol{\omega}$ *by means of Fermi-transport is dynamically adequate* and corresponds to the Newtonian definition of 'absolute' rotation: The Coriolis and centrifugal terms are absent if and only if $\boldsymbol{\omega}=0$. (Fermi-transported directions can also be characterized *optically*: A light beam moving initially along the axis of a thin, cylindrical tube and reflected by mirrors at the ends of the tube, will continue to move without hitting the walls of the tube if and only if the axis of the tube is Fermi-propagated, see refs. [23] and [52].) It also shows that a quasi-Galilean coordinate system can, for practical purposes, be identified with a nonrelativistic, Galilean frame of reference as defined in Subsection 1.2, and that *a quasi-Galilean frame with non-rotating axes along a geodesic c* (the coordinates are then called *geodesic Fermi coordinates*) *corresponds* indeed *to a local inertial frame of Newtonian theory*. Moreover, a quasi-Galilean frame with $\dot{U}^a=0$, $\omega^a=0$ in flat spacetime *is* an inertial frame. These statements can again be used in order to reduce the interpretation of general relativity to that of special relativity or Newtonian theory by local correspondence, as was done above in connection with the exponential map. Our considerations also clarify the meaning of the term 'accelerated frame' in special relativity.

It is possible to work out expansions similar to (2.38) for the g_{ab}'s associated with a quasi-Galilean coordinate system, using the method explained in connection with (2.38). We list here only some formulae for the case $\dot{U}^a=0$, $\omega^a=0$: on c, one has:

$$\Gamma^a_{bc} = 0\,, \quad \Gamma^a_{bc,4} = 0\,, \\ \Gamma^a_{b4,c} = R^a{}_{bc4}\,, \quad \Gamma^a_{\beta\gamma,\delta} = -\tfrac{2}{3} R^a{}_{(\beta\gamma)\delta}\,. \tag{2.51}$$

(These formulae hold for any geodesic Fermi-coordinate system in a manifold with a symmetric, linear connection, independently of a metric.)

Applying (2.50) to a geodesic 'observer' c with nonrotating axes we get

$$\frac{d^2x^\alpha}{d\tau^2} = R^\alpha{}_{44\beta}x^\beta . \tag{2.52a}$$

x^α is the position vector of a neighbouring particle relative to c. We may also call it the *connection vector* joining c to a neighbouring geodesic, drawn orthogonal to c (compare Figure 6). To emphasize its vector character and independence of any particular coordinate system, let us rename it V^a. Since the Γ^a_{bc}'s vanish along c, the derivatives in (2.52a) are, in fact, absolute derivatives; moreover, $U^a = \delta^a_4$ is the tangent vector of c. Hence we may rewrite (2.51) covariantly as

$$\frac{D^2V^a}{d\tau^2} = R^a{}_{bcd}U^bU^cV^d , \tag{2.52b}$$

which is called the *equation of geodesic deviation*. Its fundamental importance for gravitation theory has already been emphasized; it exhibits the role of the curvature tensor as *differential gravitational field.*

According to the derivation given here, (2.52) is an approximate equation whose relative error tends to zero if r and v, the distance and velocity of the second test particle relative to the first one, tend to zero, and which will be useful if r is much smaller than a typical radius of curvature of spacetime along c, and if $v \ll 1$. This is indeed the way in which (2.52) is used in connection with physical interpretations. It is, however, also used as a purely mathematical tool in establishing, e.g., singularity theorems, and therefore one will wish to have an interpretation and proof of (2.52) which is free of any approximations. Such a proof can easily be given; see, e.g., ref. [23]. The derivation here has the advantage of exhibiting more clearly the physical meaning of (2.52).

Equation (2.52) indicates a method for *measuring the curvature tensor* $R^a{}_{bcd}$ by means of freely falling test particles.

To see more clearly what has to be measured, we split V^a into a magnitude and a unit vector,

$$V^a = rE^a .$$

Then

$$V_a\frac{D^2V^a}{d\tau^2} = r\ddot{r} + r^2E_a\frac{D^2E^a}{d\tau^2} .$$

The function $E^a(\tau)$ describes the 'proper motion' of the second test particle on the 'celestial sphere' of an observer sitting on the first test particle. If this motion is judged relatively to Fermi-propagated axes, the term $E_a(D^2E^a/d\tau^2)$ can easily be identified as $-\dot{\alpha}^2$, where $|\dot{\alpha}|$ is the magnitude of the proper motion (referred to proper time), as in Newtonian kinematics. Combining these results with (2.52) and using the definition

$$K(U \wedge V) = \frac{R_{abcd}U^aV^bU^cV^d}{U^2V^2 - (U\cdot V)^2}$$

of the *sectional curvature* K associated with the 2-direction $U \wedge V$, we obtain

$$r^{-1}\ddot{r} - (\dot{\alpha})^2 = K(U \wedge V).$$

Thus in order to measure K one has to measure the distance $r(\tau)$, e.g., by radar, sufficiently well so that $\ddot{r}$ can be computed, and one has to measure the proper motion $\dot{\alpha}$ relatively to Fermi-propagated axes – which requires, as we have seen, additional tools like gyroscopes or photons.

There is, however, a simpler ideal measurement of curvature which I wish to discuss now.

Consider, in a manifold with symmetric linear connection, a geodesic triangle as shown in Figure 20.

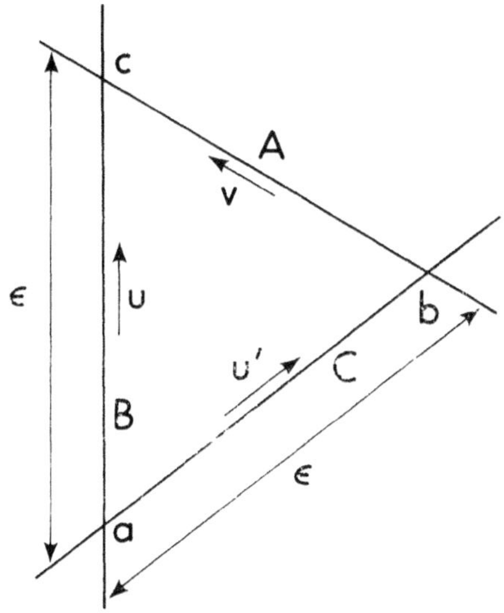

Fig. 20.

Take a to be fixed, and let u be an affine parameter on B, u' one on C; $u(a)=u'(a)=0$. Let b correspond to $u'=\varepsilon$, c to $u=\varepsilon$, and think of ε as being small. The sides of the triangle (a, b, c) can be described by the vectors **ab**, **ac**, **bc** (recall the definition (2.31)). Let **bc** be parallely displaced from b to a along C, call the result $(\mathbf{bc})_a$. In a flat space, $(\mathbf{bc})_a = \mathbf{ac} - \mathbf{ab}$. In a curved space, however, assuming the connection to be of class C^2,

$$(\mathbf{bc})_a = \mathbf{ac} - \mathbf{ab} + \tfrac{1}{3}R(\mathbf{ab}, \mathbf{ac})\cdot(\mathbf{ac} - \tfrac{1}{2}\mathbf{ab}) + 0(\varepsilon^4). \tag{2.53}$$

Here we have used the notation $R(U, V)$ for the linear map $W^a \rightarrow (R^a{}_{bcd}U^cV^d)W^b$, also called the *curvature operator*; the $\cdot$ indicates application of this map.

The vectors **ac** etc. are of order ε, the third term on the right-hand-side of Equation (2.53) is of order ε^3.

To prove (2.53) we use a geodesic Fermi coordinate system (x^a) along C, with origin $a=(0, ..., 0)$, such that C is given by $x^2 = ... = x^m = 0$, B is given by $x^1 = x^3 = = ... = x^m = 0$, $b = (\varepsilon, 0, ...0)$, $c = (0, \varepsilon, ..., 0)$. (To obtain such a system, take vectors

$e_3, \ldots, e_n$ at a which, together with the tangents of B and C, form a basis. Parallel transport of these vectors along C gives, besides the tangent of C, $m-1$ vectors which span an $(m-1)$-direction S. Within each hypersurface generated by the geodesics tangent to S introduce normal coordinates, as defined above, relative to the $m-1$ vectors; call them $x^2, \ldots, x^m$. Take $x^1 = u'$, then $(x^1, \ldots, x^n)$ has the desired properties.) Let the geodesic A be given by $x^a(v, \varepsilon)$, with $x^a(0, \varepsilon) = (\varepsilon, 0, \ldots, 0)$, $x^a(\varepsilon, \varepsilon) = (0, \varepsilon, \ldots, 0)$. Then the tangent to A is $\partial x^a/\partial v$, and we have

$$\frac{\partial^2 x^a}{\partial v^2} = -\Gamma^a_{bc} \frac{\partial x^b}{\partial v} \frac{\partial x^c}{\partial v}. \tag{2.54}$$

Using a Taylor expansion in v we get

$$\frac{1}{\varepsilon}\left(x^a(\varepsilon, \varepsilon) - x^a(0, \varepsilon)\right) = \delta^a_2 - \delta^a_1 = \frac{\partial x^a}{\partial v} + \frac{\varepsilon}{2}\frac{\partial^2 x^a}{\partial v^2} + \frac{\varepsilon^2}{6}\frac{\partial^3 x^a}{\partial v^3} + 0(\varepsilon^3) =$$
$$= \frac{\partial x^a}{\partial v} - \frac{\varepsilon^2}{6}\frac{\partial}{\partial v}\left(\Gamma^a_{bc}\frac{\partial x^b}{\partial v}\frac{\partial x^c}{\partial v}\right) + 0(\varepsilon^3),$$

where all derivatives are to be taken at $(0, \varepsilon)$ and (2.54) has been used. Expanding again with respect to ε and taking now derivatives at $(0, 0)$, we continue:

$$\delta^a_2 - \delta^a_1 = \frac{\partial x^a}{\partial v} + \varepsilon \frac{\partial^2 x^a}{\partial \varepsilon \partial v} + \frac{\varepsilon^2}{2}\left(\frac{\partial^3 x^a}{\partial \varepsilon^2 \partial v} - \tfrac{1}{3}\Gamma^a_{bc,d}\frac{\partial x^d}{\partial v}\frac{\partial x^b}{\partial v}\frac{\partial x^c}{\partial v}\right) + 0(\varepsilon^3).$$

Hence, at a,

$$\frac{\partial x^a}{\partial v} = \delta^a_2 - \delta^a_1, \quad \frac{\partial^2 x^a}{\partial \varepsilon \partial v} = 0,$$
$$\frac{\partial^3 x^a}{\partial \varepsilon^2 \partial v} = \tfrac{1}{3}\left(\Gamma^a_{22,2} + \Gamma^a_{11,2} - 2\Gamma^a_{12,2}\right),$$

where we have used that $\Gamma^a_{bc,1} = 0$ along C. Remembering the formulae (2.51) and taking into account that the role of x^4 in (2.51) is now played by x^1 we infer

$$\frac{\partial^3 x^a}{\partial \varepsilon^2 \partial v} = \tfrac{1}{3}\left(R^a{}_{121} + 2R^a{}_{212}\right),$$

whence

$$\frac{\partial x^a}{\partial v}(0, \varepsilon) = \delta^a_2 - \delta^a_1 + \frac{\varepsilon^2}{6}\left(R^a{}_{121} + 2R^a{}_{212}\right) + 0(\varepsilon^3),$$

i.e.

$$(\mathbf{bc})^a = \varepsilon\delta^a_2 - \varepsilon\delta^a_1 + \frac{\varepsilon^3}{3}\left(R^a{}_{212} - \tfrac{1}{2}R^a{}_{112}\right) + 0(\varepsilon^4).$$

Since parallel transport along C does not change vector components in our coordinate system and **ab** has components $\varepsilon\delta_1^a$, **ac** has components $\varepsilon\delta_2^a$, the last formula is nothing but the component version of (2.53), which has thus been established.

Applying (2.53) to spacetime we obtain immediately

$$\mathbf{bc}^2 = \mathbf{ac}^2 + \mathbf{ab}^2 - 2\mathbf{ab}\cdot\mathbf{ac} - \tfrac{1}{3}R\,(\mathbf{ab}, \mathbf{ac}, \mathbf{ab}, \mathbf{ac}) + 0\,(|\mathbf{ab}|^5)\,. \qquad (2.55)$$

(Here we have used the standard notation $R(U, V, W, Q) = R_{abcd}U^aV^bW^cQ^d$.) A different proof of this result is given in ref. [23].

The simplest application of (2.55) is as follows. Suppose two particles P and Q start freely falling from the event a with relative speed v. At b, particle Q sends a light signal to P which arrives at c (see Figure 21)

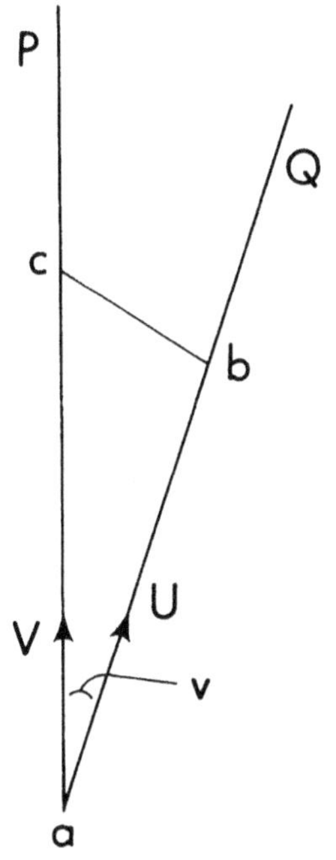

Fig. 21.

Then the proper times τ_{ab}, τ_{ac} and the speed v are related to the curvature by

$$\tau_{ab}^2 + \tau_{ac}^2 - 2\frac{\tau_{ab}\,\tau_{ac}}{(1 - v^2)^{1/2}} = -\tfrac{1}{3}\tau_{ab}^2\tau_{ac}^2 R_{abcd}U^aV^bU^cV^d\,, \qquad (2.56)$$

where U, V are the 4-velocities of P and Q at a, and the error term has been omitted. The relation (2.56) offers probably the simplest way to measure, in principle, the sectional curvature in the timelike 2-direction defined by two departing particles. If the construction of Figure 20 is combined with the construction of Subsection 2.6 one recognizes that the sectional curvature can be measured, with arbitrarily large accuracy, by means of a finite number of timelike and null geodesics and counting certain intersection events. This illustrates Einstein's claim that measurements can ultimately be reduced to the observation of coincidences and counting.

One can similarly specialize (2.55) to a triangle with three timelike sides and interpret the result as a curvature determination.

It appears from Equations (2.52) and (2.55) that by the methods considered only the tensor $R_{a(bc)d}$ can be found. However, the well known symmetries

$$R^a{}_{b(cd)} = 0 , \quad R^a{}_{[bcd]} = 0 , \tag{2.57}$$

valid for the curvature tensor of any symmetric linear connection, are equivalent to

$$R^a{}_{bcd} = \tfrac{2}{3}\left(R^a{}_{(bc)d} - R^a{}_{(bd)c}\right), \tag{2.58}$$

so that the full curvature tensor can be obtained by these procedures.

Having considered some ideal measurements, let us now ask how *actual measurements* are to be represented in curved spacetime.

The most important measurable quantities in astrophysics and cosmology are *times*, *angles*, and *energies* (of various kinds of radiation). Times are expressed as invariant integrals $\int d\tau = \int (-g_{ab} dx^a dx^b)^{1/2}$ over sections of worldlines representing clocks. Angles between directions given by unit vectors **e**, **d** contained in an observer's rest space (i.e. tangent to S_p in Figure 19) are computed from $\cos\phi = \mathbf{e}\cdot\mathbf{d} = g_{ab} e^a d^b$. If, e.g., two light rays reach the observer at event e, then the angle between them is found by projecting the tangents to the light rays at e into the observer's rest space and taking the angle between the so obtained spatial directions. If the energy of some kind of radiation which passes through the opening of some telescope is to be represented mathematically, one has to know the stress-energy-momentum tensor T^{ab} of the radiation and the 4-velocity of the opening of the telescope, and one has to integrate $u_a T^{ab}$ over the timelike strip (hypersurface) S with hypersurface element* σ_a consisting of all events contained in the telescope opening during the time of observation: $\int_S u_a T^{ab} \sigma_b$. If the observations of finer details like spectra or particle species are to be described, integrals of distribution functions (to be defined in Subsection 3.5) over the appropriate frequency or proper mass intervals have to be used. The important point is that *measurable quantities are invariants*, constructed from quantities describing the object observed (e.g., T^{ab}) and quantities describing the measuring apparatus (e.g., u^a, S). It will often be convenient to find, or to evaluate, such invariants by introducing quasi-Galilean coordinates adapted to the observer and to use the correspondence with special relativity or Newtonian theory.

* σ_a is defined in Equation (3.31)

II. *Classical Descriptions of Matter in Curved Spacetime: Kinetic Theory, Continuum Mechanics and Thermodynamics, Radiation, and Extended Bodies*

3. Kinetic Theory

3.1. Introduction

Having established a framework for the description of gravitational fields we now turn our attention to the stuff which fills spacetime. Physics is concerned with matter, and the chrono-geometrical structure of spacetime has been introduced – just like electromagnetic fields – to enable us to describe matter and its motion.

At the classical level, the most detailed and least phenomenological model of matter in curved spacetime which has been elaborated is that of the kinetic theory of gases. Its assumptions, though not without problems, are simple and rather free of arbitrariness. If several species of particles and elastic and inelastic collisions are taken into account, it can describe a wide range of phenomena. In particular it provides a common basis for the relativistic mechanics and thermodynamics of bulk matter at low density and for radiation fields, and the *form* of its thermo-mechanical laws can be maintained even for dense matter, for which constitutive equations have to be introduced separately, however.

In the following subsections we try to present a coherent, essentially self-contained and, with respect to the basic physical assumptions, concepts and laws – not with respect to applications which can be found in large parts of the other lectures contained in this volume, implicitly or explicitly – complete account of kinetic theory.

Several recent research and review articles [53, 54, 55, 56, 57] have dealt with the topics of this section, and naturally there is a considerable overlap between what follows and some of these articles. Nevertheless, to make these notes selfcontained, I have described these things here again, with some improvements (hopefully).

3.2. Phase space and phase flow

The motion of a freely falling test particle in a gravitational field (M, g_{ab}) can be characterized by the *Langrangian*

$$\mathscr{L}(x^a, \dot{x}^a) = \tfrac{1}{2} g_{ab}(x^c)\, \dot{x}^a \dot{x}^b\,; \tag{3.1}$$

for the associated Euler-Lagrange equations are equivalent to the equations (2.46) of a geodesic. (The appearance of g_{ab} in $\mathscr{L}$ illustrates the role of the metric coefficients as potentials of the combined inertial-gravitational field: In the Newtonian limit the g_{ab} give rise both to the inertial term and the gravitational term of the standard Lagrangean $\tfrac{1}{2}\dot{\mathbf{x}}^2 - U$.)

Introducing canonical momenta

$$p_a = \frac{\partial \mathscr{L}}{\partial \dot{x}^a} = g_{ab}\dot{x}^b \tag{3.2}$$

we can pass to the *Hamiltonian*

$$H(x^a, p_b) = \tfrac{1}{2} g^{ab}(x^c)\, p_a p_b, \tag{3.3}$$

and thus obtain canonical equations for the geodesics. Thus the computation of geodesics can be regarded as a problem of analytical mechanics.

Since the Hamiltonian H does not depend explicitly on the (affine) parameter of the geodesic which plays formally the role of the time, the value of H, $\frac{1}{2}p_a p^a$, is constant on each extremal. The tangent vector p^a is called the *4-momentum*, the constant $m = \sqrt{-p_a p^a} \geqslant 0$, the *mass* of the test particle. If $m > 0$, one can write

$$p^a = m u^a; \tag{3.4}$$

the unit vector u^a is the 4-velocity of the particle. We shall always take worldlines to be future-directed, so that p^a and u^a are future-directed vectors.

The Lagrangian (3.1) or, equivalently, the Hamiltonian (3.4) determine the worldlines of freely falling particles of arbitrary masses. Two solution curves which pass through the same event $x_0{}^a$ with the same direction differ in their parametrizations only; they represent particles with different masses. This fact expresses the universality of free fall.

It is easy to generalize the preceding statements to charged particles in a combined gravitational and electromagnetic field (see, e.g., [55]), but here we shall restrict our attention to the purely gravitational case.

In order to survey the totality of all possible test particle motions in a given spacetime it is useful to introduce the concept of a (one-particle) *phase space P*. Physically, P is the collection of all possible *instantaneous states* of particles which we may describe either as pairs (x^a, p_a) or as $(x^a, \dot{x}^a)$. Choosing the first description, we define P as the 8-dimensional manifold of all non-spacelike, future-directed, covariant vectors defined at arbitrary spacetime points. Thus P is the non-spacelike part of $T^*(M)$, the *cotangent bundle* over M. P is a submanifold-with-boundary of $T^*(M)$; its boundary $\partial P = P_0$ consists of the states of massless particles. More generally, the states belonging to a given mass $m(\geqslant 0)$ form the 7-dimensional *sub-phase space* P_m.

We shall use the symbols P_x, P_{mx} to denote the sets of all momentum vectors and of all momentum vectors of given mass m, respectively, at the event $x \in M$.

Because of (3.2) one can consider either (x^a, p_a) or (x^a, p^a) as local coordinates on P. On P_m we shall usually take (x^a, p^λ) as local coordinates. p^4 is determined by the mass condition

$$g_{ab}(x^c)\, p^a p^b = -m^2 \tag{3.5}$$

and the requirement that p^a be future directed.

All the manifolds P, P_m are *oriented* by the following prescription: (x^a, p^a) is an oriented coordinate-system on that 'cylinder' of P which 'lies over' the domain of the local coordinate system (x^a) of M; here p^a are natural vector components with respect to (x^a). (x^a, p^λ) is taken to be an oriented coordinate system on a corresponding part of P_m provided x^4 is a timelike coordinate, increasing towards the future.

The free fall trajectories define on P a congruence of curves; i.e., through each point of P there passes precisely one such trajectory. The tangent vector field associated with this *phase flow* can be inferred from (3.3):

$$L = \frac{\partial H}{\partial p_a}\frac{\partial}{\partial x^a} - \frac{\partial H}{\partial x^a}\frac{\partial}{\partial p_a} =$$

$$= p^a \frac{\partial}{\partial x^a} - \tfrac{1}{2} g^{bc}{}_{,a}\, p_b p_c \frac{\partial}{\partial p_a}. \tag{3.6}$$

An equivalent representation follows from (3.1):

$$L = p^a \frac{\partial}{\partial x^a} - \Gamma^a_{bc} p^b p^c \frac{\partial}{\partial p^a}. \tag{3.7}$$

In accordance with statistical mechanics we call L the Liouville vector or *Liouville operator*.

(Since spacetime carries a metric there is a canonical correspondence between contravariant vectors and covariant vectors as exemplified by (3.2). Hence there is a canonical diffeomorphism between the tangent bundle $T(M)$ and the cotangent bundle $T^*(M)$ of M, and instead of considering the phase space P as part of $T^*(M)$ one may also take it to be part of $T(M)$. In particular, L as given by (3.7) may be regarded as a vector field on $T(M)$ associated with the connection Γ; with this interpretation it is called the *geodesic spray* of Γ in differential geometry. The advantage of working with $T^*(M)$ consists in the fact that it has a natural *canonical* (or 'symplectic') *structure* and therefore permits a Hamiltonian description of the phase flow (see [58]). On the other hand, $T(M)$ has a natural *Riemannian structure* which can also be used to investigate the phase space (see [59] for a concise account of this approach).

L defines on P a local one-parameter group of local transformations: Given a sufficiently small number ε and a state $(x^a, p_a) = \xi$, there is a neighbourhood N_ξ of ξ such that each phase trajectory through a point of N_ξ exists at least for a parameter interval of length ε. Therefore, each point $\eta \in N_\xi$ can be mapped into the point $\bar{\eta}$ which lies on the phase orbit through η and whose parameter distance from η equals ε. The mappings $e^{\varepsilon L}$ so obtained describe the evolution of any system of (independent) test particles.

Because of mass conservation L is tangent to each P_m. Let L_m denote the restriction of L to P_m. We then have, e.g.,

$$L_m = p^a \frac{\partial}{\partial x^a} - \Gamma^\lambda_{bc} p^b p^c \frac{\partial}{\partial p^\lambda}. \tag{3.8}$$

The relativistic phase space P_m corresponds to the augmented phase space $\{(x, t, \mathbf{p})\}$ of nonrelativistic mechanics (for particles of fixed m); $P = \bigcup_m P_m$. Analogues of the 6-dimensional $(\mathbf{x}, \mathbf{p})$ phase space are those hypersurfaces in P_m which 'lie over' spacelike hypersurfaces of M, and the quotient manifold P_m/L whose points are the phase orbits of L in P_m (if the quotient exists). (The points of P_m/L correspond to 'Heisenberg states' of particles.)

3.3. Vector fields and Lie derivatives

Any vector field V on an m-manifold M determines a congruence of curves, provided $V \neq 0$ (everywhere), which we assume. In a suitable neighbourhood of any point $p \in M$ there exist coordinates (x^a) such that the x^1-parameter curves are tangent to V, i.e., such that $V = \partial/\partial x^1$ or $V^a = \delta^a_1$. This is intuitively obvious and can be proven easily by means of the theory of systems of ordinary differential equations (see, e.g., [60]). The local maps

$$e^{\varepsilon V} : (x^1, x^2, \ldots, x^m) \to (x^1 + \varepsilon, x^2, \ldots, x^m)$$

are uniquely determined by V, as described in the previous subsection (in the case $V = L$, $M = P$).

We define the *Lie derivative* of any tensor field or linear connection or 'geometric object' S with respect to V as that object whose components in an adapted coordinate system (x^a) as described above are given by

$$(\mathcal{L}_V S)^{\cdot\cdot}{}_{\cdot\cdot} = \frac{\partial}{\partial x^1} S^{\cdot\cdot}{}_{\cdot\cdot}. \tag{3.9}$$

It is easy to verify that this definition is meaningful, i.e., that the choice of the adapted coordinate-system in (3.9) is immaterial. $\mathcal{L}_V S$ measures the change which the quantity S undergoes if it is 'carried along' by the flow with 'velocity' V. For a geometrical discussion of Lie derivatives see, e.g., ref. [28].

The definition of $\mathcal{L}_V$ shows that $\mathcal{L}_V$ obeys a sum and (ordinary) product rule, that it commutes with contractions, and that for a scalar f, $\mathcal{L}_V f = V^a f_{,a}$.

For a vector field W we have

$$\mathcal{L}_V W = [V, W] = (W^a{}_{,b} V^b - V^a{}_{,b} W^b) \frac{\partial}{\partial x^a}, \tag{3.10}$$

where the *Lie-bracket* notation emphasizes that $(V, W) \to [V, W]$ is a kind of product on vector fields. (With respect to + and [,], the vector fields on M form a *Lie algebra*, the Lie algebra of infinitesimal diffeomorphisms.)

An object field S – e.g., a tensor field or a linear connection – is called *invariant with respect to* V if $(\mathcal{L}_V S) = 0$. In coordinates adapted to V, invariance of S means independence of its components $S^{\cdot\cdot}{}_{\cdot\cdot}$ of x^1, the coordinate 'in the V-direction'.

3.4. Volume elements, differential forms and integration

A completely skew symmetric covariant tensor with r indices is also called an *r-form*. Such objects play a preferred role in geometry, e.g., as *volume elements*.

Precise definitions of the operations 'exterior multiplication' and 'exterior differentiation' of r-form fields can be found in Dr Schmidt's lectures (this volume, p. 296) and in refs. [60], [55], [61]; here I only put together those facts which will be used in the sequel.

Any scalar f has a gradient $\mathrm{d}f$, a 1-form. If (x^a) is a local coordinate system, the gradients $\mathrm{d}x^a$ of the coordinate functions generate the *Cartan differential algebra* of the exterior form fields.

Exterior products are related to tensor products as in the following examples:

$$\mathrm{d}x^{ab} = \mathrm{d}x^a \wedge \mathrm{d}x^b = 2!\, \mathrm{d}x^{[a} \otimes \mathrm{d}x^{b]}, \tag{3.11}$$

$$\mathrm{d}x^{abc} = \mathrm{d}x^a \wedge \mathrm{d}x^b \wedge \mathrm{d}x^c = 3!\, \mathrm{d}x^{[a} \otimes \mathrm{d}x^b \otimes \mathrm{d}x^{c]}, \tag{3.12}$$

where $\mathrm{d}x^{ab}$ etc. is to be regarded as an abbreviation for the wedge product of the differentials, and $\otimes$ indicates the tensor product. With this convention the value of $\mathrm{d}x^{ab}$ on the vectors V, W, $\mathrm{d}x^{ab}(V, W)$ equals the determinant $V^aW^b - W^aV^b$, and similarly for higher dimensions, which shows in which sense the $\mathrm{d}x^{ab}$ etc. are *volume functions* assigning to each parallelotope the signed measures of its parallel projections onto the coordinate planes (or hyperplanes).

An r-form ψ can be expanded thus:

$$\psi = \frac{1}{r!} \psi_{a_1 \ldots a_r} \mathrm{d}x^{a_1 \ldots a_r}. \tag{3.13}$$

$\psi_{a_1 \ldots a_r}$ are the (natural) components of ψ, considered as a tensor.

The exterior derivative of the r-form (3.13) is the $(r+1)$-form

$$\begin{aligned} \mathrm{d}\psi &= \frac{1}{r!} \mathrm{d}\psi_{a_1 \ldots a_r} \wedge \mathrm{d}x^{a_1 \ldots a_r} \\ &= \frac{(-1)^r}{r!} \psi_{a_1 \ldots a_r, a_{r+1}} \mathrm{d}x^{a_1 \ldots a_{r+1}}. \end{aligned} \tag{3.14}$$

An r-form ψ can be integrated over an r-dimensional oriented submanifold S. The integral $\int_S \psi$ is computed by 'inserting' into ψ the parameter representation $x^a = x^a(u^1, \ldots, u^r)$ of S,

$$\int_S \psi = \frac{1}{r!} \int_S \psi_{a_1 \ldots a_r}(x(u)) \frac{\partial(x^{a_1}, \ldots, x^{a_r})}{\partial(u^1, \ldots, u^r)} \mathrm{d}u^{12 \ldots r}, \tag{3.15}$$

and by evaluating the resulting r-fold integral in 'u-space' in the standard manner. (This may require decomposition of M and/or S into coordinate patches.) The integral $\int_S \psi$ is independent both of the coordinates x^a on M and those, u^λ, on S, as indicated by the notation.

A very important theorem called *Stokes' theorem* which generalizes classical formulae of Gauss and Stokes to arbitrary dimensions and frees it of any metric or affine concepts says that if S is an oriented, compact r-submanifold with boundary ∂S

in an m-manifold M and if ψ is a continuously differentiable $(r-1)$-form field on M, then

$$\int_S \mathrm{d}\psi = \int_{\partial S} \psi . \tag{3.16}$$

For a proof, see, e.g., [62].)

An r-form ψ can be *contracted* with a vector V, to give an $(r-1)$-form $\varphi = V \cdot \psi$. In components,

$$\varphi_{a_2 \ldots a_r} = V^{a_1} \psi_{a_1 a_2 \ldots a_r} . \tag{3.17}$$

(The dot in the index free notation $V \cdot \psi$ corresponds to the summation over a pair of adjacent indices.) Remembering that tensors are defined as multilinear functions of vectors we may rewrite (3.17) as

$$\begin{aligned} &\varphi(B_2, \ldots B_r) = \psi(V, B_2, \ldots, B_r), \\ \text{i.e.,} \quad &(V \cdot \psi)(B_2, \ldots, B_r) = \psi(V, B_2, \ldots, B_r), \end{aligned} \tag{3.18}$$

which shows that the function $V \cdot \psi$ of $r-1$ vector variables arises from the function ψ of r variables by substituting the (fixed) vector V for the first variable.

Suppose that $\Omega \neq 0$ is an m-form on an m-manifold. Then the $(m-1)$-form $\omega = V \cdot \Omega$ can be characterized, except for a nonvanishing scalar factor, by the property: $\omega(B_1, \ldots, B_{m-1}) = 0$ if and only if the vector V depends linearly upon $(B_1, \ldots, B_{m-1})$. Therefore, the hypersurface volume element defined by ω is the suitable one for the computation of fluxes through hypersurfaces of quantities carried by the flow with velocity V.

Let, again, Ω denote a nonvanishing m-form on an m-manifold M, and let V be a vector field. Then $\mathrm{d}(V \cdot \Omega)$ is an m-form, since contraction with V decreases, exterior differentiation increases the degree of a form by one. Since the m-forms form a 1-dimensional linear space, $\mathrm{d}(V \cdot \Omega)$ is proportional to Ω; whence we may define a *divergence operator* on vector fields relative to Ω by

$$\mathrm{d}(V \cdot \Omega) = (\mathrm{Div}\, V)\, \Omega . \tag{3.19}$$

If, in local coordinates, $\Omega = \vee\, \mathrm{d}x^{1 \ldots m}$, $V = V^a (\partial / \partial x^a)$, one obtains

$$\mathrm{Div}\, V = \frac{1}{\vee} (\vee\, V^a)_{,a} , \tag{3.20}$$

which generalizes a well-known formula of Riemannian geometry.

The three (metric independent) operations $\mathscr{L}_V$, $V \cdot (\ \ )$, d on form fields are related by the useful identity

$$\mathscr{L}_V \psi = V \cdot \mathrm{d}\psi + \mathrm{d}(V \cdot \psi), \tag{3.21}$$

valid for any form field ψ. This formula can be verified easily by computing, relatively to a coordinate system adapted to V $(V^a = \delta^a_1)$, the components of the three terms of the equation.

Applying (3.21) to a non vanishing m-form Ω, so that $\mathrm{d}\Omega=0$, and using (3.19) we get

$$\mathscr{L}_V\Omega = \mathrm{d}\omega = (\mathrm{Div}\, V)\,\Omega\,, \tag{3.22}$$

where we have put

$$\omega = V\cdot\Omega\,. \tag{3.23}$$

Note that

$$V\cdot\omega = 0\,, \tag{3.24}$$

which expresses the fact that the ω-measure of a hypersurface Σ, $\int_\Sigma\omega$, vanishes if V is tangent to Σ, as remarked already earlier.

The product rule for $\mathscr{L}_V$ and $\mathscr{L}_V V=0$ further imply that

$$\mathscr{L}_V\omega = V\cdot\mathscr{L}_V\Omega\,. \tag{3.25}$$

Remembering the definition of invariance with respect to a vector field V given at the end of Subsection 3.3 we infer from (3.22) and (3.25):

Ω is V-invariant if and only if Div $V=0$ or, equivalently, $\mathrm{d}\omega=0$. If Ω is V-invariant, then so is ω defined by (3.23).

Also, it is clear that if Ω is V-invariant and f is a scalar, then $f\,\Omega$ is V-invariant if and only if $\mathscr{L}_V f = V^a f,_a=0$, i.e., if f is constant on the orbits of V.

Next let us verify that the formal definition of invariance actually corresponds to the property of remaining unchanged under a map, in the case of forms Ω and ω, as one would expect.

As we have seen above, V defines local diffeomorphisms $e^{\varepsilon V}$ on M, and these appear analytically as translations along the x^1-direction if expressed in coordinates adapted to V. A sufficiently small, compact oriented domain $D\subset M$ is mapped by $e^{\varepsilon V}$ onto another domain $D(\varepsilon)$. The Ω-volume of D changes according to

$$\left(\frac{\mathrm{d}}{\mathrm{d}\varepsilon}\int\limits_{D(\varepsilon)}\Omega\right)_{\varepsilon=0} = \int\limits_D \mathscr{L}_V\Omega\,, \tag{3.26}$$

as is seen by using adapted coordinates and remembering the definition (3.9) of $\mathscr{L}_V$. Similarly, for a hypersurface Σ and its image $\Sigma(\varepsilon)=e^{\varepsilon V}\Sigma$ we have

$$\left(\frac{\mathrm{d}}{\mathrm{d}\varepsilon}\int\limits_{\Sigma(\varepsilon)}\omega\right)_{\varepsilon=0} = \int\limits_\Sigma \mathscr{L}_V\omega\,. \tag{3.27}$$

These formulae show that invariance of Ω or ω under V means, in fact, that the Ω or ω-measures of domains and hypersurfaces of M remain unchanged under the maps generated by V.

We end these general remarks by noting, for later use: If Ω is an m-form in an m-manifold M, f a scalar field, and V a vector field, then

$$\mathrm{d}f\wedge(V\cdot\Omega) = V(f)\,\Omega\,. \tag{3.28}$$

To prove this, introduce local coordinates and rewrite the formula in component notation, obtaining

$$mf_{,[a_1}V^b\Omega_{|b|a_2\dots a_m]} = V^b f_{,b}\Omega_{a_1\dots a_m}.$$

(The index b is to be excluded from antisymmetrization, as indicated by vertical bars.) This equation needs to be verified only for $(a_1, \dots, a_m)=(1, \dots, m)$, and that is easily done.

The preceding remarks all apply to an arbitrary manifold M, irrespective of whether M has a connection or a metric. Let us now specialize first to spacetime, and then to phase space.

In spacetime there is defined the *Riemannian volume element*

$$\eta = \sqrt{-g}\,\mathrm{d}x^{1234} = \frac{1}{4!}\eta_{abcd}\,\mathrm{d}x^{abcd}, \tag{3.29}$$

with $g=\det(g_{ab})$, a 4-form. (Here we make use of our assumption that spacetime is oriented and the (x^a) form an oriented coordinate system, see Subsection 2.8.) It can be used to form *invariant integrals* $\int f\eta$, like action integrals. By means of a vector field V in spacetime and a hypersurface Σ we can form the *flux of V through Σ*,

$$\int_\Sigma V\cdot\eta = \int_\Sigma \tfrac{1}{6}V^a\eta_{abcd}\,\mathrm{d}x^{bcd} = \int_\Sigma V^a\sigma_a, \tag{3.30}$$

where

$$\sigma_a = \tfrac{1}{6}\eta_{abcd}\,\mathrm{d}x^{bcd} \tag{3.31}$$

is a covariant vector-valued 3-form, the *hypersurface element*. In similar fashion, a 2-contravariant tensor field can be integrated over a 2-surface, etc.

The tangent space M_x of an event x is itself a (flat) pseudoriemannian manifold, and thus has its own volume element analogous to (3.29). For it we introduce the notation

$$\pi = (\sqrt{-g})_x\,\mathrm{d}p^{1234}. \tag{3.32}$$

Physically it is used as a *volume element in 4-momentum space.*

As an example take a point observer with quasi-Galilean coordinates (x^λ, x^4) as discussed in 2.10, and choose a hypersurface $x^4=\text{const.}$ (see Figure 19). Then at the observer, i.e., at $x^\lambda=0$, we obtain for the hypersurface element (3.31), since $\mathrm{d}x^4=0$:

$$\sigma_a = u_a\,\mathrm{d}x^{123} \tag{3.33}$$

where $u_a=-\delta_a^4$, the 4-velocity of the observer.

Passing on to phase space P we introduce the *canonical Lebesgue measure* given by

$$\begin{aligned}\Omega &= -\,\mathrm{d}x^{1234}\wedge\mathrm{d}p_{1234} = \\ &= -\,\mathrm{d}x^{1234}\wedge d(g_{1b_1}p^{b_1})\wedge\cdots\wedge\mathrm{d}(g_{4b_4}p^{b_4}) = \\ &= -\,g_{1b_1}\cdots g_{4b_4}\,\mathrm{d}x^{1234}\wedge\mathrm{d}p^{b_1\dots b_4} = \\ &= \eta\wedge\pi,\end{aligned} \tag{3.34}$$

where we have used (3.29) and (3.32). Contracting Ω with the Liouville vector L (Equation (3.7)) we obtain the 7-form

$$\omega = L\cdot\Omega = p^a\sigma_a \wedge \pi - \tfrac{1}{6}\eta_{abcd}\Gamma^a_{ef}p^e p^f \,\mathrm{d}p^{bcd} \wedge \eta\,. \tag{3.35}$$

The hypersurfaces of P of physical interest are of the following kind: There is a hypersurface S in spacetime (which may be associated in some way with an observer), and at each point $q\in S$ a region D_q of 4-momenta is assigned (which may be the region of sensitivity of a counter at q); then the points (q, p) of P with $q\in S$, $p\in D_q$ form a hypersurface Σ in P. If ω is restricted to such a 'special' hypersurface, we have simply

$$\omega = p^a\sigma_a \wedge \pi \tag{3.36}$$

since the restriction of η to such a Σ vanishes. In the particular case where S is a hypersurface of constant time of an observer using quasi-Galilean coordinates we can combine (3.33) and (3.36), obtaining

$$\omega = -\,E\,\mathrm{d}x^{123} \wedge \mathrm{d}p^{123} \wedge \mathrm{d}E\,, \tag{3.37}$$

where $E = -u_a p^a =$ (at the observer) p^4 is the energy of the particle with momentum p^a relative to the observer.

3.5. Liouville's Theorem; Distribution Functions, Collision Density, Collisionless Bolzmann Equation

Having assembled the mathematical tools we can proceed to develop the relativistic kinetic theory of gases.

So far we have introduced on the 8-dimensional phase space $P=\{(x^a, p_a)\}$ the Liouville vector L, the forms Ω and $\omega = L\cdot\Omega$, and the subspaces P_m of constant mass.

Applying the general formulae (3.19), (3.20) to L and Ω as defined by (3.6) and (3.34), we see that

$$\operatorname{Div} L = \frac{\partial^2 H}{\partial x^a\,\partial p_a} - \frac{\partial^2 H}{\partial p_a\,\partial x^a} = 0\,. \tag{3.38}$$

Hence, the results of the preceding subsection given below Equation (3.25) show: Ω and ω are invariant with respect to the phase flow (*Liouville's theorem*),

$$\mathscr{L}_L\Omega = \mathrm{d}\omega = 0\,, \quad \mathscr{L}_L\omega = 0\,. \tag{3.39}$$

The two basic properties of ω — $\mathrm{d}\omega = 0$ and $L\cdot\omega = 0$ — which imply via (3.21) the third, $\mathscr{L}_L\omega = 0$, enable one to consider ω as a volume measure on P/L, the space of phase orbits, since they guarantee that all cross sections of a tube of phase orbits have the same ω-volume, which may therefore be regarded as measuring the 'number of orbits' in such a tube. Physically, $\int_\Sigma \omega$ may be regarded as the 'number' of states intersecting Σ.

Let us now assume that a spacetime region is filled with a system of many particles. These particles may be atoms or ions, elementary particles, stars, or galaxies, depending on the situation of interest. Suppose that the interactions between the particles can be divided into *long range forces* and *short range forces*, and that the system is so *dilute* that the mean free path between successive short range interactions is much larger than the range of those interaction .Then it seems reasonable to account for the long range forces in terms of an external or self-consistent *mean field*, in which each particle moves like a test particle except during collisions, and to idealize the short range interaction as *point collisions* governed by special relativistic scattering theory. *The typical, average behaviour of such systems treated under the above simplifying assumptions* (and thus bypassing the difficulties of a more realistic many particles statistical mechanics) *is the object of the kinetic theory of gases.*

The only long range force which we shall take into account in these lectures is gravity. (The extension to electromagnetic forces is straightforward, see, e.g., [55]).

Let, then, g_{ab} denote the potential of the mean gravitational field. The associated phase flow then describes all possible 'free' motions of particles between collisions.

An individual gas is described in spacetime M as a *broken complex of worldlines* (Synge [63]) as indicated in Figure 22a. The vertices represent collisions, the lines between them are geodesic segments.

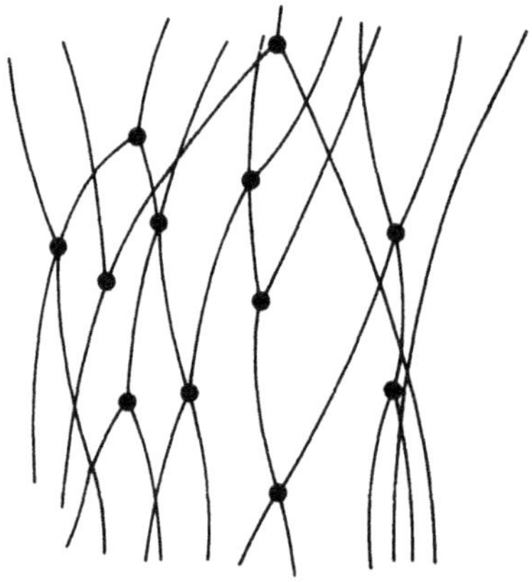

Fig. 22a. Spacetime picture of a relativistic gas.

In phase space P a *gas state* (or 'history') is represented as a collection of segments of phase orbits, the (free fall) states occupied by particles between collisions (see Figure 22b.) We do not assign phase orbits to particles during collisions; a collision corresponds to jumps of the participating particles from old to new phase orbits.

The state of the gas in a region of P is fully characterized by the *occupation functional* $\Sigma \to N[\Sigma]$ which assigns to any compact oriented hypersurface Σ the number of occupied orbit segments intersecting Σ. (The intersection of an orbit k with Σ is counted positively (negatively) if, at the event of intersection, the vector basis $(L, A_1, \ldots, A_7)$ has the same (opposite) orientation as an oriented frame of P, $(A_1, \ldots, A_7)$ being an oriented basis tangent to Σ.)

If D is a region in P with boundary ∂D then $N[\partial D]$ is the *number of collisions* in D, if creations are counted positively, and annihilations negatively (see Figure 22b).

An individual gas state is too complicated to be computed in detail, it also is not observable in practice. Therefore, one imagines a Gibbs-ensemble of such states representing a *macrostate* of a gas, and studies the *average distribution* of states,

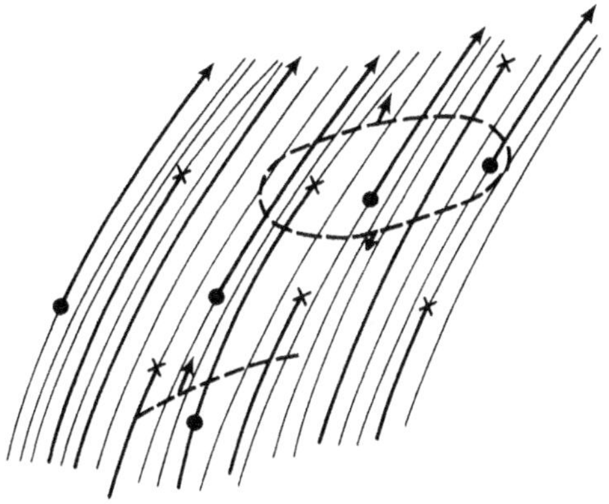

Fig. 22b. Phase-space picture of a gas. Occupied states fat. Creations are shown as dots, annihilations as crosses. The dashed lines represent open and closed hypersurfaces respectively.

given, say, by $\Sigma \to \bar{N}[\Sigma]$. (The averaging may have the additional merit that it disposes of such all-too-classical features of the gas model like precisely localized worldlines and collisions events. The average properties may well provide an approximate macroscopic description of a gas whose particles obey quantum laws; see, e.g., [64] for the nonrelativistic case.)

The averaged occupation functional $\bar{N}[\Sigma]$ shares with the volume $\int_\Sigma \omega$ the property that it vanishes if L is tangent to Σ. We assume that the average distribution of occupied states on any hypersurface Σ can be described in terms of a continuous density function f_Σ (with respect to ω):

$$\bar{N}[\Sigma] = \int_\Sigma f_\Sigma \omega .$$

We also assume that the average distribution of collisions in P can be described by means of a continuous collision density g:

$$\bar{N}[\partial D] = \int_D g\Omega .$$

These two regularity assumptions (which are usually not stated, but made implicitly as a matter of course) imply that the value of f_Σ at some phase point is independent of Σ, i.e.,

$$f_\Sigma|_{\Sigma\cap\Sigma'} = f_{\Sigma'}|_{\Sigma\cap\Sigma'}$$

for any two hypersurfaces. This is intuitively plausible since the number of collisions 'between' Σ and Σ' is of higher order of smallness than $\bar{N}[\Sigma]$, for small Σ; for a formal proof see [55]. Thus, *there exists a unique nonnegative function f on P such that*

$$\bar{N}[\Sigma] = \int_{\Sigma} f\omega . \tag{3.40}$$

f is called the (one particle) *distribution function* of the gas. Equation (3.37) shows that for any local observer the number of particles contained in d^3x having 3-momenta in d^3p and energies in dE is $Ef\, d^3x\, d^3p$; here particles of all rest masses are counted. According to the preceding discussion $f = f(x^a, p^a)$ is a *scalar* on P; i.e., its value is independent of the observer who happens to measure it.

The description of a gas state in terms of $f(x, p)$ disregards the distribution of particles with respect to internal degrees of freedom (like spin). If one wanted to include such fine details, one would have to consider a more complicated phase space and correspondingly more complicated distribution functions (see refs. [65], [66], [67]). We shall stick to a description which disregards such details, and assume tacitly spin-averaging in those cases where the existence and value of spin is taken into account for statistical reasons.

Assuming henceforth that f is continuously differentiable we obtain from (3.40) the net number of collisions in D as $\int_{\partial D} f\omega$. Using Stokes' theorem (3.16), the product rule for exterior differentiation, $d\omega = 0$, and the identity (3.28) one gets

$$\bar{N}[\partial D] = \int_D d(f\omega) = \int_D df \wedge \omega = \int_D df \wedge (L \cdot \Omega) = \int_D L(f)\, \Omega ,$$

i.e.,

$$L(f) = g \tag{3.41}$$

is the *collision density* in P with respect to Ω.

If there are no collisions, f satisfies *Liouville's equation* or the *collisionless Boltzmann equation*

$$L(f) = 0 . \tag{3.42}$$

More generally, this equation is the necessary and sufficient condition for *detailed balancing*, i.e., equality of the number of creations and annihilations of a state (x^a, p^a).

According to the meaning of L, $L(f)$ equals the rate of change of $f(x^a, p^a)$ along a free fall orbit. Hence *f satisfies Liouville's equation* (3.42) *if and only if it is a first integral of the geodesic equation*, a useful remark for solving that equation.

So far we worked in the 8-dimensional phase space P, using the idealization of a continuous mass spectrum which is adequate for macroscopic particles, e.g. stars. If the particles are atomic or subatomic they have effectively discrete masses m_A, if A is used as a label to distinguish the particle species, and then it is more appropriate to work with the discrete collection $\{P_{m_A}\}$ of 7-dimensional phase spaces. Let us adapt the preceding formulae to this case.

To obtain an analog of Ω on P_m, observe that for any function h, $h(p^2)\, \Omega$ is also an L-invariant measure on P. Replacing h by a δ-function we get in $\delta(p^2 + m^2)\, 2H(p)\, \Omega$

(where $H(p)=1$ if p is future directed and $H(p)=0$ otherwise) an L-invariant measure concentrated on P_m. Using coordinates (x^a, p^λ, p^4) on P and 'integrating' $\delta(p^2+m^2)\,2H(p)\,\Omega$ over p^4, we obtain the 7-form

$$\Omega_m = \eta \wedge \pi_m \tag{3.43}$$

on P_m. As before, η is the spacetime volume element, and

$$\pi_m = \frac{\sqrt{-g}\,\mathrm{d}p^{123}}{-p_4} \tag{3.44}$$

is a 3-form (measure) on the mass shell P_{mx}. For 'special' hypersurfaces in P_m – defined analogously to those on P which were considered in connection with Equation (3.36) – one finds

$$\omega_m = L_m \cdot \Omega_m = p^a \sigma_a \wedge \pi_m\,. \tag{3.45}$$

For a point observer using Galilean coordinates, (3.37) is replaced by

$$\omega_m = -\,\mathrm{d}x^{123} \wedge \mathrm{d}p^{123}, \tag{3.46}$$

which shows that ω_m *reduces locally to the ordinary volume element in* $(\mathbf{x}, \mathbf{p})$ *phase space*. (The sign results from our orientation conventions.)

As in the case of P, so also on P_m we have *Liouville's theorem*,

$$\mathscr{L}_{L_m}\Omega_m = \mathrm{d}\omega_m = 0\,, \quad \mathscr{L}_{L_m}\omega_m = 0\,, \tag{3.47}$$

so that all kinematical considerations can easily be transferred from P to P_m.

For particles of mass m there is a distribution function f_m on P_m such that

$$\bar{N}_m[\Sigma] = \int_\Sigma f_m \omega_m \tag{3.48}$$

gives the mean number of occupied particle states intersecting any (6-dimensional) hypersurface Σ of P_m. Because of (3.46) and (3.48), f_m *coincides*, for any local observer, *with the ordinary distribution function* in his infinitesimal $(\mathbf{x}, \mathbf{p})$-phase space.

The *collision density* in P_m is

$$L_m(f_m) = g_m, \tag{3.49}$$

and *Liouville's equation* on P_m is

$$L_m(f_m) = 0\,. \tag{3.50}$$

If *several species* of particles occur in the gas, we shall write m_A, P_A, f_A, Ω_A, ... for their masses, phase spaces, distribution functions, measures etc. It is then useful to think of $\bigcup_A P_A$ as *the* phase space, with components P_A.

For applications it is useful to express the volume element π_m defined in (3.44) in terms of the energy E and the direction angles ϑ, φ of some local Minkowskian observer. One gets

$$\pi_m = \sqrt{E^2 - m^2}\, \mathrm{d}E \wedge \sin\vartheta\, \mathrm{d}\vartheta \wedge \mathrm{d}\varphi\,, \tag{3.51}$$

so that (3.46) can also be written

$$|\omega_m| = E\sqrt{E^2 - m^2}\, \mathrm{d}V \wedge \mathrm{d}E \wedge \sin\vartheta\, \mathrm{d}\vartheta \wedge \mathrm{d}\varphi\,. \tag{3.52}$$

The definition (3.48) of the distribution function then shows that

$$E\sqrt{E^2 - m^2}\, f_m(x^a, E, \vartheta, \varphi) \tag{3.53}$$

is the number of particles per unit volume, per unit energy range and per unit solid angle observed at x^a with energy near E and direction near (ϑ, φ). This last expression is therefore what is often called the *spectral density of particles* (e.g., in cosmic ray physics). Whereas the value of f_m for given x^a, p^a is independent of the velocity of the observer, that of the spectral density is not; this is the reason why f_m is more useful for general considerations.

Specializing (3.53) to massless particles (photons, neutrinos) and using that they move with speed $c=1$ we infer that

$$E^3 f_0(x^a, E, \vartheta, \varphi)$$

is the energy flux density per unit energy range per unit solid angle, so that (because of $E=h\nu$ and our convention $h=1$),

$$I_\nu = (2\pi)^4 \nu^3 f_0(x^a, 2\pi\nu, \vartheta, \varphi) \tag{3.54}$$

is the *specific radiation intensity*. Hence, I_ν/ν^3 *is an observer-independent scalar.*

If radiation propagates through space without interaction with matter, Liouville's equation (3.50) holds, so that $\nu^{-3} I_\nu(x^a, p^a)$ has the same value along a light ray $x^a(v)$ with tangent (momentum) $p^a(v)$ for all observers. This statement is useful for discussing theoretically *optical observations*, particularly in cosmology. If we use indices o, s to denote quantities evaluated at the observer and the source, respectively, we get

$$I_{\nu_0} = \frac{I_{\nu_s}}{(1+z)^3}, \tag{3.55}$$

where

$$z = \frac{\nu_s}{\nu_0} - 1 = \frac{\lambda_0 - \lambda_s}{\lambda_s} \tag{3.56}$$

is the *redshift* between source and observer. Equation (3.55) is fundamental for the derivation of the *m-z-relation* and the formula for *angular sizes* of distant galaxies as well as for the discussion of the diffuse *background radiation*; its validity is not restricted by any symmetry assumptions about spacetime. The three factors $(1+z)$ in (3.55) can be understood intuitively as due to the energy change of each individual photon, the change of the rate with which photons are received compared to the rate with which they were emitted (thinning effect), and the dependence of the angular size of a source on the motion of the observer (aberration).

A particularly interesting application of the photon Liouville equation to cosmology is provided by the following *theorem* (ref. [68]): If a spacetime (M, g_{ab}) permits a solution f_0 to the $m=0$ Liouville equation $L_0(f_0)=0$ which is locally isotropic with respect to some geodesic, timelike, expanding unit vector field u^a, i.e. which has the functional form $f_0(x, p)=g(x, u_a p^a)$, then (M, g_{ab}) is necessarily a Robertson-Walker spacetime. Since the 3 K background radiation seems to have been collisionless for some time (see Sachs' lectures) and is highly isotropic with respect to us, this theorem gives a fairly good reason for assuming the 'late' universe to be (nearly) Robertson-Walker, provided the implication $\{f_0 \text{ isotropic}\} \Rightarrow \{g_{ab} \text{ is Robertson-Walker}\}$ is stable against small perturbations, for which there are some indications (see the appropriate remarks in ref. [69]); a general quantitative proof of stability is probably difficult to obtain.

Equation (3.55) shows: If the radiation emitted in a particular direction has a Planck-distribution with temperature T_s, the observer will also receive a Planck-distribution with temperature $T_0=(1+z)^{-1}\,T_s$.

3.6. Current densities, conservation laws, balance equations, and mean velocities

Let us rewrite (3.48) for a special hypersurface Σ of P_m, i.e., one whose projection into spacetime is a hypersurface G. Using (3.45) we get

$$\bar{N}_m[\Sigma] = \int_G \sigma_a \left\{ \int_{K_x} f_m p^a \pi_m \right\}, \tag{3.57}$$

where K_x is the part of the mass shell P_{mx} which is contained in Σ. This is the average number of particles with mass m passing through G with a 4-momentum in K_x. Examples where (3.57) (or its analog for a continuous mass spectrum) applies are measurements of cosmic rays and galaxy counts. In the first case G represents the history of the sensitive area of the counter, and K_x is determined by the energy and angular ranges registered by the counter. In the second case, G is a part of the past light cone of the observer, and $K_x=P_x$.

(3.57) shows that

$$N_m^a(x) = \int f_m p^a \pi_m \tag{3.58}$$

is the *particle 4-current density*. (We shall always write $\int \pi_m \ldots$ if the integration refers to the whole mass shell P_{mx}.) For any smooth distribution function $f_m \geqslant 0$, N_m^a is a timelike, future-directed vector field in spacetime. (Only if $m=0$ and if one permits f_0 to be a distribution concentrated on one generator of the null cone is N_0^a lightlike; this corresponds to the idealized case of a stream of photons without angular dispersion, associated with a locally plane wave.)

The spacetime tensor field

$$T_m^{ab}(x) = \int p^a p^b f_m \pi_m \tag{3.59}$$

is the *kinetic stress energy momentum tensor* of particles with mass m. This follows from the meaning of the distribution function and the characterization (1.39) of the matter tensor.

We assume always that all distribution functions vanish at infinity on the mass shell, i.e., for large energy, sufficiently fast so that integrals like (3.58), (3.59) exist.

If f_m is smooth (and not identically zero), then

$$(T_m)_{ab} V^a V^b > 0 \tag{3.60}$$

for all non-spacelike vectors, as is obvious from (3.59). This strong *energy inequality* implies, according to a theorem of Synge's (ref. [70], p. 99), that there exists a unique decomposition

$$\left.\begin{array}{l} T_m^{ab} = \varrho u^a u^b + p^{ab} \\ \text{with } u^a \text{ future directed, } u_a u^a = -1,\ p_{ab} u^b = 0\,. \end{array}\right\} \tag{3.61}$$

(In the case of a singular photon distribution mentioned before (3.59), $T_0^{ab} = k^a k^b$ with $k_a k^a = 0$, as for a plane wave.) Kinetic stress energy tensors map timelike vectors into timelike vectors and satisfy the various inequalities that have been used in singularity theorems.

If one considers a mixture of interacting particles of different species (A) carrying charges – electric e, baryonic b, leptonic L_e, L_μ, hypercharge Y – one can introduce associated currents like

$$B^a = \sum_A b_A N_A^a\,. \tag{3.62}$$

For *conserved charges* – e, b, L_e, L_μ and, for strong interactions, Y – these currents satisfy *conservations laws* like

$$B^a{}_{;a} = 0\,, \tag{3.63}$$

as follows from the fact that, according to (3.57) and (3.58), the flux $\int_{\partial D} B^a \sigma_a$ through any closed hypersurface ∂D in M must then vanish, which is equivalent to (3.63).

Although the quantities listed above seem to be the only ones which are strictly and generally conserved, it is important for applications that other currents are approximately conserved under appropriate circumstances; then one has more relations of the type (3.63). This holds for chemical and nuclear reactions where energy thresholds determine which quantities are practically conserved in given temperature ranges.

The current densities (3.58), (3.59) are special cases of moments of the distribution function in 4-momentum space. The *r-th order moment* is

$$\int p^{a_1} \dots p^{a_r} f_m \pi_m\,. \tag{3.64}$$

To obtain, in general, *balance equations* and, in particular, conservation laws, we have to compute the covariant divergence of the spacetime tensor field (3.64). To do this, introduce temporarily local inertial coordinates at x, and set up a one-to-one correspondence between P_{mx} and the mass shells of infinitesimal nearby events $x+\mathrm{d}x$ by means of parallel displacement of 4-momenta. At x, the operation $(\ )_{;a}$ then equals $(\ )_{,a}$, and because of the volume-preserving correspondence between P_{mx} and $P_{m(x+\mathrm{d}x)}$ the only x-dependence of (3.64) is the one through $f_m(x^a, p^a)$. Hence, at x,

$$\left(\int p^{a_1} \dots p^{a_r} f_m \pi_m\right)_{;a_1} = \int p^{a_2} \dots p^{a_r} L_m(f_m)\, \pi_m, \tag{3.65}$$

since at x, $L_m = p^a(\partial/\partial x^a)$ (see (3.8)). But, both sides of (3.65) being tensors, the equation is true independently of any coordinate condition.

Applying (3.65) to N^a_m we obtain the *balance equation for particles with mass m*.

$$N^a_{m;a} = \int L_m(f_m)\, \pi_m, \tag{3.66}$$

giving the number of m-type particles created per unit spacetime volume because of inelastic collisions (reactions).

Applying (3.65) to the kinetic matter tensor (3.59) we get

$$T^{ab}_{m;b} = \int p^a L_m(f_m)\, \pi_m. \tag{3.67}$$

This is the 4-*momentum balance* for particles with mass m. The right-hand side represents the 4-*force density* acting on the component of the gas consisting of particles with mass m due to collisions with particles of other species. An example for the latter is the force exerted on an electron gas by photons due to Compton scattering.

For a *simple gas*, consisting of only one species of particles, both expressions (3.66), (3.67) vanish. For a mixture, 4-*momentum conservation* during collisions requires

$$\sum_A T^{ab}_{A;b} = \sum_A \int p^a L_A(f_A)\, \pi_A = 0, \tag{3.68}$$

where one has to sum over all species. Similar equations involving $L_A(f_A)$ follow from scalar conservation laws like (3.63). (If an electromagnetic field is present, its energy momentum tensor has to be added on the left hand side of (3.68).)

If no collisions occur, Liouville's Equation (3.50) implies the conservation laws (3.63), $N^a_{A;a}=0$ (compare (3.66)), (3.68)). If collisions do occur, (3.63) (and similar equations for e, L_e, L_μ, Y) and (3.68) have to be postulated; these laws impose restrictions on the physically permissible variations of distribution functions which have to be respected by all kinetic equations which, like Boltzmann's equation, give a schematic account of collisions.

For a macroscopic,hydrodynamical description of matter a *mean* 4-*velocity* has to be defined. This can be done in several ways, and thus in order to prevent confusion it should always be stated which mean velocity is used.

For a simple gas, there are two sensible ways to define a mean 4-velocity. One can either use the fact that the particle current N_m^a is timelike and define a *kinematic mean velocity* u_k^a through

$$N_m^a = nu_k^a, \quad (u_k)^2 = -1; \tag{3.69}$$

or one can use the decomposition (3.61) of T_m^{ab} and define the $u^a = u_D^a$ which occurs therein as the *dynamical mean velocity*. An observer travelling with u_k^a will see no particle flux in his rest frame, one with u_D^a will measure no momentum density of the surrounding gas. Since in relativity $\mathbf{p} = m\gamma\mathbf{v}$ with γ depending on $|\mathbf{v}|$, these mean velocities differ. The observer with u_D^a measures the smallest energy density of the gas. – A process is *adiabatic*, i.e., free of a heat flux, if there is an observer who finds neither a particle flux nor an energy flux in his rest frame. This is the case if and only if $u_k^a = u_D^a$, which is quite a complicated condition on the distribution function.

For a *multicomponent gas* one can use the *total kinetic stress energy tensor*

$$T^{ab} = \sum_A T_A^{ab}, \tag{3.70}$$

decompose it as in Equation (3.61), and get a dynamical mean velocity u_D^a. Other reasonable choices are the *barycentric mean velocity*, defined as proportional to the total (rest) mass current $\sum_A m_A N_A^a$, and the *baryonic mean velocity* u_B^a, defined by

$$B^a = bu_B^a, \tag{3.71}$$

provided B^a is timelike (as it usually is). Other choices may be useful for special cases. (In practice, the choice of u^a does not matter much, but conceptually one should be aware of what one is talking about.)

With respect to *any* mean 4-velocity u^a one can decompose the total matter tensor (3.70) uniquely according to the scheme

$$T^{ab} = \varrho u^a u^b + 2u^{(a}q^{b)} + ph^{ab} + \pi^{ab}, \tag{3.72}$$

where

$$h_b^a = \delta_b^a + u^a u_b \tag{3.73}$$

projects the tangent space M_x onto the 3-space orthogonal to u^a, and where

$$u_a q^a = u_a \pi^{ab} = \pi_a^a = 0. \tag{3.74}$$

ϱ is the *mean energy density*, q^a the *mean energy flux density*, p the *mean kinetic pressure*, and π^{ab} is the *shear pressure* of the gas *relative to u^a*. (If $u^a = u_D^a$ then $q^a = 0$).

If, for some u^a, $q^a = 0$ and $\pi^{ab} = 0$, the system behaves like an *ideal gas* (*perfect fluid*); if all terms in (3.72) except the first one are neglected one speaks of *dust*.

3.7. Collisions; Boltzmann's Equation

In order to simplify many of the following equations it is advantageous to *change*, at this stage, *the meaning of the symbols π_A, f_A* slightly, if they refer to non-classical particles. Since a one-particle quantum state (with definite linear momentum and spin) corresponds asymptotically to a region of volume $h^3/w=(2\pi)^3/w$ in classical $(\mathbf{p}, \mathbf{q})$ phase space, if w is the *spin weight* of the particle species ($w=2$ for neutrinos, photons and gravitons, $w=2s+1$ for massive particles with spin s), $[(2\pi)^3/w_A]f_A(x, p)$ gives approximately the *average occupation number* of simple quantum states which are nearly eigenstates of p, have a definite, but arbitrary, spin and are localized near x. We therefore write, henceforth, f_A instead of $[(2\pi)^3/w_A]f_A$. Also, we write π_A instead of $[w_A/(2\pi)^3]\pi_A$. Then the new f_A's will be dimensionless, equations like (3.48), (3.58), (3.59) remain unchanged, and in formulae like (3.53) adjustments are easily made. For classical particles one may keep the old meaning and omit those terms in later equations which are of quantum origin. Note that then

$$f_A \leqslant 1 \quad \text{for fermions}. \tag{3.75}$$

In order to obtain time evolution equations for distribution functions in the presence of collisions it is necessary to make assumptions about the frequencies of various collisions. We consider a gas mixture containing species A, B, ..., and wish to take into account elastic and inelastic collisions. Let

$$(x;\, p_A, p_B, \ldots \rightarrow p_C, \ldots) \tag{3.76}$$

indicate a collision in which particles of species A, B,... with 4-momenta $p_A, p_B, \ldots$ collide at $x \in M$ and produce particles C,... with $p_C, \ldots$. (If several incident particles belong to the same species, one has to write p_A, p'_A etc. instead of p_A, p_A, of course; this is tacitly assumed here and in the sequel.)

We assume that the spacetime density of collisions (3.76) with 4-momentum ranges $K_A, K_B, \ldots, K_C, \ldots$ at the event x is expressible as

$$\int_U V(x;\, p_A, p_B, \ldots;\, p_C, \ldots)\, \delta(\Delta p)\, \pi_{AB\ldots C\ldots} \tag{3.77}$$

where $U = K_A \times K_B \ldots \times K_C \times \ldots$ and where we have used the abbreviation

$$\pi_{AB\ldots C\ldots} = \pi_A \wedge \pi_B \wedge \cdots \wedge \pi_C \wedge \cdots$$

which will be employed henceforth. V is a nonnegative, measurable function depending on the (macro) state of the gas, symmetrical with respect to the momenta of incoming or outgoing particles of the same species. The δ-function (on R^4) of the momentum change

$$\Delta p = p_A + p_B + \cdots - p_C - \cdots \tag{3.78}$$

in (3.77) expresses that collisions conserve 4-momentum. Hence, the significant domain of definition of V is given by $\Delta p=0$.

In general, V depends not only on the one-particle distribution functions $f_A, f_B, \ldots$ of the system, but also on the pair distributions $f_{AB}(x, p_A;\, x', p'_B)$ and higher order correlations; its determination is a complicated problem of many-body theory. We shall be content with the simple *Boltzmann approximation* (as modified by the requirements of quantum statistics, if the particles are atomic or subatomic) which consists in assuming

$$V(x;\, p_A, p_B, \ldots;\, p_C, \ldots) = f_A f_B \ldots f^C \ldots R^{C\ldots}_{AB\ldots}, \tag{3.79}$$

where we have used the abbreviations

$$\begin{aligned} f_A &= f_A(x, p_A),\ f_B = f_B(x, p_B), \ldots, \\ f^C &= 1 \pm f_C(x, p_C), \ldots . \end{aligned} \tag{3.80}$$

Here and in the remainder, the *upper sign* in $\pm$ or $\mp$ refers to *bosons*, the *lower* one to *fermions*.

The factors $f^C \ldots$ associated with the out-states take care of stimulated emission or scattering in the case of bosons, and of the suppression of processes with occupied out-states in the case of fermions, in accordance with the Pauli principle. (In the case of classical (macroscopic) particles, these factors f^C are to be omitted.) The *transition probability* per unit spacetime volume and unit collision volume $(\delta(\Delta p)\pi_A \wedge \ldots \pi_C \wedge \ldots)$, $R^{C\ldots}_{AB\ldots}$, is a Lorentz invariant function of the momenta $p_A, p_B, \ldots;\, p_C, \ldots$ (in that order), symmetrical with respect to 'in' or 'out' momenta of the same species.

The simple Ansatz (3.79) is plausible and, to a certain extent, justifiable from first principles under the following conditions (besides the ones formulated on p. 64):

(1) The interaction ranges are small in comparison with (a) the mean radius of curvature of spacetime and (b) the scale on which the distribution functions change appreciably with x.

(2) The gas is in a sufficiently disordered state so that particles which are about to collide have uncorrelated momenta.

(3) The evolution of the system is essentially due to collisions each involving a few particle states only, and each being almost unaffected by the presence of particles in states not occurring in that collision.

No attempt to derive (3.79) will be made here. A sketch of a derivation which shows how the above assumptions enter has been given in ref. [55]. For a more complete derivation in the nonrelativistic case see, e.g., ref. [64].

If the mean occupation numbers are all much smaller than 1, (3.79) reduces practically to Boltzmann's original Ansatz (with $f^C \to 1$).

The transition probabilities $R^{C\ldots}_{AB\ldots}$ in (3.79) have to be found experimentally or deduced from collision theory. In the case of two incoming particles and, say, q outgoing ones,

$$(x;\, p_A, p_B \to p_{C_1}, \ldots, p_{C_q}), \quad (q \geqslant 1)$$

it is usual to express the collision frequencies in terms of *cross sections*. For collisions with momentum ranges K_A, K_B, $K_{C_1}, \ldots, K_{C_q}$ the particle number densities of the incident beams in the locally Minkowskian rest frame of an observer with 4-velocity u are, according to (3.33), (3.45) and (3.48),

$$n_A = \int\limits_{K_A} f_A \left|u \cdot p_A\right| \pi_A, \quad n_B = \int\limits_{K_B} f_B \left|u \cdot p_B\right| \pi_B .$$

Choosing u such that the incident particles move collinearly before the encounter, $u=\lambda p_A+\mu p_B$, we get for the relative speed of these particles in the u-frame

$$v_{AB}=|\mathbf{V}_B-\mathbf{V}_A|=((u\cdot p_A)(u\cdot p_B))^{-1}P_{AB},$$

where

$$P_{AB}=((p_A\cdot p_B)^2-(m_Am_B)^2)^{1/2}. \tag{3.81}$$

Hence the *incident flux* for small ranges $K_A,\ldots$ is

$$\phi=n_An_Bv_{AB}=\int_{K_A\times K_B} f_Af_BP_{AB}\pi_{AB}.$$

On account of (3.77) and (3.79) the spacetime density of collisions of the specified kind, for unpopulated final states (as in scattering experiments), for small momentum ranges, is given by

$$\phi\int P_{AB}^{-1}R_{AB}^{C_1\ldots C_q}\delta(\Delta p)\,\pi_{C_1\ldots C_q}.$$

Therefore, by definition, the (invariant relativistic) cross section for scattering from the state (p_A,p_B) into the range $K_{C_1}\ldots K_{C_q}$ is given by the last integral, so that the *differential cross section* is

$$Q_{AB}^{C_1\ldots C_q}=P_{AB}^{-1}R_{AB}^{C_1\ldots C_q}\,\delta(\Delta p)\,\pi_{C_1\ldots C_q}. \tag{3.82}$$

If $q\geqslant 2$, $Q_{..}^{..}$ is, for fixed (p_A,p_B), a differential form of degree $3q-4$ on the submanifold of $P_{C_1}\times\cdots\times P_{C_q}$ defined by $\Delta p=0$, because of the δ-factor. If $q=1$, which corresponds to an absorption or fusion process, the expression (3.82) gives, on integration over P_C, the (total) cross section

$$Q_{AB}=\int_{P_C}Q_{AB}^C=\sigma_{AB}^C\,\delta\left(\frac{m_A^2+m_B^2-m_C^2}{2p_A\cdot p_B}-1\right) \tag{3.83}$$

for absorption of B-particles with p_B by A-particles with p_A, where the constant

$$\sigma_{AB}^C=\frac{w_C}{2\pi^3}\cdot(m_C^2-m_A^2-m_B^2)^{-1}\times$$
$$\times\,(m_A^4+m_B^4+m_C^4-2[m_A^2m_B^2+m_B^2m_C^2+m_C^2m_A^2])^{-1/2}\,R_{AB}^C \tag{3.84}$$

(which looks much simpler if $m_B=0$) is the cross section per A-particle for absorption of any B-particle such that a C-particle results. (More about the evaluation of expressions like (3.82) is found in ref. [54] and in books on scattering theory.)

The transition probabilities $R_{..}^{..}$ can be derived from the scattering operator. The latter's unitarity implies (ref. [71]) that

$$\int(R_{AB\ldots}^{C\ldots}-R_{C\ldots}^{AB\ldots})\,\delta(\Delta p)\,\pi_C\wedge\cdots=0, \tag{3.85}$$

a relation between a collision and its inverse. The integration refers to the out-states of (3.76).

If the scattering operator is invariant with respect to the spacetime reflection PT, the stronger symmetry

$$R^{C\ldots}_{AB\ldots} = R^{AB\ldots}_{C\ldots} \tag{3.86}$$

holds (see, e.g., [72]).

We now set up a *collision balance* for an arbitrary state (x, p_A), to obtain Boltzmann's equation. Combining the statements contained in equations (3.49), (3.77) and (3.79) we conclude

$$L_A(f_A) = \sum_{\text{coll.}} \frac{r_A}{r_{AB\ldots C\ldots}} \int (f_C \ldots f^A f^B R^{AB\ldots}_{C\ldots} - \\ - f_A f_B \ldots f^C \ldots R^{C\ldots}_{AB\ldots})\, \delta(\Delta p)\, \pi_{B\ldots C\ldots}, \tag{3.87}$$

where

$$r_{AB\ldots C\ldots} = r_A!\, r_B! \ldots r_C! \ldots,$$

an abbreviation to be used also in the sequel. The summation extends over all kinds of collisions in which A-particles participate. The integers r_A, r_B, ... indicate how many particles of species A, B,... occur on each side of the collision (3.75). The left-hand side of (3.87) is to be evaluated at (x, p_A); the arguments on the right-hand side are those specified in (3.80) and below. The factorials $(r_A - 1)!r_B!$ etc. are needed since without them the number of collisions with several indistinguishable particles on one side would be overestimated. Care must be taken to count each relevant contribution to the production rate (3.87) exactly once.

As an example of (3.87) consider photons (species A) absorbed by a nondegenerate medium (ground state B, excited state $C, f^C \approx 1$). Then, using (3.87) and (3.81), (3.82), (3.83), one gets

$$\begin{aligned} L_A(f_A) &= - f_A \int f_B R^C_{AB}\, \delta(\Delta p)\, \pi_{BC} = \\ &= - f_A \int_B f_B \pi_B P_{AB} \int_C Q^C_{AB} = \\ &= - f_A \sigma^C_{AB} \int f_B \pi_B \,|(p_A \cdot p_B)|\, \delta\left(\frac{m_B^2 - m_C^2}{2 p_A \cdot p_B} - 1\right). \end{aligned}$$

Assuming that the medium is nonrelativistic ('cold') with a mean 4-velocity u, $p_B \approx \approx m_B u$, we obtain approximately

$$L_A(f_A) \approx - f_A \sigma^C_{AB} |p_A \cdot u|\, \delta\left(\frac{m_B^2 - m_C^2}{2 m_B (u \cdot p_A)} - 1\right) n_B,$$

where n_B is the mean number density of the medium. Thus

$$L_A(f_A) \approx - f_A n_B \kappa, \tag{3.88}$$

where the *invariant opacity*

$$\kappa = |u \cdot p_A| \, \sigma = |u \cdot p_A| \, \sigma^C_{AB} \, \delta \left(\frac{m_B^2 - m_C^2}{2m_B (u \cdot p_A)} - 1 \right) \tag{3.89}$$

depends on the photon energy relative to the medium only and is sharply peaked at the transition energy ($\approx m_C - m_B$). By integrating (3.88) along a light ray one can compute absorption corrections to (3.55), e.g. (3.88) and (3.89) give, in a notation obvious from Figure 23,

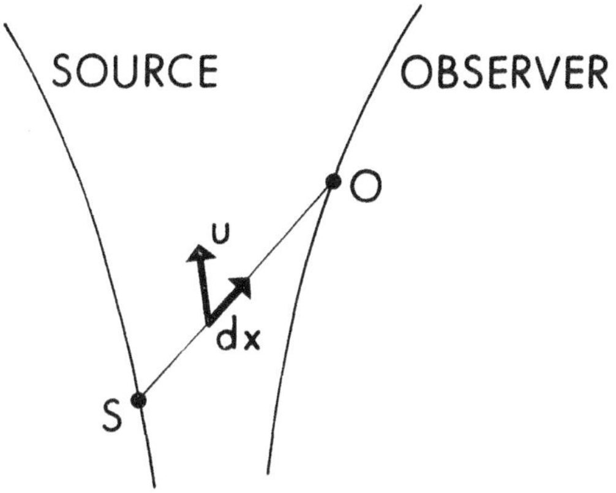

Fig. 23.

$$f_A(\text{Obs.}) = f_A(S)\, e^{-L},$$

where

$$L = - \int_S^O \sigma n_B u_a \, \mathrm{d}x^a \tag{3.90}$$

is the *optical depth* of the medium, i.e., the mean number of scattering particles contained in a tube of cross sectional area σ along the beam. Similarly one can include scattering and emission by a medium into a *photon transport equation* of the type (3.88) (see [59]).

The conservation laws considered in Subsection 3.6 follow from the Boltzmann equations, provided the transition probabilities or cross sections restrict the collisions accordingly. Since these conservation laws are more fundamental than the Boltzmann equation, their compatibility with the latter lends support to the Boltzmann approximation. Other kinetic equations are subject to the same test.

If the Boltzmann Equation (3.87) is combined with the particle balance equation (3.66) for species A, there results

$$N^a_{A;a} = \sum_{\text{reactions}} \bar{r}_A v \, . \tag{3.91}$$

Here

$$v = r^{-1}_{AB..C..} \int (f_A f_B \ldots f^C \ldots R^{C.}_{AB..} - {}_{C..} f^A f^B \ldots R^{AB..}_{C..}) \, \delta(\Delta p) \, \pi_{AB..C..}, \tag{3.92}$$

the *rate* of the reaction $A+B+\ldots \to C+\ldots$, is the difference of the spacetime densities of collisions of the type $A+B+\ldots \to C+\ldots$ and the inverse type $C+\ldots \to A+B+\ldots$; and $\bar{r}'_A$ the stöichiometric coefficient of A in the (first) reaction, is the difference between the number of A-particles created and the number of A-particles destroyed in that reaction. If several reactions occur, one has to select a system of linearly independent reactions (as judged in terms of the 'vectors' $(\bar{r}_A, \bar{r}_B, \ldots)$) and sum over them, as indicated in (3.91). (In the case of PT-invariance, the relation (3.86) can be employed to simplify the integrand in (3.92) slightly.)

In all the preceding considerations it did not matter whether g_{ab} was considered as an external field or as the field generated by the gas itself. In the first case the system of Boltzmann equations (3.87) governs the evolution of the (test) gas, whereas in the much more complicated second case the coupled *Boltzmann-Einstein system of equations*,

$$\begin{aligned} G^{ab} + \Lambda g^{ab} &= 8\pi \sum_A \int f_A p^a p^b \pi_A, \\ L_A(f_A) &= \text{coll. integr. } (3.87), \end{aligned} \tag{3.93}$$

determines 'selfconsistently' the evolution of field and matter variables, $(g_{ab}, f_A, f_B, \ldots)$.

This system of equations may be considered as the *basis of a large part of the macroscopic theory of gravity* of which other systems, involving dust or perfect or viscous fluids as sources, are approximations. A model similar to (3.93) results if one treats the source as a quantum gas obeying quasi-free field equations (with g_{ab} replacing the Minkowski metric in quantized field equations) which generates the classical mean field g_{ab} selfconsistently via the expectation value of its stress-energy tensor (ref. [73]). Other models involve phenomenological stress-energy tensors, with equations of state taken from local physics, as in the case of neutron star theory. (As far as experience is concerned, it seems that one may even hold the position that descriptions of gravity as a selfconsistent mean field are correct *in principle*, though this author is not aware of a demonstration that this point of view can be combined consistently with a quantum description of matter. Though this point of view does not appear very convincing, it represents a challenge.)

It is very remarkable that Y. Choquet-Bruhat and D. Bancel have been able to establish rigorously that the Equations (3.93) do in fact define a dynamical system, i.e., that the corresponding Cauchy initial value problem has physically unique solutions (locally) depending continuously on the initial data [74, 75]. This result, for which it is essential (though formally trivial) that $T^{ab}{}_{;b}=0$ follows *separately* from the Einstein equation and the Boltzmann equations, puts the theory on a firm mathematical basis.

A few exact solutions of the coupled equations (3.93) with vanishing collision integrals are known. They represent either collisionless systems, or collision-dominated systems in equilibrium. So far, no solutions with non-vanishing collision integrals seem to have been found, although some general theorems about isotropic solutions (as defined on p. 69) have been derived, and the system (3.93) has been analyzed in considerable detail. For a recent review of these questions see refs. [69], [76].

3.8. Growth of entropy, equilibrium

In the general theory of relativity, an *instantaneous state* of a spatially extended physical system is the collection of Cauchy data associated with a spacelike hypersurface G in spacetime. According to this general concept, an instantaneous state of a gas with various species $A, \ldots$ of particles is obtained, in the approximation of kinetic theory, by constructing the hypersurfaces Σ_A in the phase spaces P_A which 'lie over' G (i.e., $\Sigma_A = \{(x, p) \mid x \in G, p \in P_{x,A}\}$), and giving the distribution functions f_A on Σ_A.

The *entropy* for any physical system is a functional defined on the set of (statistical) states of that system which measures the degree of uncertainty (or randomness) ascribed to that state. In the case of kinetic theory, the value of the entropy S for the state (G, f_A) is defined by

$$S[G, f_A, \ldots] = -\sum_A \int_{\Sigma_A} (f_A \log f_A \mp f^A \log f^A)\, \omega_A. \tag{3.94}$$

This choice of the entropy functional can be justified by using the correspondence between the classical gas model used here and the quantum description of a gas. (Recall the remarks at the beginning of the previous subsection.) If the integrals in (3.94) are approximated by sums over cells each of size $\int \omega_A = 1$ corresponding to simple quantum states, (3.94) reduces to a sum containing average occupation numbers f_A of one-particle states: this sum is the quantum mechanical expression for the entropy of the statistical state specified by those mean occupation numbers, an expression which in turn can be justified on very general grounds information theoretically. (See ref. [77]). In the case of classical particles ($f^A \log f^A = 0$) a more elementary derivation in the manner of Boltzmann's counting argument can be given (see ref. [63]).

Accepting (3.94) and recalling the expression (3.45) for ω_A, one recognizes that the total entropy can be expressed as a flux through the spacetime hypersurface G,

$$S[G, f_A, \ldots] = \int_G S^a \sigma_a, \tag{3.95}$$

where

$$S^a = \sum_A S^a_A \tag{3.96}$$

and

$$S^a_A = -\int (f_A \log f_A \mp f^A \log f^A)\, p^a \pi_A. \tag{3.97}$$

The vector field S^a_A on M is called the *entropy current density* of species A. With respect to an observer, S^4_A is the entropy density, and S^λ_A is the entropy 3-current.

The formula (3.95) shows that the total entropy of an adiabatically isolated system will have the property never to decrease in the course of time (i.e., if G is pushed into the future) if and only if

$$S^{a}{}_{;a} \geqslant 0 \tag{3.98}$$

always. We shall now prove that this *entropy inequality* (Boltzmann's H-theorem) follows from the Boltzmann Equations (3.87), *provided we assume the PT-symmetry* (3.86) to hold for all collisions. (For nondegenerate gases, i.e. when all f^C-factors can be replaced by 1, the weaker condition (3.85) suffices, as is shown in ref. [57]. Whether this is true even in the general case I do not know.)

The divergence of S^a_A can be computed by the method explained below (3.64) and gives

$$S^{a}_{A;a} = \int \log\left(\frac{1}{f_A} \pm 1\right) L_A(f_A)\, \pi_A . \tag{3.99}$$

Inserting $L_A(f_A)$ from the Boltzmann equations (3.87) and summing over all species one obtains, after some rearranging of terms and using (3.86).

$$S^{a}{}_{;a} = \sum_{\text{coll.}} \frac{1}{r_{AB..C..}} \int \pi_{AB..C..}\, \delta(\Delta p)\, R^{C..}_{AB..} \times$$
$$\times \log \frac{(f_A f_B ..)(f^C ..)}{(f_C ..)(f^A f^B ..)} \cdot \{(f_A f_B ..)(f^C ..) - (f_C ..)(f^A f^B ..)\} . \tag{3.100}$$

The sum is to be extended over all pairs consisting of a collision (3.76) and its inverse. If a collision type is identical with its inverse, only the first term needs to be written in { }; but even then it is convenient to write both terms and compensate this by an extra factor $\frac{1}{2}$ in front of the integral.

Since $(a-b)\log a/b \geqslant 0$ for positive a, b, with equality holding only if $a=b$, we conclude from (3.100) not only (3.98), but that *each pair of inverse collisions contributes a nonnegative part to the entropy production density* $S^{a}{}_{;a}$.

Moreover, if we assume that the transition probabilities are strictly positive almost everywhere (with respect to $\delta(\Delta p)\pi_{AB..C..}$), (3.100) shows that *the contribution of the collisions of a particular type and its inverse to* $S^{a}{}_{;a}$ *vanishes* at an event x *if and only if* at that event

$$f_A f_B \ldots f^C \ldots = f_C \ldots f^A f^B \ldots . \tag{3.101}$$

A glance at the Boltzmann equation (3.87) reveals that *the last condition is* in turn *necessary and sufficient in order that the collisions* in question *do not contribute to* $L_A(f_A)$.

Combining the preceding three results with (3.99) we conclude:

At an event x, *the three following statements are equivalent*:

(a) $S^{a}{}_{;a} = 0$,

(b) $L_A(f_A) = 0$ for all A and all p_A, (3.102)

(c) Relations (3.101) hold for all collisions.

In view of these results and the physical interpretations of them we define: A gas is in *local equilibrium* at x if any (and hence all) of the conditions (a), (b), (c) holds. *Partial equilibrium* with respect to certain types of collisions holds if, for those collisions, the Equations (3.101) are satisfied. In particular, if (3.101) is satisfied for all elastic collisions in the system, we shall say the system is in *thermal* (though not necessarily 'chemical') *equilibrium*.

These definitions are really satisfactory from a theoretical point of view only if one could prove that, under suitable circumstances, a system has a tendency to relax into an equilibrium state. Although the last assertion is very plausible in view of the structure of the Boltzmann equation, the entropy inequality and the preceding remarks, no exact theorems seem to have been established in the relativistic case, nor are counterexamples (in the case of gravitational collapse?) known to me. These are interesting questions of principle.

For elastic binary collisions the equation (3.101) can be rewritten as

$$\log\left(\frac{1}{f_A} \pm 1\right) + \log\left(\frac{1}{f_B} \pm 1\right) = \log\left(\frac{1}{f_{A'}} \pm 1\right) + \log\left(\frac{1}{f_{B'}} \pm 1\right),$$

which has to hold for all momenta on the mass shells satisfying $\Delta p = 0$. The general solution is known to be (refs. [78], [79], [56] contain proofs)

$$f_A = [\exp(-\alpha_A - \beta_a p^a) \mp 1]^{-1}, \tag{3.103}$$

where α_A is a scalar, β^a a future directed timelike vector, and

$$\alpha_A \leqslant \beta m_A \quad \text{for bosons} \tag{3.104}$$

since $f_A \geqslant 0$. If $A \neq B$, a similar equation holds for f_B, and β^a in f_A and f_B is the same vector. We put

$$\beta^a = \beta u^a, \quad \beta > 0, \quad u^2 = -1. \tag{3.105}$$

Assuming that all species in a mixture are, directly or indirectly, coupled to each other by elastic binary collisions, we conclude that *thermal equilibrium at x is characterized by (3.103).*

If there is thermal equilibrium, and if inelastic collisions (reactions) are also possible in the system, then (3.103) and (3.101) show that *chemical equilibrium is characterized by the equations* ('law of mass action')

$$\alpha_A + \alpha_B + \cdots = \alpha_C + \cdots. \tag{3.106}$$

Since all collisions conserve baryon number b, electric charge e, and the lepton numbers L_e, L_μ, the conditions (3.106) are certainly satisfied if one puts

$$\alpha_A = \lambda_e e_A + \lambda_b b_A + \lambda_{L_e} L_{e_A} + \lambda_{L_\mu} L_{\mu A} \tag{3.107}$$

and similarly for $\alpha_B, \ldots$ (with the same $\lambda_e, \ldots$); e_A is the electric charge of one A-particle etc. If, in certain energy ranges, the actually occurring reactions can be characterized by the universal conservation laws together with some other (approximate) ones, the general solution of (3.106) can be written in a form analogous to (3.107), with additional terms corresponding to the additional conserved quantities.

Since photons seem never to be restricted by conservation laws (of the scalar type here relevant), one has, in accordance with (3.107), $\alpha_\gamma=0$; indeed then (3.103) and (3.54) give Planck's formula. Other illustrations of the preceding remark are the law for the distribution of atoms and ions over excited and ionization states (Saha equation), and the distribution of nuclei in the early universe (see Sachs's lectures for details).

For an observer travelling with the *kinetic mean 4-velocity* u^a defined in (3.105) the equilibrium distribution functions (3.103), or

$$f_A = [\exp(-\alpha_A + \beta E) \mp 1]^{-1}, \tag{3.108}$$

depend on the *thermal energy* $E = -u \cdot p$ only. Working in a local inertial frame with $u^a = \delta^a_4$ and using (3.51), (3.58), (3.59), (3.97), one obtains from (3.108) for each particle species the results (3.61), (3.69) with $u^a_K = u^a$, and

$$S^a = s u^a, \tag{3.109}$$

as one would expect. Moreover, the *particle density* n, *energy density* ϱ, *pressure* p and *entropy density* s in those formulae are given by the well known expressions

$$\left.\begin{aligned} n &= \int_m^\infty N\,\mathrm{d}E, \quad \varrho = \int_m^\infty NE\,\mathrm{d}E, \\ p &= \tfrac{1}{3}\int_m^\infty N(E^2 - m^2)\,E^{-1}\,\mathrm{d}E, \end{aligned}\right\} \tag{3.110}$$

$$s = \beta(\varrho + p) - \alpha n, \tag{3.111}$$

where

$$N(E: \alpha, \beta; m, w, \pm) = \frac{w}{2\pi^2} \frac{E\sqrt{E^2 - m^2}}{\exp(-\alpha + \beta E) \mp 1} \tag{3.112}$$

is the *distribution-in-energy* of particles specified by mass, spin weight, symmetry character, in an equilibrium state given by α, β. ((3.111) is obtained from the integral representation of s by partial integration.)

The quantities n, ϱ, p, s are defined on the 2-space $\{(\alpha, \beta)\}$ of equilibrium states (α, β). Besides (3.111) they satisfy the Pfaffian equation

$$\mathrm{d}p = \frac{n}{\beta}\,\mathrm{d}\alpha - \frac{\varrho + p}{\beta}\,\mathrm{d}\beta, \tag{3.113}$$

as follows from (3.110)–(3.112). Hence, $p = p(\alpha, \beta)$ is a *thermostatic potential* from which n, ϱ and (via (3.111)) s can be computed by differentiation. Applying Legendretransformations one can pass to other potentials. In particular one can introduce $\varrho(s, n)$ and get (one version of) the *Gibbs relation*

$$d\varrho = \beta^{-1}\, ds + \alpha\beta^{-1}\, dn\,, \tag{3.114}$$

a formula more familiar in thermostatics than the preceding one. From it we see that $\beta^{-1} = T$ is the (absolute) *temperature*, and $\alpha\beta^{-1} = \kappa$ is the (relativistic) *chemical potential* of the gas component. (The relativistic chemical potential, κ, differs from the usual nonrelativistic one by the summand m.)

In the case of a mixture in thermal equilibrium, relations of the form (3.113), (3.114) hold for each component; since $\beta = T^{-1}$ is the same for all components and the (extensive) quantities ϱ, s are additive, we get from (3.114) *for any mixture in local thermal equilibrium* the *Gibbs relation*

$$d\varrho = T\, ds + \sum_A \kappa_A\, dn_A\,, \tag{3.115}$$

and from (3.111)

$$\varrho = Ts + \sum_A \kappa_A n_A - p\,. \tag{3.115'}$$

The thermostatic functions defined by (3.110)–(3.112) have been studied extensively, see ref. [54] for details and further references. As an illustration, take $m = \alpha = 0$ (photons, neutrinos). Then

$$n \propto T^3\,, \quad \frac{\varrho}{n} = \tfrac{3}{4}\sigma T\,, \quad p = \tfrac{1}{3}\varrho\,, \quad \frac{s}{n} = \sigma_{\pm} = \frac{\tfrac{3}{4}\int_0^\infty x^3\, dx/(e^x \mp 1)}{\int_0^\infty x^2\, dx/(e^x \mp 1)} = \text{const.}\,,$$

i.e. the entropy per particle is a constant σ.

For a non-degenerate (i.e., $f_A, f_B, \ldots \ll 1$) mixture in local thermal equilibrium Equations (3.92) and (3.108) give

$$v = (e^{\alpha_A + \alpha_B + \cdots} - e^{\alpha_C + \cdots})\, k_{AB..}^{C..}\,, \tag{3.116}$$

where

$$k_{AB..}^{C..} = r_{AB..C..}^{-1} \int e^{-\beta(E_A + E_B + ..)}\, \delta(\Delta p)\, R_{AB..}^{C..}\, \pi_{AB..C..} \tag{3.117}$$

is a function of the temperature. Equation (3.116) for the reaction rate v clearly shows the role of the α_A's or, equivalently, the chemical potentials $\kappa_A, \ldots$, and demonstrates how this rate is related to the deviation from the condition (3.106) for chemical equilcibium.

The ronsiderations so far referred to local equilibrium at an event x. *Equilibrium in a domain D of spacetime* demands, according to (3.102b), that the equilibrium variables $\alpha_A(x)$, $\beta^a(x)$ of (3.103) satisfy $L_A(f_A) = 0$; that is, $\alpha_A + \beta \cdot p$ has to be constant on any particle trajectory in D. That implies:

$$\alpha_A = \text{const. in } D \tag{3.118}$$

and

$$\beta_{(a;b)} = 0 \quad \text{if} \quad m \neq 0 \text{ for at least one species } A\,, \tag{3.119}$$

$$\beta_{(a;b)} = \lambda g_{ab} \quad \text{if} \quad m_A = m_B = \cdots = 0\,. \tag{3.120}$$

Equation (3.119) states that the metric tensor g_{ab} has vanishing Lie derivative with respect to β^a, as can be verified using the definition of Subsection 3.3. The metric is therefore invariant with respect to the one-dimensional group of transformations generated by β^a. In coordinates adapted to β^a (recall Subsection 3.3) we have

$$g_{ab,4} = 0\,, \quad \beta^a = \delta^a_4\,, \tag{3.121}$$

where x^4 is a timelike coordinate since β^a is timelike. Spacetimes admitting such a group are called *stationary*. The result (3.119) therefore means that, except for the special case described by (3.120), *equilibrium is possible only in a stationary spacetime domain.* (In the exceptional case (3.120) β^a defines conformal transformations. Spacetime is then said to be conformally stationary. Examples are the Robertson-Walker spacetimes of cosmology.)

(3.121) means that $g_{ab}\beta^a\beta^b = g_{44} = -\beta^2 = -T^{-2}$. Therefore, in contrast to classical thermodynamics, in general-relativistic thermodynamics the temperature T, and because of (3.118) also the chemical potentials κ_A $(=\alpha_A T)$, are not spatially constant, but vary in accordance with *Tolman's law*

$$T \propto (-g_{44})^{-1/2}\,, \quad \kappa_A \propto (-g_{44})^{-1/2}\,. \tag{3.122}$$

Since in a stationary spacetime the redshift between stationary sources and observers is given by

$$z = \frac{(\sqrt{-g_{44}})_{\text{observer}}}{(\sqrt{-g_{44}})_{\text{source}}} - 1\,, \tag{3.123}$$

the relation (3.122) can be understood physically from the requirement that the equilibrium between two elements of matter is not disturbed by the exchange of photons in spite of the redshift of the latter. The argument is simplest if one employs Wien's displacement law.

3.9. Non-equilibrium, transport processes, reactions

Whereas the equilibrium solutions of Boltzmann's equation can be written down exactly, exact solutions describing nonequilibrium processes are presently not available. (One special exact solution is known in the non-relativistic case only.) Therefore, one has to resort to approximation methods. This subsection is devoted to a brief description of such approximation methods and some of the results obtained. For details we have to refer, more than in the preceding subsections, to the original papers, since the computations are quite involved.

We restrict our attention to *near-equilibrium* situations. Then each distribution function can be written as a perturbation

$$f_A = f_A^0 (1 + g_A) \tag{3.124}$$

of a local thermal equilibrium distribution f_A^0 with parameters $\alpha_A(x)$, $\beta_a(x)$ whose spacetime dependence is to be determined, along with the 'small' perturbations $g_A(x, p)$, from the Boltzmann equation. Let the index 0 denote quantities computed by means of f_A^0, so that one can distinguish between N_A^a and N_{0A}^a etc.

In order to make the decomposition (3.124) of f_A unique one has to impose $N+4$ conditions which determine the α_A, β_a in terms of the f_A, if there are N particle species. Such conditions are called *matching conditions*. Once β^a has been specified, a kinetic temperature $T=\beta^{-1}$ and a kinetic mean 4-velocity u^a are determined by means of (3.105); we shall always use this u^a in this subsection. The quantities T^{ab}, ϱ, p, q^a, π^{ab} are then also well defined through (3.59), (3.70), (3.72), (3.73), (3.74). Analogously, we put

$$N_A^a = n_A u^a + i_A^a, \quad u_a i_A^a = 0 \tag{3.125}$$

and

$$S^a = s u^a + s^a, \quad u_a s^a = 0. \tag{3.126}$$

n_A is the mean particle density, i_A^a the diffusion flux density of species A, s and s^a are the mean entropy density and entropy diffusion flux density, respectively. These quantities satisfy conservation laws like (3.63), (3.68) and the entropy inequality (3.98). These 'macroscopic' laws hold exactly, but since the number of equations, which is $4+R'$ if there are R' scalar conservation laws like (3.63) or equivalently, $R=N-R'$ linearly independent reactions (as defined on p. 78), is smaller than the number of 'macroscopic' variables n_A, i_A^a, ϱ, p, q^a, π^{ab}, additional equations are needed to determine the evolution of these quantities. The main purpose of solving the Boltzmann equations approximately by means of an Ansatz (3.124) is to obtain such missing equations, in the form of transport or reaction rate equations. Another purpose is to understand the initial stages of relaxation of an arbitrary distribution towards a local equilibrium solution which cannot be accounted for in terms of such macroscopic transport equations.

Before embarking on procedures to determine the g_A of (3.124) approximately a few simple, general observations are useful. If the change $f_A^0 \to f_A = f_A^0(1+g_A)$ is treated as a first order variation (as if g_A were 'infinitesimal'), the changes in s, ϱ, n_A are related, according to Equations (3.96), (3.97), (3.103), (3.125), (3.126), by

$$T\,\delta s = \delta\varrho - \sum_A \kappa_A\,\delta n_A \tag{3.127}$$

and

$$T\,\delta s^a = \delta q^a - \sum_A \kappa_A\,\delta i_A^a, \tag{3.128}$$

where $T=\beta^{-1}$ and $\kappa_A=\beta^{-1}\alpha_A$, as before.

Comparison of (3.127) with (3.115) shows: *The functional dependence* $\varrho=\varrho(s, n_A)$

which holds at thermal equilibrium between the functions $\varrho(x)$, $s(x)$, $n_A(x)$, remains valid to first order in the g_A for the off-equilibrium functions determined by the f_A of (3.124).

In this sense, an equation of state exists not only *at* equilibrium, but also *near* equilibrium.

Further, since the equilibrium values of s^a, q^a, i^a_A are zero according to the previous subsection, we can restate (3.128) thus: *Near equilibrium:*

$$Ts^a = q^a - \sum_A \kappa_A i^a_A . \tag{3.129}$$

These results are useful since they allow to formulate the entropy inequality in terms of the macroscopic variables listed above, and thus can serve to guess transport equations from the form of the entropy inequality (as will be discussed in Subsection 4.3), and to test approximation methods.

Let us now turn to approximation methods to solve the Boltzmann Equation (3.87). They are all based on series expansions of the 'small' functions $g_A(x, p)$, combined with truncation and/or linearization prescriptions. To describe these methods let us consider a simple (one component), nondegenerate gas.

In the *Chapman-Enskog method* which has been adapted to relativity by Israel (refs. [80]), A. H. Taub (unpublished), de Groot *et al.* (ref. [57]), and, in a mathematically more complete way, by Marle (ref. [56]), one proceeds as follows: Replace (purely formally) the transition probability $R^{\cdot\cdot}_{\cdot\cdot}$ by $1/\varepsilon\, R^{\cdot\cdot}_{\cdot\cdot}$, and write for the g of (3.124), $g=\sum_{n=1}^{\infty} \varepsilon^n g^{(n)}$. Moreover, decompose the Liouville operator L_m of (3.8) by means of the mean 4-velocity u^a and its associated projector h_{ab} (equation (3.73)) into a 'time derivative' and a 'spatial gradient',

$$L_m = p^a \nabla_a = -\underbrace{u_b p^b u^a \nabla_a}_{\substack{\text{time} \\ \text{derivative}}} + \underbrace{h^a{}_b p^b \nabla_a}_{\substack{\text{spatial} \\ \text{gradient}}} . \tag{3.130}$$

In this equation the operator (on functions on P_m)

$$\nabla_a = \frac{\partial}{\partial x^a} - \Gamma^{\nu}_{ab} p^b \frac{\partial}{\partial p^{\nu}} \tag{3.131}$$

is the partial covariant derivative with respect to x^a (also called the horizontal basis vector on the sphere bundle P_m associated with x^a). Then determine $g^{(1)}, g^{(2)}, \ldots$ successively (in principle; actually, only $g^{(1)}$ has been calculated in the relativistic case, for special cross sections) by solving the Boltzmann equation to within order $\varepsilon^0, \varepsilon^1, \ldots$, *after* eliminating at each step the time derivatives $\dot{\alpha}$, $\dot{u}_a = u_{a;b} u^b$, $\dot{\beta}$ to the required order by means of the five (in the one-component case) conservation laws $N^a{}_{;a}=0$, $T^{ab}{}_{;b}=0$. At each step this requires solving an inhomogeneous integral equation (reducible to equations of the Fredholm type, see ref. [56]). The iteration procedure shows that $g^{(n)}(x, p)$ can be expressed in terms of the local equilibrium functions α, β, u_a, their spatial gradients (like $u_{a;b}\, h^b_c$), the spatial momentum $h^a_b p^b$, and functions of the energy $E = -u_a p^a$ and temperature β^{-1} which themselves have to satisfy particular integral equations.

The functions $f^{(1)} = f^{(0)} \sum_{n=1}^{l} \varepsilon^l g^{(l)}$ obtained by this iteration process thus depend on x only through the parameters α, β, u_a and their spatial gradients; they are, therefore, functionals of α, β, u_a or, equivalently (for given matching conditions), of the basic hydrodynamical state variables s, n, u^a, defined for arbitrary (smooth) fields s, n, u^a on spacetime. Hence the *transport terms* q^a, π^{ab} and $\pi = p - p_0$ in T^{ab} can also be expressed, in any approximation, in terms of these basic hydrodynamical variables. (π is the difference between the total kinetic pressure $p = \frac{1}{3} h_{ab} T^{ab}$ and the thermostatic pressure $p_0(s, n)$ computed from s and n as in equilibrium.) The description of the Chapman-Enskog method given above shows that the distribution functions $f^{(1)}$ satisfy the Boltzmann equation approximately if and only if (a) the procedure converges (in some sense, not to be considered here) and (b) the five conservation laws $N^a{}_{;a} = 0$, $T^{ab}{}_{;b} = 0$ are satisfied for s, n, u^a and the quantities depending on them, like q^a, π^{ab}. Exact or approximate solutions of the Boltzmann equation of this type, depending on s, n, u^a, are called *normal solutions*; they provide a *kinetic theory basis for hydro-thermodynamics*.

For a simple gas, the first approximation gives (ref. [80]) the transport equations

$$\begin{aligned} \pi_{ab} &= -2\eta\sigma_{ab}, \\ \pi &= -\xi\theta, \\ q_a &= -\lambda h_a^b (T_{,b} + T\dot{u}_b). \end{aligned} \tag{3.132}$$

In these equations, $\theta = u^a{}_{;a}$ is the *expansion rate*, $\sigma_{ab} = h_a^c h_b^d u_{(c;d)} - \frac{1}{3}\theta h_{ab}$ is the *rate of shear* (or rate of distortion) of the gas as defined in the relativistic kinematics of a continuous medium (see refs. [81], [82] and Subsection 4.1). The *transport coefficients* η, ξ, λ are nonnegative functions of the local thermostatic state (s, n) (or (n, T)). They have been computed by this method for particular cross sections by Israel [80] and by de Groot *et al.* [57]. Only the heat conduction law differs in form from the nonrelativistic one; the additional acceleration term in it produces, in an equilibrium state where $q^a = 0$, precisely the temperature variation required by Tolman's law (3.122). ξ, the *bulk viscosity coefficient*, turns out to be positive for a relativistic gas consisting of point particles with elastic collisions, in contrast to the nonrelativistic case where ξ is nonzero only for particles with internal degrees of freedom. Both in the nonrelativistic $(T \to 0)$ and the ultrarelativistic $(T \to \infty)$ limit the influence of bulk viscosity is negligible. For further discussions of this point, consult refs. [53], [54].

The *Grad method of moments* has been adapted to relativity by Marle [56] and by Anderson [83] and Stewart [53]. Its basic tool is a complete orthogonal set of functions on the real Hilbert space $\mathscr{L}^2(P_{x,m}, f^0\pi_m)$ of functions on the mass shell which are square integrable with respect to the measure $f^0\pi_m$. This set is defined as follows. Put

$$\overset{0}{H} = 1, \overset{r}{H}{}^{a_1 \ldots a_r} = p^{a_1} \ldots p^{a_r} - \sum_{s=0}^{r-1} C^{a_1 \ldots a_r}_{b_1 \ldots b_s} \overset{s}{H}{}^{b_1 \ldots b_s}, \quad (r = 1, 2, \ldots) \tag{3.133}$$

and require

$$\left\langle \overset{r}{H}^{a_1 \ldots a_r}, \overset{s}{H}^{b_1 \ldots b_s} \right\rangle = 0 \quad \text{if} \quad r \neq s, \tag{3.134}$$

where the C's are constants on the mass shell and $\langle \, , \rangle$ denotes the Hilbert inner product. Such a set exists, is unique and complete, and the tensors $\overset{r}{H}^{a_1 \ldots a_r}$ are symmetrical and trace free, as shown by Marle.

Assuming that the $g(x, p)$ of Equation (3.124) is in the Hilbert space one can expand it,

$$g(x, p) = \sum_{n=1}^{\infty} \overset{n}{a}_{b_1 \ldots b_n}(x) \overset{n}{H}^{b_1 \ldots b_n}(x, p). \tag{3.135}$$

The coefficients $\overset{n}{a}\cdots$ can be shown to be linear combinations of moments of f of orders $\leqslant n$.

The Boltzmann equation is equivalent to the infinite system of *moment equations* which are obtained by inserting into equation (3.65) for $L_m(f_m)$ the collision integral, for $r = 1, 2, \ldots$. If the expansion (3.135) is inserted into these moment equations and the $\overset{n}{a}\cdots$'s expressed in terms of moments of f, an infinite system of differential equations for the moments of f is obtained which is, at least for a large class of functions f, equivalent to the Boltzmann integrodifferential equation. This system is just as unmanageable as the original Boltzmann equation. However, it is plausible (though apparently no rigorous justification is known) to assume that if one truncates the series (3.135) after a few terms and linearizes the truncated moment equations in the moments (or the $\overset{n}{a}\cdots$'s), and then solves the resulting partial differential equations, one will obtain distribution functions obeying the Boltzmann equation approximately, as long as one considers situations close to equilibrium. This procedure is called a *moment approximation*. For brief expositions of the essentials, see [83] and [54]; for a complete elaboration consult [53].

Keeping in (3.135) only the terms with $n \leqslant 2$ gives the 14 moment approximation in which $f(x, p)$ is permitted to be a polynomial of 2nd degree in the momenta only, with a spacetime dependence specified by the 19 functions α, β, u_a, $\overset{0}{a}$, $\overset{1}{a}_b$, $\overset{2}{a}_{bc}$ $(\overset{2}{a}_b{}^b = 0)$. Five *matching conditions*, e.g.

$$N^a = N_0^a, \quad T_a^a = (T_a^a)_0, \tag{3.136}$$

reduce the number of disposable functions to 14, the number of moments N^a, T^{ab}. The truncation rule $\overset{3}{a}\ldots = \overset{4}{a}\ldots = \ldots = 0$ is equivalent to equations expressing the moments of orders $\geqslant 3$ in terms of the basic moments N^a, T^{ab}. Besides the five conservation laws for N^a, T_{ab}, the 14 moment approximation provides nine differential equations for $T^{ab} - T_0{}^{ab}$, the transport terms in T^{ab}. Among the solutions are normal solutions which, again, lead to the transport law (3.132), with integral representations for the coefficients η, ξ, λ, but also 'anormal' solutions in which the spacetime dependence of $f(x, p)$ is *not* reducible to that of the basic hydrodynamical variables.

Stewart has shown [53] that the equations governing the perturbed part of the stress energy tensor form a hyperbolic system whose characteristics are contained inside the light cone, in accordance with the requirement of Einstein causality. For a simple Boltzmann gas, the maximal speed of propagation of such thermal disturbances (relative to the gas) is $(3/5)^{1/2} \approx 0.8\, c$, whereas it is well known that the upper limit for the sound speed of such a gas is $1/\sqrt{3} \approx 0.58\, c$ (see [63]).

A third way of handling the Boltzmann equation, outlined by Tauber and Weinberg [58] and systematically elaborated by Ellis and Treciokas [76], consists in choosing a mean 4-velocity field u^a and expanding the distribution function for fixed x and $E = -u_a p^a$ in spherical harmonics with respect to the direction of the 3-momentum, $h^a{}_b p^b$. This method is similar to the moment method; its usefulness will yet have to be tested in applications.

The methods described above can be extended to *degenerate* and *multicomponent gases*, as shown in the papers quoted. In particular, de Groot and van Leeuwen have treated the case of nonreacting mixtures of isobaric Maxwellian particles, and have derived *Onsager reciprocity relations* for nondegenerate systems of this kind. Relativistic corrections to transport coefficients, all of order T/m, have been worked out. W. Israel and J. N. Vardalas have treated relativistic degenerate gases. General formulae for transport and reaction coefficients have been derived by J. Stewart.

Concluding this rather long section on relativistic kinetic theory, it may be said that the conceptual and computational framework of this theory is now well established, and that it provides a useful, though limited basis for describing bulk matter of not too high density and not too low temperature in strong gravitational fields, provided one has a clean separation between long range and short range forces.

4. Continuum Mechanics and Thermodynamics

4.1. Kinematics of a Continuous Medium

The motion of a continuous medium is described by a *congruence of timelike worldlines* in spacetime or, equivalently, by a *unit timelike vector field* u^a. Since careful expositions of the kinematical concepts and the relations between them have been given elsewhere (see in particular [81], [82])and are fairly well known, I shall review only a few basic formulae here, and treat only the concepts of *strain* and *convective derivative* explicitly, following the beautiful exposition of Carter and Quintana (ref. [84]).

The gradient of u_a can be decomposed uniquely according to

$$u_{a;b} = \omega_{ab} + \theta_{ab} - \dot{u}_a u_b\,, \tag{4.1}$$

where

$$\begin{aligned} \omega_{ab} u^b &= 0\,, \quad \omega_{(ab)} = 0\,, \\ \theta_{ab} u^b &= 0\,, \quad \theta_{[ab]} = 0\,. \end{aligned} \tag{4.2}$$

ω_{ab} is the *rate of rotation* (or vorticity tensor), θ_{ab} is the *rate of deformation*, and

$\dot{u}^a = u^a{}_{;b} u^b$ is the *4-acceleration* of the medium. The trace

$$\theta = \theta^a{}_a = u^a{}_{;a} \tag{4.3}$$

is the *rate of expansion*, and the traceless part

$$\sigma_{ab} = \theta_{ab} - \tfrac{1}{3}\theta h_{ab} \tag{4.4}$$

of θ_{ab} is the *rate of shear*. As before (cf. Equation (3.73)) $h^a{}_b$ is the *projector* orthogonal to $-u^a u_b$.

By means of these quantities the relative motion between adjacent 'particles' of the medium can be discussed in close analogy to the nonrelativistic case, and one can single out special types of motions, e.g., *irrotational* motion ($\omega_{ab}=0$) and *rigid* motion ($\theta_{ab}=0$), as explained in detail in the cited papers.

If we consider a sufficiently small spacetime region V, cylindrical with respect to u^a (a tube of streamlines between two spacelike cross sections), then the restrictions of the streamlines (u^a-lines) to V form themselves a manifold $V/(u)$ with dimension 3, whose 'points' may be thought of as the 'particles' of the medium (as far as they meet V). Let us introduce the notation B for this local *quotient manifold*. If the medium is a fluid or a gas, B is a rather artificial construct, but if the medium is a solid, we may consider the points of B as related in a definite way to the sites of the lattice contituting the solid. Therefore, we call B the *body*.

Coordinates x^a adapted to u^a in the manner explained in subsection 3.3, such that $u^a = \delta^a_4$, are *comoving coordinates*; x^α may then be considered as coordinates on the body, and x^4 measures proper time along each streamline.

The definition of B implies that there is a *projection map*

$$\mathscr{P}: V \to B \tag{4.5}$$

which takes an event of V into the point of B on which that event is situated. In comoving coordinates,

$$\mathscr{P}: (x^a) \to (x^\alpha). \tag{4.6}$$

It is useful to visualize B as a manifold in its own right, separated from V; see the illustration in Figure 24.

The diffeomorphism $\mathscr{P}$ induces a map $\mathscr{P}_*$ from each tangent space $M_x (x \in V)$ onto $B_{\mathscr{P}(x)}$ (often called 'push-forward'); in comoving coordinates

$$\mathscr{P}_*: v^a \to \bar{v}^\alpha. \tag{4.7}$$

The restriction $\mathscr{P}^{\perp}_*$ of $\mathscr{P}_*$ to the subspace of M_x orthogonal to u^a is bijective.

Similarly, $\mathscr{P}$ induces a map $\mathscr{P}^*$ from each cotangent space $\mathbf{B}^*_{\mathscr{P}(x)}$ to M^*_x ('pullback'); in comoving coordinates

$$\mathscr{P}^*: \varphi_\alpha \to (\varphi_\alpha, 0). \tag{4.8}$$

The range of this map consists of precisely those 1-forms (covariant vectors) at x

which are annihilated by u^a; $\varphi(u)=\varphi_a u^a=0$. $\mathscr{P}^*$ is injective; let $\mathscr{P}^{*1}$, defined on the range of $\mathscr{P}^*$, be the inverse of $\mathscr{P}^*$.

If we call a tensor S of arbitrary rank at x *u-orthogonal* if its contractions with u^a and u_a all vanish, then it is evident that by means of $\mathscr{P}^1_*$ and $\mathscr{P}^{*1}$ one can associate with S a unique tensor at $\mathscr{P}(x)$, a *material tensor* (say).

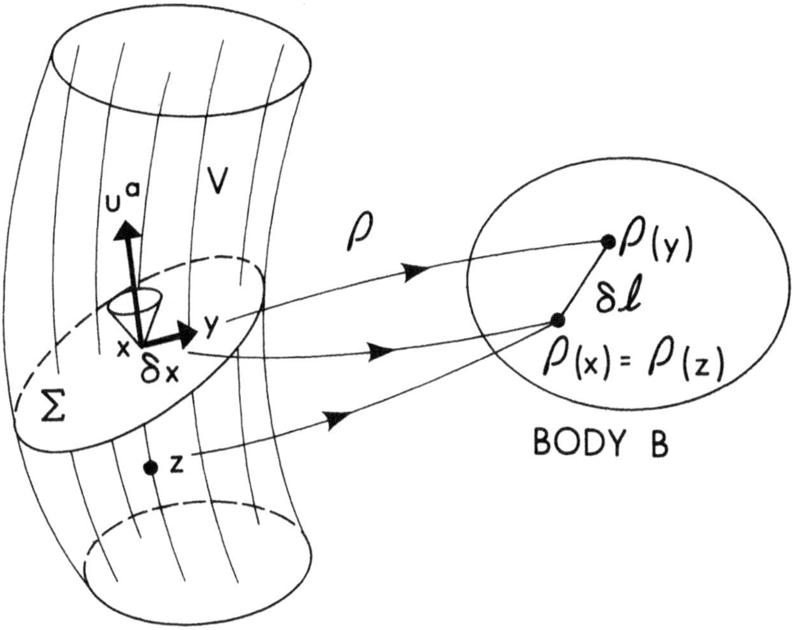

Fig. 24. The local projection from spacetime to a body.

In fact, one obtains in this way a *natural isomorphism* $\mathscr{P}^1_x$ from the tensor-subalgebra of u-orthogonal tensors of M at x onto the (full) tensor algebra of B at $\mathscr{P}(x)$. We write

$$\mathscr{P}^1_x : S \to \bar{S}, \tag{4.9}$$

and we shall also put $S \leftrightarrow \bar{S}$ since $\mathscr{P}^1_x$ is bijective. For example,

$$h^a{}_b \leftrightarrow \delta^\alpha{}_\beta, \quad h_{ab} \leftrightarrow \bar{h}_{\alpha\beta}. \tag{4.10}$$

In comoving coordinates, $\bar{S}^{\alpha\beta\cdots}{}_{\gamma\cdots} = S^{\alpha\beta\cdots}{}_{\gamma\cdots}$, and those components of S which contain an index 4 are trivially determined by the 'spatial' components and the conditions of u-orthogonality. One can therefore freely switch back and forth from material tensors on B to u-orthogonal spacetime tensors, under preservation of the tensor algebraic operations; but it must be kept in mind that this map relates tensors at $x \in M$ to tensors at $\mathscr{P}(x) \in B$; if $x \neq y$ and $\mathscr{P}(x) = \mathscr{P}(y)$ we have two different maps $\mathscr{P}^1_x$, $\mathscr{P}^1_y$ with the same range.

If x and $x+\delta x$ are two adjacent events in V, then $\delta_{\perp}x^a = h^a_b \delta x^b$ connects x with the streamline through $x+\delta x$ and is orthogonal to the streamline through x (see Figure 24), and according to the preceding discussion we have for the proper length δl of the spacelike vector $\delta_{\perp}x^a$

$$\delta l^2 = h_{ab}\delta x^a \delta x^b = \bar{h}_{\alpha\beta}\delta_{\perp}x^{\alpha}\delta_{\perp}x^{\beta}. \tag{4.11}$$

Hence, $h_{ab}(x)$ or, equivalently, $\bar{h}_{\alpha\beta}(\mathscr{P}(x), \tau)$ specifies the *proper distances* between the particle $\mathscr{P}(x)$ and all neighbouring particles at proper time τ. (Let τ always denote a smooth *proper time function* along the streamlines in the chosen region V.)

If Σ is a spacelike cross section of V – 'instant of time' – then $h_{ab}|_{\Sigma}$ defines, via (4.11), a positive definite Riemannian metric $\bar{h}_{\alpha\beta}$ on B.

These remarks and the physical interpretation of Equation (4.11) justify the following *definition*: A *state of strain* of the body B is a Riemannian metric $\bar{h}_{\alpha\beta}$ on B; $\bar{h}_{\alpha\beta}$ is also called the strain tensor.

Notice that *a strain tensor is primarily a material tensor*, defined on the body manifold B; it is an assignment of spatial distances to 'places' (lattice sites), with no reference to 'time' or spacetime. *If* a motion of B, i.e., a congruence (u^a) in M and a projection map $\mathscr{P}$ from (part of)M to B, is given, then any 'instant' Σ in M defines a unique state of strain on B, as explained above, and if Σ is considered as variable, the state of strain changes, and is given by $\bar{h}_{\alpha\beta}(x^{\gamma}, \tau)$ with respect to any coordinate system on B and proper time function τ ('Lagrangean description'). Equivalently we can, with respect to a motion (u^a), describe the strain in terms of the spacetime tensor $h_{ab}(x^c)$ ('Eulerian description').

If a u-orthogonal tensor field S is given in V, then the image $\bar{S}$ of S under $\mathscr{P}^1$ is a one-parameter family of tensor fields on B, parametrized by a proper time function: $S(x) \leftrightarrow \bar{S}(\mathscr{P}(x), \tau)$. We can form the proper time derivative $(\partial/\partial\tau)\bar{S}(\mathscr{P}(x), \tau)$ at a fixed place $\mathscr{P}(x)$ of the body, and define the *convective derivative* of $S(x)$ at x by

$$\mathring{S}(x) = \left(\frac{\partial}{\partial\tau}\bar{S}\right)(\mathscr{P}(x), \tau(x)). \tag{4.12}$$

This convective derivative can be computed directly on M by means of the Lie derivative defined in Subsection 3.3. Using comoving coordinates one can easily verify that

$$\mathring{S}^{a_1\ldots a_r}_{b_1\ldots b_s} = h^{a_1}{}_{c_1}\ldots h^{a_r}{}_{c_r}\mathscr{L}_u S^{c_1\ldots c_r}_{b_1\ldots b_s}, \tag{4.13}$$

which holds in any coordinate system (or tetrad frame). Note that for a covariant tensor field the convective derivative coincides with the Lie derivative, whereas for a contravariant or mixed tensor an additional projection is necessary to obtain $\mathring{S}^{\cdot\cdot}{}_{\cdot\cdot}$ from $\mathscr{L}_u S^{\cdot\cdot}{}_{\cdot\cdot}$.

The convective derivative measures the change of S relative to the body.

As an example needed later on, we note that

$$\frac{d\bar{h}_{\alpha\beta}}{d\tau} \leftrightarrow \mathring{h}_{ab} = \mathscr{L}_u h_{ab} = 2\theta_{ab}. \tag{4.14}$$

Thus the rate of deformation, θ_{ab}, is in fact half the rate of strain, and with a slight abuse of language we shall just also call it the *rate of strain.*

It is now clear that a motion is rigid if and only if the associated strain is materially constant, i.e., has zero convective derivative. In this case, and only in this case, can one assign a *metric* $\bar{h}_{\alpha\beta}$ unambiguously to B.

If two states of strain, $\bar{h}_{\alpha\beta}$ and $\bar{k}_{\alpha\beta}$, are given on B, one can introduce the *deformation tensor*

$$\bar{e}_{\alpha\beta} = \tfrac{1}{2}(\bar{h}_{\alpha\beta} - \bar{k}_{\alpha\beta}) \tag{4.15}$$

as a measure for the (finite) change of strain associated with the two states. Recalling (4.11) one recognizes that, for small deformations, $\bar{e}_{\alpha\beta}$ reduces to the elementary deformation tensor of the classical theory of infinitesimal deformations. Again one can pass to the Eulerian quantity e_{ab}, if a motion has been assigned. If $\bar{k}_{\alpha\beta}$ is a fixed reference state, the convective derivative of e_{ab} is indeed θ_{ab}, as anticipated by the terminology.

4.2. Balance Equations, Entropy Inequality, Conservation Laws

We assume that the macroscopic description of *any* kind of matter in bulk involves a *symmetric stress energy momentum tensor* T^{ab} which satisfies the *energy-momentum balance equation*

$$T^{ab}{}_{;b} = 0, \tag{4.16}$$

provided the matter interacts only with the gravitational field g_{ab}.

We also assume, in analogy to nonrelativistic physics (see, e.g., refs. [85] and [86] for general accounts) and in accordance with the kinetic theory (see particularly Subsections 3.8 and 3.9), that there always exists an *entropy 4-current density* S^a which satisfies the *entropy inequality*

$$S^a{}_{;a} \geqslant 0 \tag{4.17}$$

which generalizes the Clausius-Duhem inequality to relativity.

There is no generally valid relationship between the 14 variables T^{ab}, S^a; rather, the relations (4.16), (4.17) provide a *general framework* into which all specific models of material behaviour are supposed to fit.

With respect to an *arbitrary* mean 4-velocity field u^a one can decompose T^{ab} and S^a according to the equations

$$T^{ab} = \varrho u^a u^b + 2u^{(a}q^{b)} + p^{ab} \tag{4.18}$$

and

$$S^a = su^a + s^a, \tag{4.19}$$

where

$$u_a q^a = 0, \quad p_{ab}u^b = 0, \quad u_a s^a = 0, \tag{4.20}$$

as in Equations (3.72) and (3.126).

Equation (4.16) is equivalent to the system $u_a T^{ab}{}_{;b}=0$, $h_{ab}T^{bc}{}_{;c}=0$, which can be rewritten, by means of the definitions (4.1) and (4.18), as

$$\dot{\varrho} + \varrho\theta + p_{ab}\theta^{ab} + q^a{}_{;a} + q^a \dot{u}_a = 0\,, \tag{4.21}$$

$$(\varrho + p)\,\dot{u}_a + h^b_a\,(p^c_{b;c} + \dot{q}_b) + (\omega_{ab} + \theta_{ab} + \tfrac{1}{3}\theta h_{ab})\,q^b = 0\,. \tag{4.22}$$

Similarly, we rewrite (4.17), using (4.19), as

$$\dot{s} + s\theta + s^a{}_{;a} \geqslant 0\,. \tag{4.23}$$

Equation (4.21) is the *local energy balance* relative to an observer travelling with 4-velocity u^a. Only the last term in that equation has no analogue in the nonrelativistic theory. Formally it results from the fact that the infinitesimal 3-spaces orthogonal to the observer's worldline are not parallel to each other, but relatively tipped because of the curvature of that line. Physically this last term is due to the inertia of heat; it represents the power per unit volume needed to accelerate the heat energy.

Equation (4.22) is the relativistic analogue of the *Cauchy equation of motion* of the material. The q^a-terms in that equation are due to the fact that in relativity theory one cannot separate mechanical and thermal equations as in the nonrelativistic scheme, since any energy current also represents a momentum density.

If additional conservation laws hold they can be formulated in a way similar to (4.21)–(4.23). *Baryon conservation*, e.g., would be accounted for by writing

$$B^a{}_{;a} = (bu^a + b^a)_{;a} = \dot{b} + b\theta + b^a{}_{;a} = 0\,. \tag{4.24}$$

4.3. Constitutive Equations for Simple, Isotropic, Visco-Elastic Solids and Fluids

To specify a particular material one has to add to the general laws formulated in the previous subsection *constitutive equations*. In conjunction with the general laws they should give *time evolution equations* for the independent state variables of the matter under consideration.

We shall restrict our attention here to *simple materials*, by which we mean materials with a uniform composition in which no internal diffusion and no reactions take place, and which have no internal degrees of freedom like spin densities etc. To obtain equations of motion for a class of such materials we proceed in two steps. First we require only an incomplete set of constitutive equations which permits us to re-reformulate the entropy inequality in a transparent way, and on the basis of that we then complete the set of constitutive equations in several different ways, thus obtaining simple models of materials.

Definition. A *simple material* is one for which there exists a (preferred) mean 4-velocity u^a such that, with respect to this u^a, the energy density ϱ is a function of the entropy density s and the strain tensor $\bar{h}_{\alpha\beta}$,

$$\varrho = f(s, \bar{h}_{\alpha\beta})\,, \tag{4.25}$$

and for which the entropy 3-current density s^a is given by the classical (Carnot) relation

$$s^a = q^a/T\,, \tag{4.26}$$

where

$$T = \frac{\partial f}{\partial s} > 0 \tag{4.27}$$

is the temperature.

In this definition essential use is made of the kinematical concepts of Subsection 4.1. (4.25) means that for any motion of such a material (in any instantaneous configuation), the functions ϱ, s, $\bar{h}_{\alpha\beta}$, defined on the body B as functions of proper time, are functionally related according to (4.25), with f given once and for all; also, (4.26) has to hold for all motions. (Notice that the Equations (4.25)–(4.27) are equivalent to a single equation expressing S^a in terms of u^a, ϱ and $\bar{h}_{\alpha\beta}$.)

As an aside, I should like to point out that the entropy inequality (4.23), combined with (4.26), assumes the form

$$(su^a)_{;a} \geqslant -\left(\frac{q^a}{T}\right)_{;a}. \tag{4.28}$$

Integration of this inequality over a section of a world tube of streamlines bounded by two spacelike cross sections Σ_1, Σ_2 tangential to q^a and use of Gauss's theorem gives, in an obvious notation,

$$\int\limits_{\Sigma_2-\Sigma_1} su^a\sigma_a \geqslant -\int\limits_{\substack{\text{timelike}\\ \text{boundary}}} T^{-1}q^a\sigma_a\,. \tag{4.29}$$

If T is constant on the boundary of the matter element considered, (4.29) states that the entropy increase ΔS of the element is never less than the reduced influx of heat, $\delta Q/T$,

$$\Delta S \geqslant \delta Q/T\,.$$

This is the – without further explanation somewhat vague – formulation of the second law of thermodynamics in general relativity in Tolman's version (ref. [87]). Tolman's entropy flow 4-vector is to be identified *not* with S^a, but with su^a, and his version is somewhat less general than (4.17), even for simple materials as defined here.

Continuing the study of simple materials, we set

$$\frac{\partial f}{\partial \bar{h}_{\alpha\beta}} = -\tfrac{1}{2}\left(\bar{p}^{\alpha\beta}_{\mathrm{el}} + \bar{h}^{\alpha\beta}\{\varrho - Ts\}\right) \tag{4.30}$$

and call $\bar{p}^{\alpha\beta}_{\mathrm{el}}$ the *elastic pressure tensor* associated with the state $(s, \bar{h}_{\alpha\beta})$ of the body. (4.30) is the (generalized) *stress-strain relation*.

Consider now an arbitrary motion of the material, so that ϱ, s, u^a,... are functions on spacetime, and $\bar{h}_{\alpha\beta}$ is a well defined function of proper time at each particle of the body B. Differentiation of (4.25) along a streamline gives, with (4.27) and (4.30),

$$\dot{\varrho} = T\dot{s} - \tfrac{1}{2}\left(\bar{p}_{\mathrm{el}}^{\alpha\beta} + \bar{h}^{\alpha\beta}\{\varrho - Ts\}\right)\frac{\mathrm{d}\bar{h}_{\alpha\beta}}{\mathrm{d}\tau}. \tag{4.31}$$

Using the isomorphism $\mathscr{P}^1$ between material tensors and u-orthogonal tensors considered in Subsection 4.1 and Equation (4.14), we obtain from the last equation

$$\dot{\varrho} + \varrho\theta = T(\dot{s} + s\theta) - p_{\mathrm{el}}^{ab}\theta_{ab}.$$

Eliminating $\dot{\varrho}$ by means of (4.21) we get an expression for $\dot{s}$. If this and q^a from (4.26) are put into the inequality (4.23), there results

$$S^a{}_{;a} = -\frac{1}{T}\left(p^{ab} - p_{\mathrm{el}}^{ab}\right)\theta_{ab} - \frac{1}{T^2}q^a\left(T_{,a} + T\dot{u}_a\right) \geqslant 0. \tag{4.32}$$

A trivial way to satisfy this inequality is to require, as additional constitutive equations,

$$p^{ab} = p_{\mathrm{el}}^{ab}, \quad q^a = 0; \tag{4.33a}$$

the ideal material defined in this way is characterized by the conservation of entropy in all processes, it is called a *perfectly elastic, non-heat-conducting material*. For more details concerning such ideal materials, see ref. [84], [119], [120].

The next simple possibility to satisfy the entropy inequality (4.32) is to require that the viscous stress tensor, defined by

$$p_{\mathrm{vis}}^{ab} = p^{ab} - p_{\mathrm{el}}^{ab}, \tag{4.33b}$$

and the heat flow vector q^a are linear transforms of the 'thermodynamic forces' θ_{ab} and $h_a^b(T_{,b} + T\dot{u}_b)$, such that the quadratic form in p_{vis}^{ab} and q^a which then represents the entropy production density $S^a{}_{;a}$, is positive semi-definite. This then defines a *linearly viscous* and *linearly heat conducting material.*

In order to characterize such an idealized material completely one has to specify the thermostatic potential f of Equation (4.25) and the tensors which transform θ_{ab} and $h_a^b(T_{,b} + T\dot{u}_b)$ into p_{vis}^{ab} and q^a, or their material images on B. Here we shall consider only the cases of fluids, and isotropic solids in the Hookean approximation (see [84]).

Formally we define a *fluid* by the following two properties: (a) The energy density depends on the state of strain only via the Riemannian volume element $\sqrt{\bar{h}}\,\mathrm{d}^3x$ on B, where $\bar{h} = \det(\bar{h}_{\alpha\beta})$, so that, with respect to any coordinate system on B,

$$\varrho = f(s, \bar{h}) \tag{4.34}$$

or equivalently, if we assume that each part of the fluid consists of a definite, conserved number of particles $\int n\sqrt{\bar{h}}\,\mathrm{d}^3x$,

$$\varrho = g(s, n), \tag{4.34'}$$

n being the number density. (b) The relations determing p_{vis}^{ab} and q^a in terms of θ_{ab}

and $h_a^b(T_{,b}+T\dot{u}_b)$ are, in any state of strain, invariant with respect to all rotations in the (tangent) 3-space orthogonal to u^a.

The two properties are intended to express that there are no preferred directions in any state.

Equation (4.34′) generalizes the thermostatic equation which was derived in Subsection 3.9 for simple gases close to equilibrium. There the function g was derived from microscopic considerations, here it remains undetermined. Equations (4.34), (4.30) imply $p_{\text{el}}^{\alpha\beta}=p_0\bar{h}^{\alpha\beta}$, and the pressure p_0 can be obtained from (4.34′) just as in the case of gases from Equations (3.115), (3.115′)

The second requirement, (b), implies that the *transport equations must have the form* (3.132) as for gases (with π^{ab} and π the traceless part and trace of p_{vis}^{ab}, again with functions $\eta(s,n)$, $\xi(s,n)$, $\lambda(s,n)$ which remain unspecified in the phenomenological continuum theory.

In the case of fluids, the body manifold B plays no essential role. The basic state variables are s, n and u^a, and the five equations of motion are $N^a{}_{;a}=\dot{n}+n\theta=0$, $T^{ab}{}_{;b}=0$, where T^{ab} is expressed in terms of s, n, u^a by means of the constitutive equations.

To characterize an *isotropic, visco-elastic solid*, it is convenient to switch from the variables ϱ, s, $\bar{h}_{\alpha\beta}$ of Equation (4.25) to $u=\varrho/n$, $\sigma=s/n$, $\bar{h}_{\alpha\beta}$ where n is the density of a conserved particle number so that $n\sqrt{\bar{h}}=\text{const}$. Then u, σ are the *energy and entropy per particle*. Equations (4.25), (4.27) and (4.30) are equivalent to

$$\mathrm{d}u = T\mathrm{d}\sigma - \frac{1}{2n}\,\bar{p}_{\text{el}}^{\alpha\beta}\,\mathrm{d}\bar{h}_{\alpha\beta}\,.$$

Furthermore, by Legendre-transformation we go over from u, σ to $\varphi=u-T\sigma$, T, and use the *free energy per particle* $\varphi(T,\bar{h}_{\alpha\beta})$ as the thermodynamic potential, with

$$\mathrm{d}\varphi = -\,\sigma\mathrm{d}T - \frac{1}{2n}\,\bar{p}_{\text{el}}^{\alpha\beta}\,\mathrm{d}\bar{h}_{\alpha\beta}\,. \tag{4.35}$$

For an isotropic solid we require, in the Hookean approximation, that

$$\varphi(T,\bar{h}_{\alpha\beta}) = m(T) + \psi(T)\,(\bar{e}^{\alpha}{}_{\alpha})^2 + \chi(T)\,\bar{e}^{\alpha}{}_{\beta}\bar{e}^{\beta}{}_{\alpha}\,, \tag{4.36}$$

where

$$\bar{e}_{\alpha\beta} = \tfrac{1}{2}\left(\bar{h}_{\alpha\beta} - l^2(T)\,\bar{k}_{\alpha\beta}\right) \tag{4.37}$$

is the deformation relative to the (temperature-dependent) strain tensor $l^2(T)\bar{k}_{\alpha\beta}$ of a relaxed *natural state*, and $\bar{k}_{\alpha\beta}$ is a fixed Riemannian metric on the body manifold. $m(T)$ is the free energy per particle at T in the relaxed state, $\psi(T)$ and $\chi(T)$ are elastic 'constants', and $l(T)$ is related to the thermal expansion coefficient. The indices are to be raised by means of the inverse $\bar{k}^{\alpha\beta}$ of the fixed body metric $\bar{k}_{\alpha\beta}$.

For isothermal processes, (4.35), (4.36) and (4.37) reduce to the classical linear relation (Hooke's approximation) between the elastic stress tensor $\bar{p}_{\text{el}}^{\alpha\beta}$ and the deformation tensor $\bar{e}_{\alpha\beta}$. – Furthermore, we require the transport equations for p_{vis}^{ab}

and q^a to be of the same form as for a fluid, with transport coefficients depending on T and n.

Using the isomorphism $\mathscr{P}^1$ of Subsection 4.1 one can translate all material tensors into u-orthogonal spacetime tensors. Thus $\bar{k}_{\alpha\beta}$ gives rise to k_{ab} with

$$k_{[ab]} = 0, \quad k_{ab}u^b = 0, \quad \mathscr{L}_u k_{ab} = 0, \tag{4.38}$$

and $\bar{k}^{\alpha\beta}$ has a spacetime image ${}^1k^{ab}$ characterized by

$${}^1k^{[ab]} = 0, \quad {}^1k^{ab}u_b = 0, \quad {}^1k^{ab}k_{bc} = h^a_c. \tag{4.39}$$

With these quantities it is straightforward to express (4.36), (4.37) and the stress-strain relations in Eulerian, spacetime language, One can then take T, n, u^a and k_{ab} as the *basic state variables* of a solid (for a given spacetime metric), express $N^a = nu^a$, $T^{ab} = \ldots$ (Equation (4.18)) in terms of them, and get a complete system of *Eulerian time evolution equations* from $N^a{}_{;a} = \dot{n} + n\theta = 0$, $T^{ab}{}_{;b} = 0$, $\mathscr{L}_u k_{ab} = 0$.

These equations are complicated and will not be written out here; our purpose was to outline how, on the basis of the kinematics of Subsection 4.1 and the general laws of Subsection 4.2, it is possible to formulate consistent laws of continuum mechanics and thermodynamics for fluids and solids in curved spacetime, so that one has a basis for approximation methods. One can use this formalism, e.g., to derive the equations for the strain waves excited in a Weber bar by a gravitational wave pulse, and, more interestingly, to set up the theory of a solid star (see ref. [88]).

Anisotropic materials can be treated similarly, and it is also straightforward to formulate equations for multicomponent fluids with reactions, thus obtaining a continuum analogue of the kinetic theory of such systems. If one defines equilibrium by $S^a{}_{;a} = 0$ as in kinetic theory, one can rederive in this framework that equilibrium requires a stationary spacetime, that Tolman's law (3.122) holds, etc.

From the point of view taken in this section the kinetic theory is a method to obtain consitutive equations for a particular class of materials, i.e., dilute gases. However, kinetic theory is less phenomenological and gives additional microscopic information, e.g., momentum distributions or specific intensities. It is sometimes useful to combine the continuum and kinetic approaches, treating part of the matter as a continuum and another one, e.g., radiation, as a particle system.

These considerations show again that as long as the scale given by the curvature is large in comparison with the scale at which short range interactions and thermodynamics operate, the combination of general-relativistic kinematics with ordinary, local laws of material behaviour poses no serious problems*; the role of the concepts of temperature and entropy is clear; and the long range effects of gravity are accounted for simply by the way in which the connection enters through covariant derivatives. The difficult question of how matter behaves and interacts with gravity in extremely strong gravitational fields – in a collapse near a singularity, or immediately after a big bang – is a mysterious one, and no attempt will be made here to discuss it.

* Apart from causality difficulties in the Cauchy problem due to viscosity terms.

5. The Transition from Maxwell's Equations to Geometrical Optics

Electromagnetic fields in empty space are governed by the covariantly generalized Maxwell equations

$$F_{[ab,c]} = 0, \quad F^{ab}{}_{;b} = 0 \tag{5.1}$$

obtained by means of the strong principle of equivalence (Subsection 2.10). These equations correspond to the Lagrangian

$$L = -\tfrac{1}{4} F_{ab} F^{ab}, \quad F_{ab} = 2A_{[b,a]}, \tag{5.2}$$

from which the stress-energy-momentum tensor

$$T^{ab} = F^{ac} F^{b}{}_{c} - \tfrac{1}{4} g^{ab} F_{cd} F^{cd} \tag{5.3}$$

is deduced according to the variational formula (2.28).

On the other hand, we have described radiation in the framework of kinetic theory as a collection of massless particles with null geodesic orbits, and we have constructed an energy tensor with the help of the distribution function; recall Equation (3.58). Moreover, the identification of light rays with null geodesics played a fundamental role in the motivation for and construction of the pseudoriemannian spacetime metric.

The purpose of this subsection is to establish a connection between the two descriptions of radiation which shows that they are compatible.

A *locally approximately plane wave* is a solution of (5.1) which is approximately of the form $e^{iS} A_{ab}$ with an *amplitude* A_{ab} which varies very slowly in spacetime in comparison with the real, scalar *phase* S; here and in the sequel it is to be understood that the physical field is the real part of the complex field. (It is in the nature of the problem under consideration that this definition is not mathematically precise.)

For computational purposes it is convenient to write

$$F_{ab} = e^{i\varepsilon^{-1} S} (A_{ab} + \varepsilon B_{ab} + 0(\varepsilon^2)) \tag{5.4}$$

and to require that this expression should satisfy the Maxwell equations asymptotically for $\varepsilon \to 0$; for very small ε we shall then indeed have $F_{ab} \approx e^{i\varepsilon^{-1} S} A_{ab}$, and $\varepsilon^{-1} S$ will vary rapidly in comparison with A_{ab}. At the end, one can simply substitute $\varepsilon^{-1} S \to S$, and get a locally nearly plane wave. (For more complete and rigorous discussions, see [89], [90].)

From (5.4),

$$F_{ab;c} = e^{i\varepsilon^{-1} S} \left(\frac{i}{\varepsilon} A_{ab} S_{,c} + i\, B_{ab}\, S_{,c} + A_{ab;c} + 0(\varepsilon) \right). \tag{5.5}$$

Validity of (5.1) for $\varepsilon \to 0$ requires and is implied by

$$A_{[ab}\, k_{c]} = 0, \quad A_{ab}\, k^b = 0 \tag{5.6}$$

and

$$i\, B_{[ab}\, k_{c]} = -\, A_{[ab,c]}, \tag{5.7a}$$

$$i\, B_{ab}\, k^b = -\, A_{ab;}{}^{b}, \tag{5.7b}$$

where we have put

$$k_a = S_{,a}\,. \tag{5.8}$$

The Equations (5.6) imply

$$k^c(A_{ab}k_c + A_{bc}k_a + A_{ca}k_b) = A_{ab}k_c k^c = 0$$

and, if it is assumed that A_{ab} vanishes at most on hypersurfaces, that k^a is a null vector:

$$S_{,a}\,S^{,a} = 0\,. \tag{5.9}$$

Moreover, $k_{a;b}k^b = S_{;ab}k^b = S_{;ba}k^b = k_{b;a}k^b = \frac{1}{2}(k_b k^b)_{,a} = 0$. That is, S defines a *family of null hypersurfaces* $S = \text{const.}$, and the k^a-curves which generate these hypersurfaces are *null geodesics*. Equation (5.9) is the *eikonal equation* of geometrical optics.

By means of any vector p^c such that $k_c p^c = 1$ we get from (5.6)

$$p^c(A_{ab}k_c + A_{bc}k_a + A_{ca}k_b) = 0\,,$$

or

$$A_{ab} = 2\,k_{[a}\,q_{b]}\,,\quad q_a k^a = 0\,, \tag{5.10}$$

where we have set $A_{ab}p^a = q_b$. Thus A_{ab} is a *null bivector* with principal null vector k^a.

Conversely, (5.8), (5.9) and (5.10) imply (5.6). We now determine the conditions which q_a has to satisfy in order that Equations (5.7) have a solution B_{ab}. Inserting (5.10) into (5.7a) and remembering that k_a is a gradient one obtains $(iB_{[ab} + 2q_{[a,b})k_{c]} = 0$ or, equivalently, $iB_{ab} + 2q_{[a,b]} = 2k_{[a}p_{b]}$ for some p_a (which is not related to the p_a used above). Equation (5.7b) then demands $2\dot{q}_a + q_a k^b{}_{;b} = k_a(p_b k^b + q^b{}_{;b})$, where use has been made of Equation (5.10).

The two amplitudes A_{ab} and B_{ab} and their representation in terms of k_a, q_a and p_a remain unchanged under the gauge transformation $q_a \to q_a + \lambda k_a$, $p_a \to p_a + \lambda_{,a}$ with an arbitrary scalar λ. λ can be chosen such that

$$p_a\,k^a = -\,q^a{}_{;a}\,. \tag{5.11}$$

Then Equation (5.7b) is equivalent to the propagation equation

$$\dot{q}_a = -\,\tfrac{1}{2}q_a\,k^b{}_{;b}\,. \tag{5.12}$$

Thus, any locally approximately plane wave (5.4) is obtained by solving the eikonal Equation (5.9) whose solution determines a hypersurface orthogonal null geodesic congruence of *rays*, and determining an amplitude vector q_a which propagates along the rays according to (5.12). Then

$$F_{ab} = \Re\,(2\,k_{[a}\,q_{b]}\,e^{i\varepsilon^{-1}S}) + 0\,(\varepsilon)\,. \tag{5.13}$$

(The main correction to the amplitude is obtained by choosing any p_a satisfying (5.11) and putting $iB_{ab} = -2q_{[a,b]} + 2k_{[a}p_{b]}$. For a continuation of the procedure, see ref. [90].)

The energy tensor (5.3) of the wave (5.13) is

$$T^{ab} = k^a k^b\,(\tfrac{1}{2}q_c\bar{q}^c + \tfrac{1}{2}\Re\,[q_c q^c e^{2i\varepsilon^{-1}S}]) + 0\,(\varepsilon)\,. \tag{5.14}$$

This formula shows that *the energy propagates along the rays.* If a radiation field is represented by an ensemble of locally approximately plane waves, the average energy tensor obtained from (5.14) will have the form (3.58) used in kinetic theory, provided the distribution function is properly related to the expectation value of $\frac{1}{2}q_c\bar{q}^c$ depending on the k^a-dependent random variable q_a.

It is useful to associate with the ray congruence of a phase function S a positive scalar r such that

$$\dot{r} = \tfrac{1}{2} r k^a{}_{;a} . \tag{5.15}$$

Just as in hydrodynamics $u^a{}_{;a}$ measures the rate of expansion of a fluid element, so $k^a{}_{;a}$ measures the relative rate of change, with respect to an affine parameter associated with k^a, of the area of a small cross section of a slender pencil of rays contained in a hypersurface $S = \text{const.}$ (see ref. [91] or [92]). Hence along a ray r^2 varies like the area of such a cross section.

(5.12) and (5.15) imply $(rq^a)^{\cdot} = 0$, i.e. $q^a \propto r^{-1}$ along a ray, whence from (5.13) and (5.14)

$$\left.\begin{aligned} F_{ab} &\propto r^{-1} \\ T_{ab} &\propto r^{-2} \end{aligned}\right\} \begin{aligned} &\text{along} \\ &\text{a ray:} \end{aligned} \tag{5.16}$$

moreover, the direction of q^a is constant along a ray.

If F_{ab} is the field of a point source (in the radiation zone), r can be chosen so that it equals the radar distance for observers close to the source. r is then uniquely determined everywhere, and the Equations (5.16) have an obvious physical interpretation which justifies the name *luminosity distance* for r.

The (circular) frequency measured by an observer with 4-velocity u^a is $\omega = -\mathrm{d}S/\mathrm{d}\tau$ where the phase S is restricted to the world line of the observer with proper time τ. (Here we have put $\varepsilon = 1$.) That is, $\omega = -S_{,a}u^a = -k_a u^a$. Therefore, the *redshift* z is given by

$$1 + z = \frac{(k \cdot u)_{\text{source}}}{(k \cdot u)_{\text{observer}}} . \tag{5.17}$$

The results (5.16), (5.17) can be used to rederive some of the statements deduced at the end of Subsection 3.5; now we could also make statements about polarization, however.

For more on geometrical optics and its use in relativistic astrophysics see, e.g., refs. [93], [82], [23].

6. Note on the Dynamics of Extended Bodies

The description of motions of bodies as a whole is of great conceptual as well as practical importance in any macroscopic mechanics. The assumption that freely falling test particles have geodesic world lines was fundamental in setting up the pseudoriemannian spacetime geometry inductively. However, once the field Equation (2.23) and a model of matter, i.e., a model of T^{ab} and equations for matter variables on which T^{ab} depends, have been specified, there is no room for an additional 'geodesic hypothesis' any more. We encounter here the familiar problem of the consistency of the final theory with the preliminary assumptions made in motivating it. In the present case the problem is to deduce that the world tubes of those matter distributions which can reasonably be called test particles do indeed, in a sufficient approximation, contain timelike geodesics. Moreover, the theory should provide corrections to the simple geodesics law which apply to heavy bodies or bodies with a significant multipole structure, and it should give definitions and evolution equations for structure parameters such as the intrinsic angular momentum. (An important application of the latter is the theory of the gyroscope experiment proposed by Schiff and being prepared by Fairbank, Everitt *et al.*, see [6].)

Setting up such a theory on the basis of the 4-momentum balance Equation (2.25) in curved spacetime is an old and formidable problem. It will not be treated here in detail; instead only the main ideas of the theory will be outlined in this section, with references to some pertinent papers.

The difficulties encountered in setting up equations of motion for extended bodies in curved spacetime are due to the non-integrability of the linear connection. Because of it, there are no 'free' (i.e., position-independent) vectors (like the total linear momentum in Newtonian or special relativistic mechanics) and no 'plane' hypersurfaces of constant time; integration of vectors or tensors over spacelike hypersurfaces requires the specification of a transport law; there is no obvious position vector; and Equation (2.25) is not a true conservation law. (In the Newtonian theory these difficulties are absent; for the physical connection introduced in Equation (1.18) is integrable if restricted to spacelike vectors.) In spite of these difficulties, exact and general equations of motion have been established by Dixon (refs. [94]) whose work 'rigorizes' and generalizes important earlier contributions by Mathisson, Papapetrou, B. and W. Tulczyjew, Taub and others; in the remainder of this section we follow essentially Dixon's work. A simplified version of this approach which gives a fair idea of its working can be found in Schild's lectures (ref. [11]).

An isolated, extended body is idealized as a timelike world tube $\mathfrak{T}$ outside of which the matter tensor T^{ab} vanishes; $\mathfrak{T}$ may be taken to be the support of T^{ab}. In order to obtain general-relativistic analogues of the total 4-momentum and spin of $\mathfrak{T}$ one defines, for events z near $\mathfrak{T}$ and spacelike hypersurfaces Σ containing z,

$$p^a(z, \Sigma) = -\int_\Sigma K^a{}_{b'} T^{b'c'} \sigma_{c'}, \tag{6.1}$$

$$S^{ab}(z, \Sigma) = -2\int_\Sigma r^{[a} H^{b]}{}_{c'} T^{c'd'} \sigma_{d'}. \tag{6.2}$$

In these equations primed indices refer to quantities at the (variable) integration event $x \in \Sigma$. $\sigma_{a'}$ is the hypersurface element defined in Equation (3.31); r^a is the position vector of x relative to z, which is by definition equal to **zx** in the notation of Equation (2.31). $K^a{}_{b'}(z, x)$ and $H^a{}_{b'}(z, x)$ are maps from M_x to M_z ('bitensor fields), to be characterized below. It is assumed that $\Sigma \cap \mathfrak{T}$ is small enough so that $r^a(z, x)$ and the 'propagators' K and H are well defined.

If spacetime were flat and both K and H were replaced by the parallel propagator, (6.1), (6.2) would reduce to the standard definitions of linear momentum and spin (with respect to z) in special relativity. (The $-$ signs are needed because of our sign convention (3.33).) In this case, (2.25) would imply that p^a is independent of z and Σ, and that S^{ab}, while independent of Σ, depends on z in the obvious, trivial way, so that a center-of-mass line L_0 can be defined by the condition that $z \in L_0$ if and only if

$$S^{ab} p_b = 0 \,. \tag{6.3}$$

In curved spacetime it might appear that one should also choose the parallel propagator in place of K and H, simply by analogy. This (and other) choices have indeed been made, and from a practical point of view and in the context of weak-field approximations the proposed choices of propagators are probably equivalent. In principle, however, the choice of K and H is all-important, and the only convincing choice (to this author) which is based on dynamically significant requirements is the one proposed by Dixon (*loc. cit.*). It is obtained as follows. If spacetime admits a Killing vector field ξ^a, then Equations (2.25) and Killing's equation $\mathscr{L}_\xi g_{ab}=0$ or, equivalently, $\xi_{(a;b)}=0$, imply the local conservation law

$$(T^{ab}\xi_b)_{;a} = 0 \,. \tag{6.4}$$

Hence, because of Gauss's theorem,

$$\mathscr{E}(\xi) = \int_\Sigma T^{a'b'} \xi_{b'} \sigma_{a'} \tag{6.5}$$

is independent of Σ, i.e., is a constant of the motion for any body $\mathfrak{T}$. In the case of flat spacetime one has $\xi_{b'} = c_{b'} + F_{b'c'} r^{c'}$ with constants c, F, so with (6.1), (6.2) it follows that

$$\mathscr{E}(\xi) = p^a \xi_a(z) + \tfrac{1}{2} S^{ab}(z)\, \xi_{[a,b]}(z) \tag{6.6}$$

for arbitrary z; i.e., the constant $\mathscr{E}(\xi)$ can be expressed bilinearly in terms of the integral quantities p^a, $S^{ab}(z)$ of the body and the values of ξ_a, $\xi_{[a,b]}$ at an arbitrary event z. In view of the importance of the connection between symmetries and conservation laws in dynamics it is reasonable to ask whether, in the general case of curved spacetime, the propagators K and H in (6.1), (6.2) can be chosen such that whenever g_{ab} admits a Killing field ξ^a, the conserved quantity (6.5) can again be written in the 'integrated' form (6.6), for arbitrary z and Σ, and with p^a and S^{ab} as given by (6.1), (6.2). As Dixon has shown, this is indeed the case. The propagators turn out to be simply expressible in terms of the *world function* of Ruse (see Synge [23]) (They are naturally related to the propagation properties of Jacobi fields, i.e., solutions of the geodesic deviation Equation (2.52), and therefore to those of Killing vector fields, and may therefore be useful outside the theory of motion.) We shall not give the formulae here.

Once K and H have been specified p^a and S^{ab} are well defined functions of z and functionals of Σ. To get rid of the arbitrariness of Σ it is useful henceforth to choose, at z, an arbitrary timelike unit vector n^a, and to define $\Sigma(z, n)$ as the hypersurface generated by the geodesics starting at z in directions orthogonal to n^a. Then $p^a(z, n) = p^a(z, \Sigma(z, n))$ and $S^{ab}(z, n) = \ldots$ are functions of (z, n) (maps from the sphere bundle over M into the appropriate tensor bundles). To eliminate the n-dependence, one will wish to make a 'best choice' of n for each event z. By applying fixed point theorems Beiglböck [95] and Madore [96] have established that, under mild and physically reasonable restrictions about g_{ab}, T^{ab}, there exists one and only one vector field $n^a(z)$ near $\mathfrak{T}$ such that $p^a(z) = p^a(z, n(z))$ is proportional to n^a. This vector field n^a is a general-relativistic analogue of the 'field of observers at rest in the C-M-frame' of special relativity, and the corresponding $p^a(z)$ is the *total 4-momentum vector of* $\mathfrak{T}$ *at* z. Analogously, $S^{ab}(z) = S^{ab}(z, n(z))$ is the *total angular momentum bivector of* $\mathfrak{T}$ *at* z.

Beiglböck and Madore (*loc. cit.*) have also shown that there exists a unique, smooth, timelike line L_0 contained in the convex hull of $\mathfrak{T}$ such that, for $z \in L_0$, $p^a(z)$ and $S^{ab}(z)$ are related by the center-of-mass condition (6.3); this L_0 is called the *center-of-mass world line* of the body. (In the local rest frame of p^a Equation (6.3) reduces to the classical characterization of the center of mass.)

The appropriateness of these concepts is strengthened by the fact (Dixon [94]) that in a spacetime of constant curvature, L_0 is necessarily a geodesic, p^a and S^{ab} are parallely transported along L_0, and p^a is tangent to L_0, in consequence of (2.25).

In analogy to Newtonian mechanics one might expect that there exists a sequence of *reduced multipole moments* of an extended body with respect to an arbitrary world line L (in particuar, L_0) which, together with L, characterize the tensor field T^{ab} uniquely, and which satisfy a finite system of ordinary differential equations along L, equivalent to $T^{ab}{}_{;b} = 0$. This is indeed true. The reduced set of moments consists of p^a, S^{ab} and a series $J^{c_1 \ldots c_n abcd}$ $(n \geqslant 0)$ of 2^{2+n}-pole moments, symmetric in the c_i's and with the curvature tensor symmetries with respect to the last four indices; moreover, $p_{c_1} J^{c_1 \cdots c_n abcd} = 0$. The *moment equations* are (Dixon [94])

$$\frac{Dp^a}{ds} = -\tfrac{1}{2} R^a{}_{bcd} v^b S^{cd} - \tfrac{1}{6} R_{bcde;}{}^a J^{bcde} + \cdots, \tag{6.7}$$

$$\frac{DS^{ab}}{ds} = 2p^{[a} v^{b]} - \tfrac{4}{3} R_{cde}{}^{[a} J^{b]ecd} + \cdots. \tag{6.8}$$

The dots indicate higher multipole terms, s is a parameter of the line L such that $dz^a/ds = v^a$ satisfies

$$-p_a v^a = +(-p_a p^a)^{1/2} = M, \tag{6.9}$$

where M is the mass of the body.

If, in particular, L is chosen to be the center-of-mass line L_0, one has in addition Equation (6.3). If (6.3) is differentiated along L_0 and (6.7), (6.8) are inserted, the resulting equation determines the 'kinematical velocity' v^a as a function of the reduced moments, so one has

$$\frac{dz^a}{ds} = v^a (p^b, S^{bc}, J^{bcde}, \ldots). \tag{6.10}$$

Given the J's the Equations (6.7), (6.8) and (6.10) determine L_0 and the evolution of p^a, S^{ab} – this is precisely as in Newtonian mechanics. Within this framework it is possible to develop a general-relativistic theory of dynamically rigid bodies, and to discuss the inductive energy transfer by gravity between deformable and rotating bodies.

Under most, if not all, circumstances p^a is very nearly equal to Mv^a, as follows from the explicit form of (6.10). Moreover, for small bodies in a slowly varying external field, the forces and torques in Equations (6.7), (6.8) will be small, and will decrease rapidly with increasing multipole order. In view of present limits of observational accuracy, it appears that for planets (and even more for smaller, freely falling bodies) one may safely replace (6.7) by

$$\frac{Dp^a}{ds} = 0;$$

this combined with the very nearly correct equation $p^a = Mv^a$ implies *geodesic motion*. Applied to the Earth, the second term on the right hand side of (6.8) produces the lunisolar precession of the Earth's axis. Applied to a precisely spherical, rapidly spinning, small body (such as the ones to be used in the Stanford gyroscope experiment) Equation (6.8) implies that the rotation-axis is Fermi-propagated along L_0, whether L_0 is geodesic or not.

So far we have considered the metric g_{ab} as given, and drawn conclusions from $T^{ab}{}_{;b}=0$. Actually, g_{ab} is linked to T^{ab} by the field equation, and so the preceding equations should be supplemented by relations giving g_{ab} in terms of body parameters. This can so far be done only approximately in terms of formal iteration methods which will be mentioned in Section 7. One may, however, wish to treat the motion of a small body in an external field by taking the actual metric as consisting of a given background contribution and a small perturbation produced by the body. For the body one has then $T^{ab}{}_{;b}=0$, where the covariant derivative refers to the actual, perturbed linear connection. If the latter is split into the background part and a difference tensor Δ^a_{bc} one gets

$$T^{ab}{}_{;b} = - \Delta^a_{bc} T^{bc} - \Delta^b_{bc} T^{ac}, \tag{6.11}$$

where now ; refers to the background connection. The 4-vector on the right-hand side of this equation may be called the gravitational *self-force* density.

If the non-gravitational internal forces acting within the body in question (which appear explicitly in (6.11) if the left-hand side is written out in detail in the manner of Equation (4.22)) are much larger than the gravitational self forces, as is the case for artificial satellites or laboratory-sized objects, one may neglect the right-hand side in Equation (6.11) and apply the previous theory to obtain equations of motion for such test bodies in terms of a given background spacetime metric. This reasoning, however, is inapplicable if one wants to justify the usual elementary treatment of the motion of a planet in the 'external' field of the Sun; for in that case the self force in (6.11) is roughly equal to the pressure gradient term from the left hand side. A justification of the geodesic law (with respect to the background geometry!) in this case requires a demonstration that the total, integrated self force which appears on the right-hand side of the modified Equation (6.7) is negligibly small. This can roughly be inferred from the fact that within the world tube of the planet the background (Schwarzschild) geometry is very nearly flat, and that in the slow motion, weak-field approximation the self force density reduces essentially to the Newtonian one which is known to contribute a vanishing total force.

Different approaches to planetary motion consist in determining the total, perturbed gravitational potential g_{ab} along with the motion of the bodies by means of iteration procedures; such methods will be indicated in Section 7.

The point to be emphasized is that the main problem still requiring further work in connection with the derivation of the geodesic law and its generalizations is the splitting off of the self field and the determination of useful approximations for the total self force (the gravitational radiation reaction force).

Another open problem is whether there exist suitable weak solutions of the field equation (in the sense of distribution theory) representing *point particles*. The Schwarzschild-Kruskal geometry of the spherically symmetric empty spacetime shows that such solutions representing particles without directional structure and with positive mass do not exist, and leads one to conjecture that point particles are incompatible with the Einstein field equation.

An ingenious approach to equations of motion which avoids the introduction of source terms and characterizes the structure of a particle in terms of the surrounding geometry is due to Newman and Posadas [97]; see also [98]). It requires a special formalism not to be presented here. It appears to be unknown whether this approach can be used to describe the motions of several interacting particles.

III. *Approximation Methods and Examples of Spacetimes*

7. Approximation Methods

7.1. Infinitesimal perturbations; the linear approximation

The set of solutions (g_{ab}, T_{ab}) of Einstein's field equation

$$G_{ab} = 8\pi T_{ab} \tag{7.1}$$

on some manifold M may be considered as a function space. Even without a rigorous formalization this viewpoint is useful for the discussion of approximation methods.

Let $(g_{ab}(x, \varepsilon), T_{ab}(x, \varepsilon))$ be a one-parameter family of solutions of Equation (7.1). It may be thought of as a *curve* in the space of solutions passing through the 'point' given by $\varepsilon=0$, which we denote simply by $(g_{ab}(x), T_{ab}(x))$. Approximation methods aim at constructing such curves for small values of ε, starting with a given solution which may, in particular, be the empty, flat solution $(\eta_{ab}, 0)$.

The 'tangent' to the curve considered at $\varepsilon=0$ is defined as the pair

$$\delta g_{ab} = \left(\frac{\partial}{\partial\varepsilon} g_{ab}\right)_{\varepsilon=0}, \quad \delta T_{ab} = \left(\frac{\partial}{\partial\varepsilon} T_{ab}\right)_{\varepsilon=0} \tag{7.2}$$

of tensor fields. In a similar way δ can be applied to other tensor fields, connections etc. defined along solution curves. $(\delta g_{ab}, \delta T_{ab})$ is also called an *infinitesimal perturbation of* $(g_{ab}, T_{ab})_{\varepsilon=0}$.

If $(\delta g_{ab}, \delta T_{ab})$ is an infinitesimal perturbation of (g_{ab}, T_{ab}), one might expect $(g_{ab}+\varepsilon\delta T_{ab}, T_{ab}+\varepsilon\delta g_{ab})$ to be an approximate solution 'near to' (g_{ab}, T_{ab}) if ε is sufficiently small. No precise justification (or even analysis of the meaning) of that statement is known to the author, but nevertheless we shall accept this plausible, physical point of view as a motivation for considering infinitesimal perturbations.

Since the identification of events of M as 'the same' for different solutions, e.g., different values of ε, is conventional and not intrinsically meaningful, one may apply to $(g_{ab}(x, \varepsilon), T_{ab}(x, \varepsilon))$ a diffeomorphism

$$\bar{x}^a = f^a(x^b, \varepsilon) \tag{7.3}$$

depending smoothly on ε and such that (7.3) equals the identity map if $\varepsilon=0$, and thus obtain a new curve $(\bar{g}_{ab}(x, \varepsilon), \bar{T}_{ab}(x, \varepsilon))$ which describes intrinsically the same family of solutions as the original one. Such a transformation is (unfortunately) called a *gauge transformation*.

By means of the generating vector field

$$\delta x^a = \xi^a = \left(\frac{\partial f^a}{\partial\varepsilon}\right)_{\varepsilon=0} \tag{7.4}$$

of (7.3) and the concepts introduced in Subsection 3.3 one proves straightforwardly that for any curve of geometrical objects (tensor fields, connections) $A^{\cdot\cdot}{}_{\cdot\cdot}(x, \varepsilon)$

$$\overline{\delta A^{\cdot\cdot}{}_{\cdot\cdot}} = \delta A^{\cdot\cdot}{}_{\cdot\cdot} - \mathscr{L}_\xi A^{\cdot\cdot}{}_{\cdot\cdot} \tag{7.5}$$

under a gauge transformation. (Sachs [99]) Hence, *the perturbation of* $A^{\cdot\cdot}{}_{\cdot\cdot}$ *is gauge invariant with respect to (7.3) if and only if* $\mathscr{L}_\xi A^{\cdot\cdot}{}_{\cdot\cdot}=0$. (This is the case for *all* gauge transformations precisely if the components of $A^{\cdot\cdot}{}_{\cdot\cdot}$ with respect to an arbitrary coordinate system are constants, as is true for δ^a_b and its analogs.) It is clear that the important object of perturbation theory is not $\delta A^{\cdot\cdot}{}_{\cdot\cdot}$ itself, but its gauge equivalence class. This class reduces to $\delta A^{\cdot\cdot}{}_{\cdot\cdot}$ itself only if $A^{\cdot\cdot}{}_{\cdot\cdot}$ is gauge invariant for all ξ's.

Let us determine the conditions imposed on $(\delta g_{ab}, \delta T_{ab})$ by the field Equation (7.1).

For brevity we put

$$\delta g_{ab} = h_{ab} \tag{7.6}$$

and use the unperturbed metric, connection etc. in formulating tensor equations for the perturbed fields.

The preceding definitions imply

$$\delta g^{ab} = - h^{ab}, \tag{7.7}$$

$$\delta \Gamma^{a}_{bc} = h^{a}_{(b;c)} - \tfrac{1}{2} h_{bc;}{}^{a}, \tag{7.8}$$

$$\begin{aligned} \delta R^{a}{}_{bcd} &= - 2\delta \Gamma^{a}_{b[c;d]} = \\ &= - h^{a}_{b;[cd]} - 2 g_{be} h^{[a\cdot\ e]}_{\ \ [c;\cdot d]}. \end{aligned} \tag{7.9}$$

($\delta \Gamma^{a}_{bc}$ is a tensor. (7.8) and (7.9) are obvious if expressed in geodesic coordinates.) By means of these formulae and with the abbreviations

$$\Box = (\ \ )_{;a}{}^{a} \tag{7.10}$$

(for arbitrary tensors) and

$$\tau_{ab} = g_{c(a}\delta\left(T^{c}{}_{b)} - \tfrac{1}{2} T \delta^{c}_{b)}\right) \tag{7.11}$$

one obtains from (7.1) the following *general form of the perturbation equation*:

$$\begin{aligned} &\left(\tfrac{1}{2}\, \delta^{(a}_{(c}\delta^{b)}_{d)} \Box + R^{(a\cdot b)}_{\ (c\cdot d)}\right) h^{cd} - \left(h^{c(a;\cdot b)}{}_{c} - \tfrac{1}{2} h^{;ab}\right) \\ &= - 8\pi\tau^{ab} \end{aligned} \tag{7.12}$$

(If the unperturbed matter tensor vanishes then $\tau_{ab} = \delta T_{ab} - \frac{1}{2} g_{ab} \delta T$.) The second term on the left-hand side of this equation vanishes if the gauge condition

$$h^{b}_{a;b} = \tfrac{1}{2} h_{,a} \tag{7.13}$$

is satisfied.

As pointed out by Sciama *et al.* [100] (7.12) has the following properties:

(a) It connects the field h_{ab} linearly and homogeneously with the source τ_{ab}.

(b) If $-h^{ab}$ is replaced by g^{ab} and τ^{ab} is replaced by $T^{ab} - \frac{1}{2} g^{ab} T$, (7.12) reduces to the unperturbed field Equation (7.1).

(c) The operator

$$D^{ab}{}_{cd} = \tfrac{1}{2} \delta^{(a}_{(c}\delta^{b)}_{d)} \Box + R^{(a\cdot b)}_{\ (c\cdot d)} \tag{7.14}$$

maps the space of smooth, symmetric tensor fields h^{ab} with compact support linearly into itself. It is hyperbolic, and selfadjoint with respect to the (nondegenerate, symmetric, indefinite) inner product

$$\langle h, k \rangle = \int h^{ab} k_{ab} \eta \,. \tag{7.15}$$

Because of these properties it is possible to replace the wave Equation (7.12) (with

(7.13)) by a generally covariant integro-differential equation of the Kirchhoff type, and because of (b) this *Sciama-Waylen-Gilman integral equation* continues to hold if the unperturbed variables g^{ab}, $T^a{}_b - \frac{1}{2}T\delta^a{}_b$ are substituted for the infinitesimal perturbations.

It reads

$$g^{a'b'}(x') = 16\pi \int_D G^{a'b'p}{}_q \left(T^q{}_p - \tfrac{1}{2}T\delta^q_p\right) \eta + \int_{\partial D} G^{a'b'p}{}_p{}^{;q}\sigma_q\,, \tag{7.16}$$

where $G^{\cdot\cdot}{}_{\cdot\cdot}$ is a Green's function (2-point tensorial distribution) relative to the operator (7.14), and primed indices refer to the field event x', whereas unprimed ones refer to the source event x. D is a (sufficiently small) compact, orientable domain with boundary ∂D. For a derivation of (7.16) from (7.12) resp. (7.1), see ref. [100], for a discussion of many interesting questions related to (7.16) see ref. [101].

Equations (7.12) (or equivalent ones) form the basis of many calculations concerning, in particular, perturbations of cosmological models; an example is treated in Prof. Sachs' lectures in this volume, p. 227. They (or Equation (7.16) combined with the form of $G^{\cdot\cdot}{}_{\cdot\cdot}$, and applied to perturbations) show clearly that perturbations propagate in a wavelike manner relative to the background, with null hypersurfaces as wave fronts.

Let us now specialize the unperturbed solution to the trivial one, $g_{ab}=\eta_{ab}$, $T_{ab}=0$. Then, with

$$\bar{h}^{ab} = h^{ab} - \tfrac{1}{2}h\eta^{ab}\,, \tag{7.17}$$

and with the harmonic gauge (7.13) we obtain from (7.12) the 'linearized Einstein equations'

$$\Box\, \bar{h}^{ab} = -\,16\pi\tau^{ab}\,, \tag{7.18}$$

$$\bar{h}^b_{a,b} = 0\,, \tag{7.19}$$

which imply the conservation law

$$\tau^{ab}{}_{,b} = 0\,. \tag{7.20}$$

According to (7.5) the gauge condition (7.19) is maintained under the restricted gauge transformation satisfying

$$\Box\, \xi^a = 0\,. \tag{7.21}$$

Whereas neither h_{ab} nor $\delta\Gamma^a_{bc}$ are gauge invariant, the linearized curvature tensor

$$\delta R^{ab}{}_{cd} = -\,2h^{[a\ b]}_{[c,d]} \tag{7.22}$$

(compare Equation (7.9)) is; it satisfies the linearized Bianchi identity

$$\delta R_{ab[cd,e]} = 0 \tag{7.23}$$

and has, of course, the usual symmetries. Moreover, the map $h_{ab} \rightarrow \delta R^{ab}{}_{cd}$ given by

(7.22) can be inverted explicitly: If δR_{abcd} has the appropriate symmetry and satisfies (7.23), then the general potential h_{ab} giving rise to the field δR_{abcd} is

$$h_{ab} = 2x^c x^d \int_0^1 \delta R(tx)_{acdb} t(1-t)\,\mathrm{d}t + \eta_{a,b} + \eta_{b,a} \tag{7.24}$$

with arbitrary functions η_a (see ref. [98]. Equations (7.22) and (7.24) show explicitly that h_{ab} is gauge equivalent to zero if and only if $\delta R^a{}_{bcd}=0$, as one would expect from the exact theory.

One may consider (7.23) and

$$\delta R^c{}_{acb} = 8\pi(\tau_{ab} - \tfrac{1}{2}\tau\eta_{ab}) \tag{7.25}$$

as field equations of the linear approximation, expressed in terms of gauge invariant quantities.

In empty spacetime the field Equations (7.18), (7.19) reduce to

$$\square\, \bar{h}^{ab} = 0\,, \quad \bar{h}^{ab}{}_{,b} = 0\,, \tag{7.26}$$

and one can use the restricted gauge freedom to arrange that, in addition,

$$\bar{h} = 0\,. \tag{7.27}$$

Then there still remain gauge transformations restricted by

$$\square\, \xi^a = 0\,, \quad \xi^a{}_{,a} = 0\,. \tag{7.28}$$

It is not difficult to define (via Fourier transformation, e.g.) on a set of gauge equivalence classes of solutions of the Equations (7.26), (7.27) the structure of a complex Hilbert space which is a representation space of a unitary representation of the Poincaré group, and thus to establish that a quantum theory of a free field belonging to the Equations (7.26), (7.27) has quanta of *mass zero* and *spin two*. (See, e.g., ref. [102)).

Solutions of the Equation (7.18), (7.19) can be obtained in close analogy to electrodynamics. For a careful exposition, see Pirani (ref. [92]). As an example of solutions of (7.26), (7.27) we may take the general *plane wave* travelling in the z-direction. In transverse tracefree gauge, its metric is

$$\mathrm{d}s^2 = (\mathrm{d}s^2)_0 + A_+(t-z)(\mathrm{d}x^2 - \mathrm{d}y^2) + 2A_\times(t-z)\,\mathrm{d}x\,\mathrm{d}y\,, \tag{7.29}$$

where $(\mathrm{d}s^2)_0$ is the flat metric in inertial coordinates (x, y, z, t) and A_+ and $A_\times$ are the two amplitudes. The gauge transformation generated by

$$\xi = \tfrac{1}{2}(xA_+ + yA_\times)\frac{\partial}{\partial x} + \tfrac{1}{2}(xA_\times - yA_+)\frac{\partial}{\partial y} + \\ + \tfrac{1}{2}\left(\tfrac{1}{2}(x^2 - y^2)A'_+ + xyA'_\times\right)\left(\frac{\partial}{\partial z} + \frac{\partial}{\partial t}\right) \tag{7.30}$$

satisfies (7.28) and transforms (7.29) into the longitudinal form

$$ds^2 = (ds^2)_0 + (\tfrac{1}{2}(x^2 - y^2) A''_+ + xyA''_\times)(dt - dz)^2 . \tag{7.31}$$

Miraculously (7.31) *satisfies the exact, nonlinear vacuum field equations*, whereas (7.29) does not, although in (7.29) one may choose the amplitudes such that the h_{ab} are uniformly as small as one wishes, which is not possible in (7.31). This reminds us of the difficulties inherent in the notion of a 'small' metric perturbation. With (7.22) one ealsiy calculates for (7.29) and (7.30)

$$\delta R_{abcd} = k_{[a} C_{b][c} k_{d]} , \tag{7.32}$$

where $k = \partial/\partial z + \partial/\partial t$ is the propagation vector of the wave and

$$C_{[ab]} = 0 , \quad C^a{}_a = 0 , \quad C_{ab} k^b = 0 , \tag{7.33}$$

so that

$$\delta R_{abcd} k^d = 0 . \tag{7.34}$$

This shows that plane waves are of Petrov type (4), in Penrose's notation. By means of Equation (7.32), (7.33) and the geodesic deviation Equation (2.52) one can easily study the effect of such waves on pairs of test particles. More details about the exact plane waves (7.31) can be found, e.g. in refs. [103] and [104].

7.2. Finite perturbations; iteration methods

It is apparent from Equation (7.20) that the linear approximation is incapable of describing the gravitational interactions between bodies; the bodies do not 'feel' the field in this approximation. Thus in order to account for such interactions, even in the case of weak fields like that of the solar system, one has to go beyond the linear approximation.

The literature on iteration or expansion methods in general relativity is enormous and not easily surveyable; no method seems to be clearly superior. This section only aims at outlining the problems as understood by the author (who is not an expert on these matters).

One good point of departure is the *Landau-Lifshitz* form

$$(\mathfrak{g}^{a[b} \mathfrak{g}^{d]c})_{,cd} = (-g)\, 8\pi (T^{ab} + t^{ab}) = \theta^{ab} \tag{7.35}$$

of the field Equation (7.1). It is obtained by introducing (in the variational principle (2.27), e.g.) the quantities*

$$\mathfrak{g}^{ab} = \sqrt{-g}\, g^{ab} \tag{7.36}$$

as the field variables and combining terms containing second derivatives as shown on the left hand side of Equation (7.35), and collecting the remaining terms in $-gt^{ab}$ whose components are quadratic forms in $\mathfrak{g}^{ab}{}_{,c}$ with coefficients depending on the $\mathfrak{g}^{ab}$'s.

* Warning: Here the symbols $\mathfrak{g}^{ab}$ has a meaning different from that in Subsection 5.3.

The form of (7.35) shows that the effective source term satisfies the (formal) conservation law

$$\theta^{ab}{}_{,b} = 0 . \tag{7.37}$$

Since θ^{ab} is symmetric, one can formulate a corresponding, formal angular momentum conservation law as in special relativity (compare section 6).

In *harmonic coordinates*, i.e., if

$$\mathfrak{g}^{ab}{}_{,b} = 0 , \tag{7.38}$$

the left-hand side of (7.35) simplifies to $\frac{1}{2}\mathfrak{g}^{cd}\mathfrak{g}^{ab}{}_{,cd} - \frac{1}{2}\mathfrak{g}^{ad}{}_{,c}\mathfrak{g}^{bc}{}_{,d}$, so that by modifying t^{ab} slightly one gets

$$\mathfrak{g}^{cd}\mathfrak{g}^{ab}{}_{,cd} = -\ 16\pi g\,(T^{ab} + \tilde{t}^{ab}) . \tag{7.39}$$

This form of the equation is useful for the discussion of the *Cauchy problem*; e.g., it shows immediately that the *characteristics* are given by

$$g^{ab}S_{,a}S_{,b} = 0 . \tag{7.40}$$

For approximation methods (7.39) is not immediately useful since the operator on the left contains the unknown field. If the aim is to find weak fields it is plausible to introduce an auxiliary flat metric η_{ab} and to put

$$\mathfrak{g}^{ab} = \eta^{ab} - \gamma^{ab} . \tag{7.41}$$

This splitting of $\mathfrak{g}^{ab}$ is largely arbitrary and a matter of computational convenience only. We use, in connection with a splitting (7.41), always coordinates x^a which are (globally) inertial with respect to η^{ab}. A transition $(\eta^{ab}, x^c) \to (\eta^{a'b'}, x^{c'})$ from one flat metric and associated coordinate system to another pair, for a given metric $\mathfrak{g}^{ab}$, induces via (7.41) a change of the functions $\gamma^{ab}(x^c)$; such changes are called *gauge transformations* in this context.

With (7.41) and (7.38) or

$$\gamma^{ab}{}_{,b} = 0 , \tag{7.42}$$

Equation (7.39) can be rewritten as

$$\Box\, \gamma^{ab} = -\ 16\pi H^{ab} , \tag{7.43}$$

where now

$$\Box = \eta^{cd}\partial_c\partial_d \tag{7.44}$$

is the D'Alembertian of η_{ab} and

$$H^{ab} = -\ g\,(T^{ab} + \tau^{ab}) . \tag{7.45}$$

τ^{ab} contains second derivatives of γ^{ab}, but all its terms are nonlinear in γ^{ab} and its derivatives, and for a suitable choice of η_{ab} and weak and slowly varying fields, one

may expect γ^{ab} and its derivatives to be uniformly small, so that Equation (7.43) appears to be useful for an iteration process.
Eqs. (7.42), (7.43) imply

$$H^{ab}{}_{,b} = 0\,. \tag{7.46}$$

If the fields γ^{ab} to be constructed would decrease sufficiently rapidly at past null infinity (i.e., on past null cones of η_{ab}), one could replace (7.43) by the integrodifferential equation

$$\gamma^{ab}(x) = 4 \int \frac{H^{ab}_{\text{ret}}}{|x - x'|}\, \mathrm{d}^3 x'\,. \tag{7.47}$$

This equation is meaningful independently of the gauge condition (7.42); provided the domain of the integral operator is sufficiently large, one could consider (7.47) as defining (implicitly) the field $\gamma^{ab}(x)$ as a functional of suitable matter variables on which T^{ab} depends. (For a perfect fluid, e.g., one may consider the (proper) energy density ϱ, the pressure p and the coordinate velocity $v^\lambda = \mathrm{d}x^\lambda/\mathrm{d}x^4 = u^\lambda/u^4$ as matter variables; T^{ab} is to be considered as a function of ϱ, p, v^λ and γ^{ab}, to be inserted in Equations (7.45), (7.47).) Thus one could associate (formally) a field with an arbitrarily moving source, as in electrodynamics, and obtain equations of motion for the matter variables by subjecting these functionals to the (hopefully) equivalent conditions (7.42) or (7.46). The point is that in such a scheme the Einstein equation is *separated* into a *truncated field equation* ((7.43) or (7.47)) which can be solved (?) for any source distribution, and (functional) *equations of motion* which can be imposed afterwards, such that at the end the full Equation (7.1) is satisfied.

Of course, one cannot solve (7.47) exactly. But one may, for given matter variables, insert η^{ab} on the right hand side of (7.47), to obtain a field $\underset{1}{\gamma}^{ab}$, then insert $\underset{1}{\gamma}^{ab}$ on the right to get $\underset{2}{\gamma}^{ab}$, etc. and stop at some stage. Using the approximate field $\mathrm{g}^{ab} = \eta^{ab} - \underset{n-1}{\gamma}^{ab}$ and the associated matter tensor $\underset{n-1}{T}^{ab}$, one can form the approximate equations of motion $\underset{n-1}{H}^{ab}{}_{,b} = 0$.

The procedure outlined can (and has) been modified by using different variables instead of g^{ab}, different coordinate conditions and truncation methods, and in particular by imposing *slowness assumptions* which amount to approximating retarded potentials arising from (7.47) by instantaneous potentials (and some of their time derivatives). Some pertinent references are [105, 106, 8, 107, 108, 109, 110, 111]. The most far reaching results along these lines are those of Chandrasekhar and his coworkers. For generalisations to other theories of gravity, see refs. [6].

In this way one can understand, modulo some conceptual and mathematical obscurities and with a considerable amount of labour, how the Newtonian theory of gravity arises as an approximation to general relativity, which post-Newtonian corrections occur, and one can even obtain radiative corrections to the equations of motion at higher orders. These considerations still appear to be in need as well as capable of simplifications and clarifications (see, e.g., ref. [112]), in spite of the enormous efforts which have already been put into them. In particular, no precise error estimates seem to be known so far. The form of the (exact) Equation (7.16) suggests that it might be used to provide a more precise iteration procedure, in particular since the propagation of the higher order fields according to the Green's function of lower order fields (and not along flat null cones) might be taken into account, as has been indicated by the inventors of that equation. The difficulties in determining the Green's function iteratively appear to have prevented such an improvement so far, however.

8. Spherically Symmetric Spacetimes

8.1. Action of SO_3 and Metric Tensor

A spacetime (M, g_{ab}) is spherically symmetric if the rotation group SO_3 acts on (M, g_{ab}) as an isometry group with simply connected, complete, spacelike, 2-dimensional orbits. (The orbit of $p \in M$ under the transformation group G is the set $O_p = G(p)$ of events into which p is mapped by the elements of G. A diffeomorphism of M is an isometry if it leaves the metric invariant.)

Theorem 8.1. (M, g_{ab}) is spherically symmetric if and only if $M = S^2 \times N$ (direct product), where S^2 is the 2-sphere, and there exists a positive scalar r on N such that $r^{-2} g_{ab}$ is the direct sum of a Lorentzian metric g_N on N and the standard metric g_S on the unit sphere. (I.e., g_{ab} is conformally decomposable.)

Sketch of proof (compare ref. [113]): The orbits under SO_3 are spheres, since they are Riemannian 2-spaces with maximal isometry groups, and simply connected and complete. – Let $r(p)$ be the radius of the orbit O_p of $p \in M$. r is a positive, smooth function on M, constant on orbits. – M consists of disjoint orbits, $M = \bigcup_p O_p$. Let $O_p^\perp$ be the surface, near p, generated by the geodesics through p and orthogonal to O_p at p. For q near p, O_p intersects $O_p^\perp$ in a point s. Any rotation i about p maps the geodesic through p and s onto the geodesic g through p and $i(s)$ which is orthogonal to O whence $g \subset O_p^\perp$ and $i(s) \in O_p$. Since also $i(s) \in O_q$, we conclude that $i(s) \in O_q \cap O_p^\perp$, and thus $i(s) = s$. I.e., $O_p^\perp$ is pointwise invariant with respect to i, and i is contained in the isotropy subgroup F_s of SO_3 of every point $s \in O_p^\perp$. Thus $F_p \subset F_s$ if $s \in O_p^\perp$. Since dim $F_p =$ dim F_s and F_p, F_s are connected, $F_p = F_s$. – Let T be any tangent vector to $O_p^\perp$ at s, and let $T_\perp$ be the orthogonal projection of T onto O_s, and let $i \in F_p = F_s$. Then $i_*(T) = T$ and, since $i(O_s) = O_s$, $i_*(T_\perp) = T_\perp$. This is possible only if $T_\perp = 0$, i.e., T is orthogonal to O_s, so $O_p^\perp$ intersects O_s orthogonally at all points s of $O_p^\perp$. Our result is that the 2-surfaces $O_p^\perp$ intersect the orbits O_q orthogonally and in isolated points. Since the family of orthogonal 2-surfaces of a given family of 2-surfaces, if it exists, is unique, it follows that $s \in O_p^\perp$ implies $O_s^\perp = O_p^\perp$. This implies in turn that $O_p^\perp$ is totally geodesic, and that each $O_p^\perp$ can be extended to a maximal element of the collection of 2-surfaces orthogonal to $\{O_p\}$. – If O_p is any (fixed) orbit we have the decomposition $M = \bigcup_{q \in O_p} O_p^\perp$ of M into disjoint 2-surfaces. If we choose a point p arbitrarily as origin, each event $r \in M$ can be characterized by its 'components' (q, s) in O_p, $O_p^\perp$, respectively, which are defined by the relations $r \in O_q^\perp$ and $\{s\} = O_p^\perp \cap O_r$. Since O_p is a sphere, this establishes that $M = S^2 \times N$, with $N = O_p^\perp$. – We assign a metric to N by taking the metric induced in $O_p^\perp$ by (M, g_{ab}). Then, since $i(O_p^\perp) = O_{ip}^\perp$, all the $O_p^\perp$'s are isometric to N. – Any two orbits O_p, O_r can be mapped onto each other as follows: $O_p \ni q \to q'$, $\{q'\} = O_q^\perp \cap O_r$. These maps f_{pr} commute with the action of SO_3. A circle about p in O_p is obtained by taking a point $q \neq p$, $q \in O_p$, and applying the isotropy group F_p to q. Since F_p is also the isotropy group of $r = f_{pr}(p)$, as we have shown above, and $f_{pr} \circ F_p = F_p \circ f_{pr}$, it follows that a circle in O_p is mapped onto a circle in O_r by f_{pr}. The ratio of the radii of these circles remains unchanged if an isometry i is applied to both of them. These results prove f_{pr} maps O_p conformally onto O_r with a constant conformal factor. – These remarks establish the theorem.

In the above definition of spherical symmetry we have excluded the occurrence of *centers*, i.e., world lines consisting of events which are fixed points of SO_3. That is artificial and was done only to keep the formulation of theorem 1 simple. The modifications required by the inclusion of centers will be left to the reader (see ref. [114]). The number of centers of a spherically symmetric spacetime with respect to a particular action of SO_3 may be 0, 1 or 2. A spacetime may be spherically symmetric with respect to several different actions of SO_3; in particular, each event may be contained in a center with respect to one of these actions (Examples: Flat spacetime, any Robertson-Walker universe.)

From now on we assume (M, g_{ab}) to be spherically symmetric, and we assume that one action of SO_3 is specified. Then r is a well defined, smooth scalar on M, positive in general and zero along centers.

In the special case $r = \text{const.} > 0$ the metric of spacetime is the direct sum of the metrics of S^2 and N, whence the curvature is also a direct sum (see, e.g., ref. [115]). Einstein's equation then implies that each vector orthogonal to the SO_3-orbits is an eigenvector of T^{ab} with eigenvalue $\varrho = (8\pi r^2)^{-1}$, and all vectors tangent to these spheres are eigenvectors with eigenvalue $p = -K/8\pi$, if K denotes the Gaussian curvature of N. Such matter tensors can be realized by electromagnetic fields (example: Levi-Civita's homogeneous magnetostatic field), but not for ordinary (e.g., stellar) matter; therefore we shall henceforth suppose that

$$\mathrm{d}r \neq 0 \tag{8.1}$$

except at special events. In particular (8.1) holds for nonflat vacuum fields.

Consider a part of N where (8.1) holds. It is covered by a congruence of curves $r = \text{const.}$ If these curves are not lightlike, there exist orthogonal curves which may be labeled by means of a scalar t, $\mathrm{d}t \neq 0$. Then the metric of N is

$$(\mathrm{d}s^2)_N = \pm\,(e^{2\lambda}\,\mathrm{d}r^2 - e^{2\phi}\,\mathrm{d}t^2), \tag{8.2}$$

with λ and ϕ depending on (r, t). (Spacetimes in which the curves $r = \text{const.}$ are lightlike do not satisfy the vacuum field equation and do not seem to have sensible physical applications; they will not be considered here.)

Combining (8.2) with theorem (8.1) we obtain that the metric of a spherically symmetric spacetime satisfying (8.1) can be written, in a region where the lines $(\vartheta, \phi, r)=\text{const}$ are not lightlike, in the form

$$ds^2 = r^2\, d\Omega^2 \pm (e^{2\lambda}\, dr^2 - e^{2\phi}\, dt^2), \tag{8.3}$$

where

$$d\Omega^2 = d\vartheta^2 + \sin^2\vartheta\, d\varphi^2 \tag{8.4}$$

is the metric of the unit sphere and the sign is $+$ $(-)$ if the t-lines are timelike (spacelike).

Henceforth we shall choose the $+$ sign in (8.3) and permit λ and ϕ to have an imaginary part $i(\pi/2)$.

If (M, g_{ab}) admits but one action of SO_3, the coordinates in (8.3) are essentially unique; for

(a) the decomposition of M into orbits and their orthogonal surfaces is intrinsic,
(b) r is an invariant scalar since $4\pi r^2$ is the area of the orbit through an event,
(c) ϑ, φ are unique up to rotations of the unit sphere,
(d) t is unique except for monotonic transformations $t = f(t')$.

Consequently, λ is an invariant scalar, whereas ϕ is intrinsic only up to the transformations

$$\phi(r, t) \to \phi(r, f(t')) + \log(f'(t')). \tag{8.5}$$

8.2. Curvature

One convenient way to compute the curvature tensor of (8.3) is this: The metric $\widetilde{ds}^2 = r^{-2}\, ds^2$ is a direct sum of two 2-dimensional parts. For a 2-surface, $R_{abcd} = K(g_{ac}g_{bd} - g_{ad}g_{bc})$ where K is the Gaussian curvature. Thus only the K of the (r, t)-part of $\widetilde{ds}^2$ has to be computed to obtain the curvature tensor of $\widetilde{ds}^2$. To get the curvature of ds^2, one applies a conformal transformation. (The metric $\widetilde{ds}^2$ used here only for computational convenience also occurs naturally if one wishes to consider the asymptotic flatness of the Schwarzschild solution by means of Penrose's conformal completion; see [32].)

In this (or any other) way, one gets for the Einstein tensor of (8.3) ($\partial/\partial r = (\ \ )'$, $\partial/\partial t = (\ \ )^{\cdot}$):

$$G^t_t = r^{-2}(-1 + [re^{-2\lambda}]'), \tag{8.6}$$

$$G^r_t = -r^{-1}(e^{-2\lambda})^{\cdot}, \tag{8.7}$$

$$G^r_r - G^t_t = 2r^{-1}e^{-2\lambda}(\phi + \lambda)', \tag{8.8}$$

whereas $G^\vartheta_\vartheta = G^\varphi_\varphi$ can be expressed in terms of the preceding components of G^a_b by means of the Bianchi identity

$$G^a_{r;a} = 0. \tag{8.9}$$

The remaining components of G^a_b are zero. We also note for later reference

$$R_{\vartheta r\vartheta r} = r\lambda'. \tag{8.10}$$

In accordance with the analysis of the Cauchy problem of Einstein's theory, G_t^t and G_r^t do not contain second time derivatives. Whereas R_{rr} contains $\ddot{\lambda}$, $R_{\vartheta\vartheta}$ does not contain any time derivatives; i.e., in view of the spherical symmetry (and the coordinate conditions adapted to it), the (ϑ, ϑ)-field equation degenerates into a constraint, and there is a single evolution equation only.

8.3. Schwarzschild-Kruskal spacetime

If the vacuum field equations $G_b^a = 0$ are imposed and the transformation (8.5) is taken into account, Equations (8.6) to (8.8) (and the remark following it) lead to the following spherically symmetric spacetime metric:

$$ds^2 = r^2\,d\Omega^2 + (1 - 2M/r)^{-1}\,dr^2 - (1 - 2M/r)\,dt^2. \tag{8.11}$$

This *Schwarzschild metric* leads to the following spacetimes:

$$M < 0, \quad 0 < r < \infty, \tag{8.12}$$

$$M > 0, \quad 2M < r < \infty, \tag{8.13}$$

$$M > 0, \quad 0 < r < 2M. \tag{8.14}$$

(In all cases, $-\infty < t < \infty$, and ϑ, φ have their usual ranges.)

The spacetime given by (8.12) is *maximal*, i.e., it cannot be extended, since any of its geodesics is either complete (can be extended indefinitely in terms of an affine parameter) or approaches $r = 0$ arbitrarily closely, in which case the curvature invariant $R_{abcd}R^{abcd}$ diverges like r^{-6}. It admits the time-translations $(\vartheta, \varphi, r, t) \to (\vartheta, \varphi, r, t + \tau)$ as isometries, and since the associated Killing vector $\partial/\partial t$ is hypersurface orthogonal, it is *static*. A discussion of test particle orbits shows that M has to be interpreted as a mass, and since $M < 0$ this spacetime appears to be unphysical.

The spacetime corresponding to (8.13) is also *static*, but its curvature invariants are bounded, so that it is not obvious whether it can be extended or not. Since $M > 0$, the subspacetimes given by $R(t) < r < \infty$ $(R(t) > 2M)$ must be interpreted as the fields surrounding spherical bodies with radii $R(t)$.

The case (8.14) represents a spacetime in which r is a time coordinate, t is a space coordinate. Since $\partial/\partial t$ is a Killing vector, the 3-spaces $r = \text{const.}$ are homogeneous (but not isotropic). The metric of these 3-spaces evolves in time r; there is an initial singularity at $r = 0$. For $r \to 2M$ the curvature behaves as in case (8.13).

The spacetimes (8.12) and (8.13) are asymptotically flat whereas (8.14) is not, as can be inferred from (8.11) and the curvature values.

According to the consideration in Subsection 8.1 any spherically symmetric empty spacetime must consist of pieces of the types (8.12)–(8.14), and these pieces must consist of full orbits, i.e., they must correspond to regions in (r, t)-space.

Without attempting to discuss the general problem of patching together such pieces, we now show that, for a given $M > 0$, two pieces of type (8.13) and two of type (8.14) can be fitted together to form a single, analytic spacetime.

Since $g^{ab} r_{,a} r_{,b} = g^{rr} = 1 - 2M/r$ and r is an invariant, it follows that if the spacetime with $r > 2M$ ($r < 2M$) can be smoothly extended, then the hypersurface $r = 2M$ in the extended spacetime must be lightlike. This explains why the previous calculation has given the pieces $r > 2M$, $r < 2M$ separately: We *assumed* the curves $r = \text{const.}$ (in N) to be nonnull. This observation suggests to introduce instead of (r, t) coordinates in N which are adapted to at least one of the two families of null geodesics. (Remark: On a 2-space with indefinite metric null lines are null geodesics. Since in our case, N, or rather the $O_p^{\perp}$'s, are totally geodesic in M, these null lines are radial null geodesics of spacetime.) The outgoing geodesics (for $r > 2M$, say) are given by

$$\mathrm{d}r = \left(1 - \frac{2M}{r}\right)\mathrm{d}t \quad \text{or} \quad \frac{t - r}{2M} - \log\left(\frac{r}{2M} - 1\right) = \text{const}.$$

If any function u of the last expression is introduced to label these outgoing, SO_3-invariant null hypersurfaces (spherical wave fronts), the metric of N reads

$$g_{uu}\,\mathrm{d}u^2 + 2g_{ul}\,\mathrm{d}u\,\mathrm{d}l. \tag{8.15}$$

Without loss of generality we require l to be an affine parameter along the outgoing null rays. That is equivalent to $g_{ul,l} = 0$. A change $u \to f(u)$ brings about $g_{ul} = 1$. Our aim is to find functions f and g such that, with

$$u = f\left(\frac{t - r}{2M} - \log\left[\frac{r}{2M} - 1\right]\right), \quad l = g(t, r),$$

the metric $(1 - 2M/r)^{-1}\,\mathrm{d}r^2 - (1 - 2M/r)\mathrm{d}t^2$ of N goes over into (8.15) with $g_{ul} = 1$ *and remains regular in as large a region as possible*, beyond the domain corresponding to $r > 2M$. A transformation of this kind has been found by Israel [121]:

$$\begin{aligned} u &= 2\,\mathrm{sgn}\,(r - 2M)\,(M\,|r - 2M|)^{1/2} \exp\frac{r - t}{4M}, \\ l &= 2\,\mathrm{sgn}\,(r - 2M)\,(M\,|r - 2M|)^{1/2} \exp\frac{t - r}{4M}, \end{aligned} \tag{8.16}$$

with inverse

$$\begin{aligned} r &= 2M + \frac{ul}{4M} \\ t &= 2M + \frac{ul}{4M} - 2M \log\left|\frac{u}{l}\right|. \end{aligned} \tag{8.17}$$

The map $(r, t) \to (u, l)$, (8.16), has the following properties:

(a) It maps the half plane $2M < r$ analytically and bijectively onto the quadrant $u > 0$, $l > 0$.

(b) It takes the strip $0 < r < 2M$ analytically and bijectively onto the region $u < 0$, $0 < l < -8M^2/u$.

(c) It maps the half plane $0 < r$ homeomorphically (but *not* differentiably at

$r = 2M$!) onto the region

$$0 < l < \begin{cases} -\dfrac{8M^2}{u} & \text{if} \quad u < 0 . \\ \infty & \text{if} \quad u \geqslant 0 . \end{cases}$$

(d) In *both* subregions $0 < r < 2M$ and $2M < r < \infty$ it transforms the Schwarzschild metric (8.11) into the desired form

$$\mathrm{d}s^2 = \left(2M + \frac{ul}{4M}\right)^2 \mathrm{d}\Omega^2 + 2\,\mathrm{d}u\,\mathrm{d}l + \frac{2l^2}{8M^2 + ul}\,\mathrm{d}u^2 . \tag{8.18}$$

This metric (due to Israel) is obviously analytic in the whole (u, l)-region described in (c), even at $r = 2M \Leftrightarrow u = 0$.

We have thus obtained the desired common extension of the two spacetimes $r > 2M$, $r < 2M$. The result shows, however, that the restriction of l given in (c) is unnecessary. The metric (8.18) admits the reflection-isometry $(u, l) \to (-u, -l)$, and thus we should use the manifold

$$M = S^2 \times \{(u, l) \mid -8M^2 < ul\} \tag{8.19}$$

as the domain of (8.18). We have established

THEOREM 8.2. (Kruskal [122], Szekeres [123]). The spacetime (M, g_{ab}) given by (8.19) and (8.18) is a common analytic extension of the two spacetimes (8.13), (8.14). It contains two regions isometric to (8.13), and two regions isometric to (8.14). These four open subspacetimes are separated by the two 'null-cylinders' $u = 0$ and $l = 0$ each of which has the degenerate 'metric' $(2M)^2 \mathrm{d}\Omega^2$, and which intersect in the spacelike 2-sphere $u = l = 0$, characterized by $r_{,a} = 0$.

The role of the null cylinders as *horizons* will be described in other lectures.

The $(r > 2M)$-parts of the Kruskal spacetime can be characterized as its *maximal static domains*.

Kruskal spacetime (whose metric is usually written in a less explicit form than (8.18)) is *maximal*, since on any incomplete, inextendible half-geodesic r approaches 0, and thus $R_{abcd}R^{abcd}$ diverges, as will not be shown here.

It is worth observing that the null-cylinders are totally geodesic, as noticed by Hajicek (personal communication).

It is probably true that any spherically symmetric vacuum spacetime with a metric of class C^1, piecewise C^3 is isometric to a sub-spacetime of either (8.12) or (8.18), (8.19), though I am not aware of a proof.

8.4. EQUILIBRIUM CONFIGURATIONS OF PERFECT FLUIDS

A spherically symmetric perfect fluid must have a 4-velocity u^a orthogonal to the SO_3-orbits, this follows from symmetry and the field equations. Thus $u^\vartheta = u^\varphi = 0$, which implies that the vorticity ω_{ab} defined in (4.1), (4.2) vanishes.

It has been established in Subsection 3.8 that for a gas in thermodynamical equilibrium u^a is collinear with a Killing vector ξ^a, and this also holds for a fluid as can be shown with the methods of Subsections 4.2, 4.3. Since r is invariant, it follows that $r_{,a}\xi^a = \xi^r = 0$, and this combined with $\xi^\vartheta = \xi^\varphi = 0$ shows that the t-lines of the standard coordinate system employed in (8.3) coincide with the streamlines of the fluid,

$$u^a = e^{-\phi}\delta^a_t . \tag{8.20}$$

That is, in equilibrium there can be no radial motion, as one would expect. This implies $T^{tr} = 0$, hence $G^{tr} = 0$, and finally because of (8.7), $\dot{\lambda} = 0$.

Assuming henceforth equilibrium we conclude that the space sections $t = \text{const.}$ (which are intrinsic according to remark (d) in Subsection 8.1) have an induced metric which is independent of t and fully determined by the (t, t) field equation. Regularity of the metric of $t = \text{const.}$ at the center $r = 0$ requires $\lambda(0) = 0$, and (8.6) with $G^t_t = 8\pi T^t_t = -8\pi\varrho$ gives

$$e^{-2\lambda} = 1 - 2m/r \tag{8.21}$$

with a mass distribution

$$m(r) = 4\pi \int_0^r \varrho(\bar{r})\,\bar{r}^2\,\mathrm{d}\bar{r} . \tag{8.22}$$

Since in equilibrium p depends functionally on ϱ it follows that $p = p(r)$, and the (r, r)-field equation gives, according to (8.8),

$$\phi' = \frac{m + 4\pi r^3 p}{r(r - 2m)} = \frac{m}{r^2}\,\frac{1 + 4\pi r^3 p/m}{1 - \dfrac{2m}{r}}, \tag{8.23}$$

the relativistic analog of the gravitational field strength. Therefore ϕ' depends on r only, and because of the gauge-freedom (8.5) we can take $\phi = \phi(r)$.

We have shown that spacetime is *static*, $\xi^a = \delta^a_t$ is (without loss of generality) a *hypersurface orthogonal Killing vector*, just as in the vacuum solution with $r > 2M > 0$.

According to the preceding considerations and the remark following Equation (8.8), all field equations are satisfied if, in addition to (8.21), (8.22) and (8.23), we demand $T^{ab}{}_{;b} = 0$, i.e., since $\dot{u}_a = \phi_{,a}$,

$$(\varrho + p)\,\phi' = -p' , \tag{8.24}$$

the relativistic condition of hydrostatic equilibrium.

Given $\varrho = \varrho(p)$, the thermostatic 'equation of state' of the fluid, one obtains $m(r)$, $p(r)$ by integrating

$$m' = 4\pi r^2 \varrho \tag{8.25}$$

and

$$p' = -\frac{(\varrho + p)(m + 4\pi r^3 p)}{r(r - 2m)}, \tag{8.26}$$

the Tolman-Oppenheimer-Volkoff equation, starting at the center with the initial values

$$m(0) = 0, \quad p(0) = p_0 > 0.$$

From $m(r)$ and $p(r)$, λ and ϕ can be computed according to (8.21) and (8.23), except for an arbitrary additive constant in ϕ.

The spacetime metric obtained is *regular at* $r=0$, provided ϱ and p are continuous in $0 \leqslant r < \varepsilon$; for the above equations then imply

$$\lambda(0) = 0, \quad \lambda'(0) = 0, \quad \phi'(0) = 0,$$

which implies that the coordinates

$$\begin{aligned} x &= l \sin\vartheta \cos\varphi \\ y &= l \sin\vartheta \sin\varphi \\ z &= l \cos\vartheta \\ \tau &= e^{\phi(0)} t, \quad l = \int_0^r e^{\lambda} \,\mathrm{d}\bar{r} \end{aligned} \tag{8.27}$$

form a local inertial system at the center.

Several analytic solutions of the preceding equations are known (e.g. [116], where one finds earlier references), and many numerical solutions with (hopefully) realistic equations of state have been worked out, as discussed by J. Hartle in this volume, p. 153.

We have assumed spherical symmetry and equilibrium, and deduced the other properties. It seems obvious physically that equilibrium and absence of a rigid rotation – say $\omega_{ab}=0$ at the center – should imply spherical symmetry for an isolated body. This, however, has not been proven yet mathematically.

8.5. Junction Conditions at the Surface

The Lichnerowicz *junction conditions* [117] require that at the surface of a body the metric should be continuously differentiable (in suitable coordinates), provided there is no surface distribution of matter. For the free surface of a body of perfect fluid these conditions imply that $p=0$ at the surface, and that the boundary hypersurface should be generated by fluid streamlines. (For a derivation, see ref. [23]).

Applying these conditions to spherically symmetric equilibrium solutions we see that the integration of (8.25), (8.26) has to be continued until $p(R)=0$; this condition defines the fluid boundary $r = R$. Let $m(R)=M$. Then (8.21), which holds for $0 \leqslant r < \infty$ according to its derivation, shows that M is the *total mass* of the 'star' which determines the exterior metric (8.11). One can now fix the additive constant in ϕ by setting

$$e^{2\phi(R)} = 1 - 2M/R, \tag{8.28}$$

as required by the continuity of g_{tt}. Now the interior and exterior metrics are completely specified, and so is the boundary, and it is clear that g_{ab} is continuous at $r = R$.

THEOREM 8.3. If λ, ϕ are determined as described above and $p(R)=0$, then the metric tensor g_{ab} is continuously differentiable with respect to the coordinate system $(\vartheta, \varphi, l, t)$ where $l = \int_0^r e^{\lambda(\bar{r})} d\bar{r}$ (for $0 \leqslant r < \infty$).

Proof $l(r)$ is C^1, $dl/dr = e^{\lambda} > 0$, whence $r(l)$ is C^1 everywhere. The metric reads

$$ds^2 = (r(l))^2 \, d\Omega^2 + dl^2 - e^{2\phi(r(l))} \, dt^2,$$

so only the continuity of $(d/dl)e^{\phi} = e^{\phi}\phi' e^{-\lambda}$ at $r = R$, i.e., that of ϕ', has to be checked. From (8.24), (8.26) and $p(R)$ we get $\phi'(R) = M/R(R-2M)$. This value equals $\phi'(R)$ as computed from the exterior metric (8.11).

This result illustrates a theorem due to Darmois, cf. [117], according to which the junction conditions hold in some coordinate system if and only if they hold in Gaussian normal coordinates with respect to the boundary hypersurface, which can also be rephrased in terms of the equality of the first and second fundamental forms of the boundary, computed from the interior and the exterior metrics, respectively.

Combining the result of Subsections 8.3, 8.4 and 8.5 we infer: For a perfect fluid configuration in equilibrium, mass M and radius $R(=$ square root of $((4\pi)^{-1} \times \text{area}))$ satisfy the inequality (Schwarzschild, 1916)

$$2M/R < 1, \tag{8.29}$$

which has been improved under various additional assumptions.

8.6. THINGS NOT TREATED

For reasons of time and energy, and partly because they are treated by other lecturers, several important topics have been omitted, among them orbits of particles and light rays, tidal accelerations, time dependent interior fields (collapse models), metrics taking into account the radiation going from a 'star' to infinity (Vaidya [118]), stability theorems, the virial theorem, quasi-static contractions, and Israel's [124] characterization of the exterior Schwarzschild solution in terms of the regularity of its event horizon. These items are all important and ought to have been included, but ... (see above).

References

1. P. Havas: *Rev. Mod. Phys.* **36**, 938 (1964).
2. A. Trautman: in B. Hoffman (ed), *Perspectives in Geometry and Relativity*, Bloomington 1966, p. 413.
3. H. P. Künzle: *Ann. Inst. Henri Poincaré* **17**, 337 (1972). This paper contains many more references.)
4. O. H. L. Heckmann and E. Schücking: 'Newtonsche und Einsteinsche Kosmologie', in S. Flügge (ed.), *Encyclopedia of Physics* **53**, 489 (1959).

5. G. Rosen: *Am. J. Phys.* **40**, 183 (1972) and references therein.
6. K. S. Thorne, C. M. Will, and W. T. Ni: in R. W. Davies (ed.), *Proceedings of the Conference on Experimental Tests of Gravitational Theories*, NASA-JPL Technical Memorandum 33–499. See also C. Will, Lectures presented at the 56th Fermi summer school, Varenna 1972.
7. H. Weyl: *Raum, Zeit, Materie*, 5th ed., Berlin 1923. H. Weyl: *Mathematische Analyse des Raumproblems*, Berlin 1923.
8. V. Fock: *Theory of Space, Time and Gravitation*, Moscow 1955, esp. § 8 and Appendix A.
9. E. Breitenberger: *Nuovo Cimento, Ser. XI* **1B**, 1 (1971).
10. J. Ehlers, R. Penrose, and W. Rindler: *Am. J. Phys.* **33**, 995 (1965).
11. A. Schild: in J. Ehlers (ed.), *Relativity Theory and Astrophysics*, Lectures in *Appl. Math.* **8**, Providence 1967.
12. R. H. Dicke: *The Theoretical Significance of Experimental Relativity*, Gordon and Breach, New York, 1964.
13. S. Deser: *General Relativity and Gravitation* **1**, 9 (1970).
14. W. Wyss: *Helv. Phys. Acta* **38**, 469 (1965).
15. R. U. Sexl: *Fortschr. Physik* **15**, 269 (1967).
16. J. Ehlers, F. A. E. Pirani, and A. Schild: in L. O'Raifeartaigh (ed.), *General Relativity*, Oxford 1972, p. 63.
17. M. Castagnino: *J. Math. Phys.* **12**, 2203 (1971).
18. R. P. Geroch: *Geometry and Gravity*, unpublished.
19. W. Kundt and B. Hoffmann: in *Recent Developments in General Relativity,* Warszawa 1962, p. 303.
20. E. H. Kronheimer and R. Penrose: *Proc. Camb. Phil. Soc.* **63**, 481 (1967).
21. R. F. Marzke and J. A. Wheeler: in H. Y. Chiu and W. F. Hoffmann (ed.), *Gravitation and Relativity,* New York 1964.
22. H. Reichenbach: *Axiomatik der relativistischen Raumzeitlehre*, Braunschweig 1924.
23. J. L. Synge: *Relativity: The General Theory*, North-Holland Publ. Comp., Amsterdam, 1960.
24. A. Trautman: *Soviet Phys. Usp.* **9**, 319 (1966).
25. N. M. J. Woodhouse: *J. Math. Phys.* **14**, 495 (1973).
26. G. R. Isaak: *Nature*, London **223**, 161 (1969).
27. F. A. E. Pirani and A. Schild: article in B. Hoffmann (ed.), *Perspectives in Geometry and General Relativity*, Bloomington 1966, p. 291.
28. J. A. Schouten: *Ricci-Calculus*, Berlin 1954.
29. P. G. Bergmann: *Introduction to the Theory of Relativity*, New York 1942.
30. A. Ostrowski: *Aequationes Mathematicae* **1**, 66 (1968).
31. R. Penrose: *Proc. Roy. Soc.* **A284**, 159 (1965).
32. R. Penrose: 'Structure of Spacetime', in C. DeWitt and J. A. Wheeler (ed.), *Battelle Rencontres*, New York, 1968.
33. F. A. E. Pirani: *Symposia Mathematica*, Rome (to be published).
34. J. Ehlers and A. Schild: *Geometry in a Manifold with Projective Structure*, Commun. Math. Phys. (in print).
35. A. A. Robb: *Geometry of Time and Space*, Cambridge 1936.
36. A. D. Alexandrov: *Commun. of the State University of Leningrad* **8**, 103 (1953).
37. A. Trautman: *Bull. Acad. Pol. Sci., Math.-Astr.-Phys. Series* **20**, 185 (1972); 503 (1972).
38. R. P. Geroch: Spacetime Structure from a Global Viewpoint, in R. K. Sachs (ed.), *General Relativity And Cosmology*, New York 1971, p. 71.
39. R. P. Geroch: *J. Math. Phys.* **9**, 1739 (1968); **11**, 343 (1970).
40. D. Lovelock: *J. Math. Phys.* **13**, 874 (1972)
41. O. Costa de Beauregard: in B. Hoffman (ed.), *Perspectives in Geometry and Relativity*, Bloomington 1966, p. 44.
42. Y. Choquet-Bruhat: in L. Witten (ed.), *Gravitation: An Introduction to Current Research*, New York 1962, p. 130.
43. A. Lichnerowicz: *Relativistic Hydrodynamics and Magnetohydrodynamics*, New York 1967.
44. E. Schrödinger: *Space-Time Structure*, Cambridge Univ. Press, 1950.
45. J. G. Fletcher: *Rev. Mod. Phys.* **32**, 65 (1960).
46. K. S. Thorne and C. M. Will: *Astrophys. J.* **163**, 595 (1971). C. M. Will: *Astrophys. J.* **163**, 611 (1971): **169**, 125 (1971). W. T. Ni: *Astrophys. J.* **176**, 769 (1972). See also C. M. Will: ref. [6].

47. W. Rindler: *Astrophys. J.* **157**, L147 (1969).
48. A. Peres: *Nuovo Cimento* **11**, 617 (1959).
49. G. F. R. Ellis and D. W. Sciama: L. O'Raifeartaigh (ed), in *General Relativity*, Oxford 1972, p. 35.
50. A. Trautman: in *Lectures on General Relativity*, Brandeis Summer Institute in Theoretical Physics, 1964.
51. F. B. Estabrook and H. D. Wahlquist: *J. Math. Phys.* **5**, 1629 (1964).
52. F. A. E. Pirani: *Bull. Acad. Polonia, Math.-Astr.-Phys. Series* **13**, 239 (1965).
53. J. M. Stewart: *Non-Equilibrium Relativistic Kinetic Theory*, Lecture Notes in Physics, Vol. **10**, Berlin 1971.
54. W. Israel: in L. O'Raifeartaigh (ed.), *General Relativity*, Oxford 1972, p. 201.
55. J. Ehlers: in R. K. Sachs (ed.), *General Relativity and Cosmology*, New York 1971, p. 1.
56. C. Marle: *Ann. Inst. H. Poincaré* **A10**, 67, 127 (1969).
57. S. R. de Groot, C. G. van Weert, W. Th. Hermens, and W. A. van Leeuwen: *Physica* **40**, 257 (1968); 581 (1969); **42**, 309 (1969). W. A. van Leeuwen and S. R. de Groot: *Physica* **51**, 1; 16; 32 (1971).
58. G. E. Tauber and J. W. Weinberg: *Phys. Rev.* **122**, 1342 (1961).
59. R. W. Lindquist: *Ann. Phys.* **37**, 487 (1966).
60. N. J. Hicks: *Notes on Differential Geometry*, New York 1965.
61. R. L. Bishop and S. J. Goldberg: *Tensor Analysis on Manifolds*, Macmillan, New York, 1968.
62. M. Spivak: *Calculus on Manifolds*, New York 1965.
63. J. L. Synge: *The Relativistic Gas*, North-Holland Publ. Co. Amsterdam, 1957
64. L. P. Kadanoff and G. Baym: *Quantum Statistical Mechanics*, New York, 1962.
65. L. Waldmann: in S. Flügge (ed.), *Encyclopaedia of Physics*, Vol. 12, Springer-Verlag, Berlin, 1962.
66. S Chandrasekhar: *Radiative Transfer*, Oxford, 1950.
67. Ch. G. van Weert: *Proc. Kon. Ned. Akad. Wet. Amsterdam* **73**, 381 (1970).
68. J. Ehlers, P. Geren, and R. K. Sachs: *J. Math. Phys.* **9**, 1344 (1968).
69. R. Treciokas and G. F. R. Ellis: *Commun. Math. Phys.* **23**, 1 (1971).
70 J. L. Synge: *Relativity: The Special Theory*, Amsterdam 1956.
71. E. C. G. Stuckelberg: *Helv. Phys. Acta* **25**, 577 (1952).
72. W. Brenig and R. Haag: 'General Quantum Theory of Collision Processes', *Fortschr. Physik* **7**, 183 (1959); reprinted in M. Ross (ed.): *Quantum Scattering Theory*, Bloomington, 1963.
73. R. Ruffini and S. Bonazzola: 'Systems of Self-Gravitating Particles in General Relativity', preprint Princeton (1969).
74. Y. Choquet-Bruhat: Annales de l'Institut Fourier (1971).
75. D. Bancel and Y. Choquet-Bruhat: *Comm. Math. Phys.* **33**, 83 (1973).
76. G. F. R. Ellis and R. Treciokas: *Comm. Math. Phys.* (to appear).
77. E. T. Jaynes: *Phys. Rev.* **106**, 620 (1957).
78. N. A. Chernikov: *Acta Phys. Polonica* **26**, 1069 (1964).
79. K. Bichteler: *Z. Physik* **182**, 521 (1965).
80. W. Israel, *J. Math. Phys.* **4**, 1163 (1963).
81. J. Ehlers: *Akad. Wiss. Mainz Abh., Math.-Nat. Kl.*, Jahrg. 1961, Nr. 11.
82. G. F. R. Ellis: in R. K. Sachs (ed.), *General Relativity and Cosmology*, New York 1971, p. 104.
83. J. L. Anderson: in M. Carmeli, S.I. Fickler, and L. Witten (ed.), *Relativity*, London 1970, p. 109.
84. B. Carter and H. Quintana: *Proc. Roy. Soc. London* **A331**, 57 (1972).
85. S. R. de Groot and P. Mazur: *Non-Equilibrium Thermodynamics*, Amsterdam 1962.
86. J. Meixner and W. Reik: in S. Flügge (ed.), *Encyclopedia of Physics*, Vol. 3, 2, Berlin 1959.
87. R. C. Tolman: *Relativity, Thermodynamics, and Cosmology*, Oxford 1934.
88. B. Carter: 'Elastic Perturbation Theory in General Relativity', Part I, Part II, preprints, Cambridge 1972; *Comm. Math. Phys.* **30**, 261 (1973).
89. M. Born and E. Wolf: *Principles of Optics*, New York 1959, esp. Chap. III.
90. J Ehlers: *Z. Naturforsch.* **22a**, 1328 (1967).
91 P. Jordan, J. Ehlers, and R. K. Sachs: *Akad. Wiss. Mainz Abh., Math.-Naturw. Kl.*, Jahrg. 1961, Nr. 1.
92. F. A. E. Pirani: in *Lectures on General Relativity*, Brandeis Summer Institute in Theoretical Physics, 1964.
93. J. Kristian and R. K. Sachs: *Astrophys. J.* **143**, 379 (1966).

94. W. G. Dixon: *Nuovo Cimento* **34**, 317 (1946); *Proc. Roy. Soc.* **A314**, 499 (1970); **319**, 509 (1970).
95. W. Beiglböck: *Commun. Math. Phys.* **5**, 106 (1967).
96. J. Madore: *Ann. Inst. Henri Poincaré* **11**, 221 (1969).
97. E. T. Newman and R. Posadas: *Phys. Rev.* **187**, 1784 (1969); *J. Math. Phys.* **12**, 2319 (1971); **11**, 3154 (1970); *GRG J.* **1**, 401 (1971).
98. I. Robinson and J. R. Robinson: in L. O'Raifeartaigh (ed.), *General Relativity*, Oxford 1972, p. 151.
99. R. K. Sachs: in C. DeWitt and B. DeWitt (ed.), *Relativity, Groups and Topology*, New York 1964, p. 523.
100. D. W. Sciama, P. Waylen, and R. Gilman: *Phys. Rev.* **186**, 1762 (1969).
101. G. Ellis and D. W. Sciama: in L. O'Raifeartaigh (ed.), *General Relativity*, Oxford 1972, p. 35.
102. J. Ehlers: in C. Cattaneo (ed.), *Relativita Generale*, Centro Internazionale Matematico Estivo, Rome, 1964.
103. H. Bondi, F. A. E. Pirani, and J. Robinson: *Proc. Roy. Soc.* **A251**, 519 (1959).
104. J. Ehlers and W. Kundt: in L. Witten (ed.), *Gravitation: An Introduction to Current Research*, New York 1962, p. 49.
105. L. Infeld and J. Plebanski: *Motion and Relativity*, New York 1960.
106. R. P. Kerr: *Nuovo Cimento, Ser. X* **16**, 26 (1960).
107. J. L. Synge: *Proc. Roy. Irish Acad.* **67A**, 47 (1969); **69A**, 11 (1970).
108. L. Landau and E. Lifshitz: *The Classical Theory of Fields*, Reading, Mass. 1951.
109. S. Chandrasekhar: in M. Carmeli, S. Fickler, and L. Witten (ed.), *Relativity*, New York 1970, p. 81, and references therein.
110. A. Papapetrou: *Proc. Roy. Irish Acad.* **A52**, 11 (1948).
111. A. Peres: *Nuovo Cimento* **11**, 617, 644 (1959); **13**, 439 (1959); **15**, 351 (1960).
112. W. G. Dixon: *GRG J.* **3**, 227 (1972), see also forthcoming continuations.
113. B. Schmidt: *Z. Naturforsch.* **22a**, 1351 (1967).
114. H. P. Künzle: *Proc. Roy. Soc.* **A297**, 244 (1967).
115. P. Jordan: *Schwerkraft und Weltall*, Braunschweig 1958.
116. H. Heintzmann: *Z. Physik* **228**, 489 (1969). See also the contribution of S. Chandrasekhar in L. O'Raifeartaigh (ed.), *General Relativity*, Oxford 1972, p. 185.
117. A. Lichnerowicz: *Théories relativistes de la gravitation et de l'électromagnetisme*, Paris 1955.
118. P. C. Vaidya: *Curr. Sci.* **12**, 183 (1943).
119. E. N. Glass and J. Winicour: *J. Math. Phys.* **13**, 1934 (1972).
120. Y. Choquet-Bruhat and L. Lamonreux-Brousse: *Compt. Rend. Acad. Sci. Paris* **276**, 1317 (1973).
121. W. Israel: *Phys. Rev.* **143**, 1016 (1966).
122. M. D. Kruskal: *Phys. Rev.* **119**, 1743 (1960).
123. G. Szekeres: *Publ. Math. Debrecen* **7**, 285 (1960).
124. W. Israel: *Phys. Rev.* **164**, 1776 (1967).

OBSERVATIONAL CONTACTS OF GENERAL RELATIVITY*

DIETER R. BRILL

Dept. of Physics and Astronomy, University of Maryland, College Park, Md., U.S.A.

1. Introduction

Not many years ago it was easy to survey in an hour the observational contacts of general relativity because there was only a handful of experiments. In recent years the advances in experimental technique have made many interesting effects measurable, and today the increasing effort and possibilities in experimental gravitation parallel the great upswing of interest in theoretical gravitation which began two decades ago.

Due to the weakness of the gravitational interaction and the necessity to use large and often uncontrollable masses, the experiments generally require great amounts of effort and money. It is therefore appropriate to look for criteria to identify important and meaningful experiments. At least three points of view have been advocated in this connection: (a) The simplest view considers on an equal footing all experiments on any effect predicted by general relativity. The significance of the experiment is to be derived from the residuals in a fit of the theory to the experimental data. This view allows no effective distinction between 'typical' general relativistic phenomena and those explained by wider classes of theories or by Newtonian gravity theory. (b) The opposite view considers it essential to compare general relativity to other complete theories of gravity; significant experiments would be those which can serve to distinguish general relativity and the alternative theories. The advances in space technology make solar system experiments a particularly important area of application of this approach. A discussion of these methods and results is given in Chapter 3. (c) An intermediate view considers it appropriate to experiment on phenomena which distinguish only large classes of gravitational theories, as long as the accuracy is very high. Such experiments tend to establish very fundamental properties of the gravitational field, common to a large class of theories. Since null experiments can usually be performed with much higher accuracy than measurements of a finite quantity, most of these experiments on 'fundamental' properties measure effects which vanish in general relativity. In the design of such experiments, the order of magnitude of possible deviations from general relativity was often estimated from general principles (such as Mach's principle), rather than from a particular alternate theory (Bertotti *et al.*, 1962). Although such estimates may be open to reinterpretation**, they can give some useful guidance about the accuracy needed for significant conclusions. I discuss some of these experiments in the next chapter.

* Supported in part by the National Science Foundation, Grant No. GP 25548.

** The history of the 'mass isotropy' experiment is an example. The experiment was motivated by the 'Machian' idea that an anisotropic distribution of large masses, such as our galaxy, surrounding an experiment, would cause an anisotropy of ordinary mass. But in a sense an absence of such an effect is even more in the spirit of Mach's principle, if the latter is taken to imply that all physics would be made anisotropic, so that the anisotropy is not measurable and can be 'transformed away'.

W. Israel (ed.), Relativity, Astrophysics and Cosmology, 127–152.

2. Fundamental Properties of Gravity

2.1. Foundations of Riemannian Geometry

Roughly speaking, a Riemannian* manifold is a space which in many respects looks locally like Euclidean or Minkowski space. Translated into physical terms, this correspondence between space with a gravitational field and Minkowski space is the principle of equivalence. What is the experimental evidence for various aspects of equivalence, and what deviations might one expect?

(a) Equivalence of differentiable structure claims that in the presence of gravity, locally the differentiable structure of space-time is the same as that of Minkowski space. There are no explicit experiments to test this (and a complete test may in principle be impossible to perform (Trautman, 1965, 1971)), but deviations would be expected only for extremely strong gravitational fields, e.g. when a description in terms of superspace (see, for example, Misner *et al.*, 1973, Chapter 43) is more appropriate than a Riemannian manifold.

(b) Equivalence of affine connections or geodesics demands that the local influence of gravity on all physical systems can be described by one affine connection (which can locally be 'transformed away' in the usual freely falling inertial system). The Eötvös and the redshift experiments are experimental evidence, and will be discussed below.

(c) Equivalence of metric structure means that space-time in the presence of gravity has properties (a) and (b) and also possesses a Riemannian metric. From (b) it follows that this metric also determines the unique affine connection. Conversely, if all the timelike and null geodesics are given, one can construct in the neighborhood of any geodesic a Riemannian metric, but a distance transferred along a closed polygon of geodesics does not necessarily return to its original value. The existence of identical particles which obey the exclusion principle but have had different histories is generally taken as strong evidence against such path dependence of length (see e.g., Marzke and Wheeler, 1964).

2.2. The Eötvös Experiment

For discussion of the Eötvös experiment it is appropriate to distinguish the weak and strong form of the equivalence principle: The weak principle demands that one affine connection** be determined by the *paths of test particles*, independent of the particles' constitution; the strong principle demands that a unique affinity determine and be determined by *all* physical processes.

An important aspect of weak equivalence can be tested by a null experiment, since general relativity predicts *zero* relative acceleration of particles with identical initial

* We shall use the term 'Riemannian' instead of the more correct 'semi-Riemannian' or 'nonsingular'.

** More exactly, test particle paths can only define a *projective* structure (parallel displacement of directions but not length). To assure an affine structure one needs either strong equivalence or (a) existence of a causal (conformal) structure, i.e. lightcones with appropriate properties *and* (b) compatibility between causal and projective structure. This latter can be provided by the redshift experiment.

position and velocity. This is the principle of the Eötvös experiment. Since ordinary test particles consist primarily of neutrons and protons, a deviation from equivalence which one may expect to detect would occur if there were a differential acceleration between neutrons and protons. The modern version of the Eötvös experiment due to Dicke (1964a) and Braginskii (1971) therefore compares test masses of very different neutron/proton ratio, such as gold and aluminum. Differential acceleration toward the Sun, with respect to which the Earth is in free fall, is measured by comparing the accelerations of the masses in a torsion balance. The masses start with identical velocity but necessarily with not quite identical position. They are therefore subject to tidal forces, and this was the main limitation of Eötvös' experiment. In the modern experiments these forces are reduced by symmetrical arrangement of the test masses; e.g. Dicke uses a triangular arrangement with two gold and one aluminum masses. The main limitation on the accuracy in this experiment is the temperature sensitivity of the torsion pendulum. The accelerations of the masses were found to be equal (as a fraction of their free-fall acceleration toward the Sun) to 10^{-11}, and a similar experiment by Braginskii (1971) yields equality to several parts in 10^{12}.

2.3. The Strong Principle of Equivalence

Two types of violation of strong equivalence which preserve weak equivalence are usually considered: (1) additional gravitational fields or an affinity different from that governing the free motion of the particles may enter the correct description of the *interaction* between particles; (2) the numerical constants entering natural laws (e.g. the fine structure constant α, or the gravitational constant G) may depend on space-time position. This would contradict strong equivalence because more than the affinity would be needed to determine these locally measurable quantities.

Violation (1) would imply that the free-fall acceleration of interaction energies would be different from that of particles. Since some fraction f of the masses in the Eötvös experiment is due to interaction energies, certain weaker conclusions can be drawn from the Eötvös experiment about the (affine connection) equivalence principle for these interactions. Table I summarizes the values of f for various constituents of typical masses. For example, unless there are unlikely fortuitous cancellations, the result of the Eötvös experiment implies that the strong binding satisfies equivalence

TABLE I

Summary of the values of f for various constituents of typical masses

constituent	f
neutron-proton difference	0.1
strong binding	10^{-4}
electrons, EM interactions	10^{-5}
virtual antiparticles	10^{-7}
weak interactions	10^{-11}
gravitational interaction	10^{-25}

to about 10^{-8}, and antiparticles satisfy equivalence to about 10^{-5} (Schiff, 1959). Present experiments do not allow significant conclusions about the weak interactions. This is one of the reasons why an Eötvös experiment is planned in a satellite in Earth orbit. Due to the larger free-fall acceleration (with respect to the force center) and other improvements, an increase in accuracy by a factor of 100 is expected (Chapman, 1972).

Violation (2), i.e. a variation of 'constants' may also show up weakly in the Eötvös experiment e.g. if the variation is described by an energy-carrying field (Dicke, 1964a). There are also direct tests, such as timing planetary motion by atomic clocks. The former depends on G whereas the latter does not, so that a time variation of G could be detected. Using optical and radar measurements of the planets since 1956, when good atomic clocks became available, Shapiro (1972) finds $|\dot{G}/G| < 10^{-10}\ \text{yr}^{-1}$. This is about the value expected from Mach's principle (Bertotti *et al.*, 1962); Brans-Dicke theory, which incorporates a form of Mach's principle predicts $\dot{G}/G \sim -3 \times 10^{-11}\ \text{yr}^{-1}$ (Dicke, 1962). If α is logarithmically related to G one might expect $\dot{\alpha}/\alpha \sim -10^{-12}\ \text{yr}^{-1}$. Such a variation is excluded by radioactive dating by comparing decays whose half-lives depend in different ways on α. For example the half-life of Re 187 (4×10^{10} yr) is very sensitive to α due to electromagnetic contributions to the energy level of the daughter, Os 187. The limit obtained under reasonable assumptions about the nuclear structure is $|\dot{\alpha}/\alpha| < 10^{-13}\ \text{yr}^{-1}$ (Peebles and Dicke, 1962; Dyson, 1967). α has also been measured from the spectrum of radiogalaxies at a distance of 2×10^9 lightyears, and agreement with the terrestial value was found to 0.2% (Bahcall and Schmidt, 1967). This limits the average change in α to $|\dot{\alpha}/\alpha| < 10^{-12}\ \text{yr}^{-1}$.

2.4. The Redshift Experiment

The Eötvös experiment does not directly test the weak equivalence principle for lightlike particles. A direct test could involve two kinds of lightlike particles, or it could compare lightlike and timelike particles. An example of the former would be a comparison of arrival times of a pulse of gravitational and electromagnetic radiation which were emitted simultaneously at the center of the galaxy. Unfortunately so far no electromagnetic pulses accompanying Weber's gravitational wave pulses have been found.

An experiment which compares the behavior of timelike and lightlike particles is the redshift experiment. This experiment is capable of various interpretations, depending on what is assumed about the timelike particles and their interactions. If we assume that natural clocks satisfy strong equivalence and treat the redshifted photons as test particles, then the redshift experiments simply tests weak equivalence for photons. But like the Eötvös experiment, the redshift experiment also has implications for strong equivalence if one considers the details of the test particles' constitution; and since the 'test particles' are a purely electromagnetic phenomenon, the redshift experiment in this sense tests the strong equivalence principle for electromagnetism. Finally, this type of experiment can be considered as a test of the strong equivalence principle for natural clocks.

For an experiment of such wide implications the present accuracy is not spectacular: the solar observations are obscured by unknown Doppler shifts due to currents in the Sun's atmosphere; the best measurements on the center of strong resonance lines (which are formed high in the solar atmosphere, above the convection) yield agreement to about 5% (Brault, 1962; Dicke, 1964b). The most accurate measurement of the redshift to date is the terrestial experiment of Pound and Snyder (1965) which uses the very narrow Mössbauer lines in Fe^{57} as the natural clocks; it yielded agreement with the equivalence principle to 1%. To reach greater accuracy, space vehicle experiments are being studied. The substantial first order Doppler shift can be measured by a two-way tracking signal, so that a correction can be applied to the one-way clock signal. By using accurate maser clocks (stability better than 10^{-13}) an overall accuracy of better than 10^{-4} for the redshift is expected (Vessot, 1972).

2.5. Experiments on Fields which can Violate Strong Equivalence

Here we consider some specific types of fields that can play a role in violating strong equivalence, without contradicting the Eötvös experiment. This experiment shows that the metric tensor plays a major role in determining gravitational interaction, but other fields may also be involved. We take the view that any field which couples to the stress-energy of matter is part of gravity. If such other fields play an intermediate role in determining the metric, but do not directly influence test particles, weak equivalence is strictly valid; if non-metric gravitational fields interact with matter in a velocity-dependent way, their effect may not show up in the Eötvös experiment, since no large velocities are involved in this experiment. It should be remarked that if other gravitational fields in addition to the metric interact with matter, the geometrical metric tensor cannot usually be uniquely defined; for example, the 'scalar-tensor theory' (Brans and Dicke, 1961) can be formulated so that the metric is solely responsible for the influence of gravity on test-particles. But another formulation uses a different metric which has other simple features (such as satisfying equations of the Einstein type), but which supplies only part of the gravitational interaction, the rest being supplied by direct interaction with the scalar field (Dicke, 1962b).

Scalar, vector, and tensor fields have been considered in addition to the metric. The latter two define preferred frames; e.g. a time-like vector field defines preferred rest systems at each point (with time axes parallel to the vector). If these fields are generated by the galaxy or by matter at cosmological distances, the preferred system would be expected to be the rest frame of the galaxy, or the cosmological comoving system. Since the velocity of the solar system with respect to this frame is not large (about 200 km s^{-1}), the effects may not show up in the Eötvös experiment, but could be detected in 'mass anisotropy' or 'ether drift' experiments.

The mass anisotropy experiments of Hughes *et al.* (1960) and Drever (1961) measured the dependence of atomic clock rates on the orientation of the clock's electron orbits with respect to the galaxy and found the rates to be constant to 10^{-20}. This puts strong restrictions on a possible second tensor component of the gravitational field (in addition to the metric), because such a second tensor field would be expected to make the electron mass anisotropic, i.e. dependent on the electron's motion (Peebles and Dicke, 1962b). However, a second tensor is not excluded by this experiment if it couples only to gravity (Will and Nordvedt, 1972).

Since the electron orbitals do not define a (polar) vector, the mass anisotropy experiment is insensitive to any vector part of the gravitational field. To test this possibility, Turner and Hill (1964) measured the dependence of nuclear (Mössbauer) clock rates on the direction of their velocity. The Mössbauer radiator was mounted on a centrifuge and its frequency measured in the four quadrants of the motion. The fractional change in frequency per dimensionless clock velocity (v/c) was found to be less than 4×10^{-5}.

2.6. The Sources of the Gravitational Field

In Einstein's theory all forms of mass and energy contribute equally to the generation of gravitational fields; however it is conceivable that the active gravitational mass of a body, which measures the amount of gravitational field produced by the body (i.e. the 'm' in the Schwarzschild solution) could depend not only on the body's inertial mass but also on its structure. To test this possiblity, Kreutzer (1968) (see also Gilvary and Muller, 1972) performed a Cavendish experiment where the gravitational force from a teflon cylinder was compared with that of an equal volume of fluid. The nuclear structure of the cylinder and the fluid was chosen as different as possible, but the (inertial) density was the same. If the active mass of the cylinder differed from that of the displaced fluid, a motion of the cylinder would entail a change of the gravitational field, which would be measured by the Cavendish balance. The null result of the experiment implies a fractional variation of gravitational mass by less than 5×10^{-5}.

TABLE II
Summary of the experimental results described in Chapter 2

Non-Einsteinian effect	Experimental upper limit	Discussed in section
Difference in inertial and gravitational mass of		
Au and Al	10^{-12}	2.2
neutrons and protons	10^{-11}	2.3
nucleons and electrons	10^{-7}	2.3
nucleons and (real) photons	10^{-2}	2.4
nucleons and (virtual) anti-particles	10^{-4}	2.3
Difference in passive and active gravitational mass	5×10^{-5}	2.6
Fractional variation of		
G	$10^{-10}\ \mathrm{yr}^{-1}$	2.3
α	$10^{-12}\ \mathrm{yr}^{-1}$	
Second tensor gravitational field	10^{-20}	2.5
Vector component of gravitational field	4×10^{-5}	2.5

2.7. Summary

Table II summarizes the experimental results described in this chapter. The smallness of the numbers measures the accuracy with which one knows the various listed non-Einsteinian effects to be absent, but for a more precise interpretation the sections above, or the original papers, ought to be consulted.

3. Alternative Theories and Solar System Experiments

3.1. Metric theories

In this Chapter, I want to consider in more detail the experimental consequences of a particular class of gravitational theories, namely metric theories. This class is selected partially for theoretical convenience, but more importantly because the experiments discussed in Chapter 2 favor very strongly this type of theory. The restriction of attention to metric theories does not exhaust the implications of all the experiments already discussed; but the class of metric theories is sufficiently narrow that it makes sense to consider definite complete theories of gravitation alternate to Einstein's. One obvious advantage of this approach is that one can calculate the magnitude of observable effects without having to resort to heuristic arguments like Mach's principle.

Any gravitational theory is a metric theory if it can be formulated so that free particles move along geodesics of a suitably defined Riemannian 4-dimensional manifold. (To describe the response of more general matter to the gravitational field one may demand that the matter satisfy the conservation law $T^{\mu\nu}{}_{;\nu}=0$ in this manifold.) Einstein's theory is unique among metric theories in its simplicity, and there appears to be no particularly unique modification of Einstein's theory. In fact, over fifty a priori more or less equally likely alternatives have been devised, and their number increases at the rate of several per year. To date, only solar system experiments have been able to some extent to distinguish between alternate theories. It is therefore efficient to classify these theories by their implications for solar system experiments. Fortunately the geometry in the solar system is nearly spherically symmetric, and can therefore be described to good accuracy by the general spherically symmetric static geometry, which contains only two arbitrary functions. These functions in turn are largely determined by the Newtonian limit, and the deviations can be characterized by the first few coefficients in a series expansion. Similarly, small deviations from spherical symmetry can be characterized by such coefficients. This scheme originated with Eddington and Robertson, and has been extended by Schiff, Nordvedt, Thorne, Will and others ('Parametrized Post-Newtonian Approximation' or PPN; see Will, 1971a, and the references cited there).

3.2. Metric for the solar system

The Eddington form of the metric is an expansion in $1/r$,

$$\mathrm{d}s_{\mathrm{E}}^2 = -(1 - 2M/r + 2\beta M^2/r^2 + \cdots)\,\mathrm{d}t^2 + (1 + 2\gamma M/r + \cdots) \times \times (\mathrm{d}r^2 + r^2\,\mathrm{d}\Omega^2). \qquad (1)$$

Here M is the solar mass; its Newtonian field determines the lowest order terms in (1). The terms of higher order in $1/r$ are multiplied by parameters β, γ which characterize different gravitational theories, and which are so chosen that their value is 1 for the Einstein theory. For most of the present day experiments it is sufficient to treat the influence of the other planets in the Newtonian approximation, and assume that, by correcting a given planet's orbit in the usual fashion for these perturbations, one can find the unperturbed, geodesic orbits in the metric (1).

The two parameters β and γ are sufficient to discuss the most important experiments which have yielded results to date. However, as the accuracy of the experiments increases, it becomes possible to see effects due to further PPN parameters. For example, if the central body rotates, two physical effects necessitate additional terms in the line element: (a) the body becomes oblate so that its Newtonian potential acquires a quadrupole component. The amplitude J_2 of this component is determined to Newtonian order by the mass distribution in the oblate body, and is not a PPN parameter. (b) The angular momentum L of the body may have the non-Newtonian, 'Machian' effect of dragging the local inertial frames along with the body (Brill and Cohen, 1966). A new parameter Δ measures the amplitude of this effect.* Thus the metric becomes

$$ds^2 = ds_E^2 - J_2 (MR^2/r^3)(3\cos^2\theta - 1)\,dt^2 - \Delta(4L\sin^2\theta/r)\,d\phi\,dt. \quad (2)$$

The number of parameters needed to describe the general metric theory continues to increase if we want to describe the gravitational field to higher and higher accuracy. However, certain physical principles (suggested e.g. by the null experiments of Chapter 2) imply algebraic relations between PPN parameters. For example, suppose we take the experimental evidence on absence of preferred frames in nature (Chapter 2, Section 5) to indicate that gravity is described by at most a scalar field in addition to the metric tensor field. This restriction on the class of allowed theories implies a restriction of the allowed parameter values in (2), so that γ and Δ are related by**

$$\Delta = \tfrac{1}{2}(1 + \gamma). \quad (3)$$

In fact, one can show that if there are no preferred frames, and if there is no violation of energy-momentum conservation, then all the PPN parameters depend only on β and γ (Will, 1971b). The measurement of β and γ is clearly an important task for the solar system experiments.

3.3. Relativistic effects on null geodesics

The behavior of null geodesics in a stationary geometry like (1) or (2) can be treated

* In some theories the dragging does not depend on the angular momentum alone but differs according as the motion is radial or transverse to the observer. To describe such theories, two parameters Δ_1, Δ_2 are needed (Will, 1971b).

** The following argument indicates why this is reasonable: In scalar-tensor theory a fraction s of the gravitational force can be considered as due to the scalar field. By its very nature, a scalar field cannot define a dragging of inertial frames, hence Δ must be reduced by this fraction s from its general relativity value 1. It is also easy to see (e.g. via conformal invariance of null geodesics) that a scalar field cannot contribute to light bending. Hence the light bending factor $\frac{1}{2}(1+\gamma)$ (see discussion) below must also be reduced by s from its general relativity value 1. Equation (3) expresses this relationship.

by geometrical optics in a three-dimensional space with an equivalent index of refraction. It is often easier to think about measurable effects in this more familiar language. To obtain this formulation consider the general stationary metric,

$$ds^2 = g_{00}\,dt^2 + 2g_{0i}\,dt\,dx^i + g_{ij}\,dx^i\,dx^j, \quad g_{\mu\nu} = g_{\mu\nu}(x^i), \qquad i = 1, \ldots, 3\,. \tag{4}$$

Any curve $x^i(t)$, $0 \leqslant t \leqslant 1$ in the three-dimensional space $t = \text{const.}$ can be 'lifted' to a *null* curve in the 4-space by supplementing it with an appropriate $x^0(t)$. This can be computed by solving the null requirement ($\cdot = d/d\tau$, $\tau = $ proper time)

$$0 = g_{00}(\dot{x}^0)^2 + 2g_{0i}\dot{x}^i\dot{x}^0 + g_{ij}\dot{x}^i\dot{x}^j \tag{5}$$

for $\dot{x}^0$ and integrating,

$$x^0(\tau) = \int \dot{x}^0(\tau)\,d\tau\,. \tag{6}$$

Suppose the $x^\mu(\tau)$ so obtained is a null geodesic. Make a variation $\delta x^i(\tau)$ of the three-dimensional curve with fixed endpoints, $\delta x^i(0) = \delta x^i(1) = 0$ and compute for the corresponding lifted curves the variation of the Action for null geodesics, $\int g_{\mu\nu}\dot{x}^\mu\dot{x}^\nu\,d\tau$. This vanishes,

$$\delta \int g_{\mu\nu}\dot{x}^\mu\dot{x}^\nu\,d\tau = 0$$

since all lifted curves are null curves. But by carrying out the variation we find

$$\delta \int g_{\mu\nu}\dot{x}^\mu\dot{x}^\nu\,d\tau = \int (\text{geodesic equation})_\mu\,\delta x^\mu\,d\tau + 2g_{\mu\nu}\dot{x}^\mu\,\delta x^\nu\big|_0^1\,.$$

Since the original curve is a geodesic, the first term on the right vanishes; also we have $\delta x^i(0) = \delta x^i(1) = 0$ but $2g_{\mu\nu}\dot{x}^\mu \neq 0$. In order that the second term on the right vanish we must therefore have

$$\delta x^0 = \delta \int \dot{x}^0\,d\tau = 0$$

as the condition for null geodesics under this kind of 'lifted' variation. The integral is independent of parametrization, so that we can replace τ by the three-dimensional arclength parameter l in any 3-dimensional metric,

$$\delta \int \dot{x}^0\,d\tau = \delta \int (dx^0/dl)\,dl = 0\,. \tag{7}$$

This has the form of a Fermat principle in a three-dimensional space, with (dx^0/dl) playing the role of the effective index of refraction. To work it out explicitly, confine attention to the physically interesting case of slow source motion, hence small g_{0i}. Then an approximate solution of the quadratic Equation (5) is

$$\dot{x}^0 = [-(g_{ij}/g_{00})\,\dot{x}^i\dot{x}^j]^{1/2} - (g_{0i}/g_{00})\,\dot{x}^i + \cdots \tag{8}$$

so that (7) becomes,

$$\delta \int \left\{ \left[-(g_{ij}/g_{00}) \frac{\mathrm{d}x^i\,\mathrm{d}x^j}{\mathrm{d}l\,\mathrm{d}l} \right]^{1/2} - (g_{0i}/g_{00}) \frac{\mathrm{d}x^i}{\mathrm{d}l} + \cdots \right\} \mathrm{d}l = 0 . \tag{9}$$

The expression in curly brackets is the equivalent index of refraction.

The index of refraction is in general anisotropic if we choose for $\mathrm{d}l^2$ the flat space metric; however, if the geometry is static ($g_{0i}=0$) and if the space part of the metric is conformally flat, $g_{ij}\,\mathrm{d}x_i\,\mathrm{d}x^i = g\,\mathrm{d}l_\flat^2$ where $\mathrm{d}l_\flat^2$ denotes the flat space line element, (9) becomes

$$\delta \int (-g/g_{00})^{1/2}\,\mathrm{d}l_\flat = 0 \tag{10}$$

and the light propagation is completely analogous to propagation in ordinary geometrical optics with an isotropic index of refraction.

Now return to the metric (1). It is written in the 'isotropic' form appropriate for definition of $\mathrm{d}l_\flat$ of (10). To lowest order the index of refraction of (10) is therefore,

$$n = (-g/g_{00})^{1/2} = 1 + \gamma^* \frac{2M}{r} \quad \text{with} \quad \gamma^* = \tfrac{1}{2}(1+\gamma) . \tag{11}$$

Thus, to lowest order all effects on light are determined by the parameter γ only, since all the deviations from flat space optics are proportional to γ^*. The radial measure r in (11) can of course, be replaced to this order by the radial distance r of the usual Newtonian solar system dynamics.

As an estimate of the deviations in the parameters one might expect, we mention the values predicted by one leading contender, the scalar-tensor Brans-Dicke theory. This theory contains a parameter ω which describes the coupling of the scalar field to the trace of the matter stress-energy; Dicke has conjectured $\omega \sim 6$, which we shall use here (the general relativity value is $\omega \to \infty$). The parameters then have the value $\gamma = (1+\omega)/(2+\omega) \sim 0.87$, $\gamma^* = (3+2\omega)/(4+2\omega) \sim 0.94$.

3.4. Experiments on Light Bending

The equivalent index of refraction which increases as $1/r$ near the Sun has as one of its consequences the well-known light bending or gravitational lens effect. The bending angle between the asymptotic directions of a light ray of impact parameter a is well defined independent of coordinates and given to lowest order by $\theta = \gamma^* \cdot 4M_\odot/a$; the general relativity value for a grazing ray is $4GM_\odot/c^2R_\odot = 1''.75$.

In eclipse experiments a starfield is photographed during a solar eclipse and the resulting plate is compared with a plate of the same starfield taken six months later, at night. The very different conditions under which the two plates are taken lead to uncertainties in the scale factor to be used in comparing the plates; the success of the experiment also depends on the star distribution around the Sun at the time of the eclipse. Some of the first eclipse measurements (in 1922) still represent some of the best data, yielding the general relativity result to about 10%. For a review, see Klüber (1960) or Bertotti *et al.* (1962).

Eclipse experiments are still of interest today, but more accurate data for bending of starlight will probably come from H. Hill's (1971a) photoelectric detection of stars in the daytime. In these experiments, the angular size of the Sun is used to define the scale factor. The experiments therefore will also yield important data on solar oblateness, and on the Earth's orbit.

The light bending experiment has also been performed at cm wavelengths, using the strong, nearly pointlike radio sources 3C279 and 3C273. The former is occulted by the Sun each October, and the latter, separated by 11°, is used as a reference. The angular separation of the sources is monitored via the phase difference of their interferometer signal. Experiments to date had to cycle between the two sources and extrapolate the phase during the off cycle; the extrapolation is more difficult for longer baselines, and for this reason, very long (4000 km) baseline interferometry has not yet yielded the high accuracy possible with this method. (This difficulty can of course be overcome by having separate radiotelescopes for the two sources). The plasma surrounding the Sun affects signals at these wavelengths and causes additional bending. Since the refraction is proportional to λ^2, corrections can be made for the plasma effect if measurements are taken at different wavelengths. In the analysis of the data, γ^* and several other parameters describing the source structure and the plasma distribution are determined by a least squares fit. It is possible to test the arrangement for several sources of systematic errors. These were found to be negligible (the most important apparently being due to the Earth's atmosphere), so that the formal errors are a reasonable indication of the accuracy. The Table II in Section 8 indicates some results obtained to date. None of the experiments have been able to test the $1/r$ dependence, or higher order (in $1/r$) corrections. of the light bending effect.

3.5. The Time Delay Experiments

A refractive medium causes not only bending of lightrays (through spatial variation of n) but also a difference in light travel time (proportional to n) compared to the case $n=1$. The relativistic time delay is therefore a different effect from the light bending but it also measures the parameter γ^*. To compute the relativistic delay to lowest order one can neglect the light bending and simply compute $\int n\,\mathrm{d}s$ along a straight line in the flat background space. In order to be able to measure the relativistic time delay one must know accurately the position of the source and receiver, so that the comparison case $n=1$ can be computed. Since in practice these objects are in solar orbit, the general relativistic effects on light cannot be totally separated from those on the particle orbits in the analysis of these experiments. However, the orbital elements are most accurately measured at inferior conjunction (source closest to Earth), whereas the time delay is largest at superior conjunction (source on opposite side of the Sun from the Earth). Thus an effective separation of the time delay is possible. The experiments use a two-way signal reflected by a planet, or by the transponder on a space probe. The typical roundtrip time is of the order of 10^3 s, and the general relativity delay is as large as 200 μs (grazing signal from Mercury). The solar plasma also contributes to the time delay, but with a different r-dependence and by a frequency-dependent amount which is usually over an order of magnitude smaller than the relativistic delay.

Radar reflections from planets have the advantage that the orbit of the reflector at superior conjunction can be very accurately calculated. The main source of systematic error is the unknown topography of the planet, and the solar corona. These errors could be eliminated, e.g. via radar measurement of the topography, but they are present in the current data. For this reason Shapiro quotes a 'realistic' error limit as well as the formal error from the least square fit. Since the orbits of the Earth and the target planet must be determined as part of the measurements, over a dozen parameters in addition to γ^* are adjusted: the perturbations of other planets and the Moon are found to sufficient accuracy from the classical analysis.

The advantage of space probes is a higher signal/noise ratio of the reflected signal from the space probe's radar transponder, and the absence of topography uncertainties. However, a space probe responds to a number of nongravitational forces which cannot all be measured independent of the time delay, and which result in systematic errors. A promising prospect appears to be the use of orbitors or landers whose position is tied closely to a planetary orbit.

Table II (in Section 8) gives a summary of the data from time-delay experiments.

3.6. Perihelion Precession Experiments

The relativistic precession of Mercury amounts to less than 1% of the precession due to perturbations from other planets, yet observational astronomy at the turn of the century was accurate enough to identify this 43″/century as a discrepancy unexplained by Newtonian physics. Pre-relativity attempts to explain it involved existence of hitherto unknown masses or force laws different from the Newtonian one, but were not successful or convincing. The Einstein theory gives a simple explanation of this effect to a few percent; this agreement has long been considered the most accurate test of general relativity, particularly when space probe measurements confirmed the accepted value for the mass of Venus (one of the important contributors to Mercury's Newtonian perihelion motion). However, in 1962 Dicke pointed out that a moderate solar quadrupole moment was consistent with the data then known, so that the apparent agreement between the experimental and (non-oblate) theoretical perihelion motion may be fortuitous and the data may be in disagreement with general relativity.

For a simple derivation of the factor λ_p by which the general relativity results must be multiplied to obtain the result for the line element (1) we compare the radial geodesic equation to the corresponding Newtonian equation. We change coordinates to Schwarzschild-type coordinates by introducing a new radial variable $r+\gamma M$ (which we again denote by r), so that

$$ds^2 = -(1 - 2M/r + 2(\beta - \gamma) M^2/r^2)\, dt^2 + (1 + 2\gamma M/r)\, dr^2 + r^2\, d\Omega^2$$

and so that the 'conservation of angular momentum' takes the Newtonian form,

$$u \cdot \partial/\partial\phi = r^2\, d\phi/d\tau = l\,,$$

where τ is proper time. Another constant of the motion is the energy at infinity,

$$-u\cdot\partial/\partial t = -g_{00}(\mathrm{d}t/\mathrm{d}\tau) = E.$$

Substitute these constants into the conserved length of the 4-velocity,

$$u\cdot u = -1$$

and expand the result to order M^2 (noting that $E^2-1\sim M^2/l^2$) to obtain the 'conservation of energy' equation

$$E^2-1=(\mathrm{d}r/\mathrm{d}t)^2+\left[-2Mu+l^2u^2+(2\beta+2\gamma-4)M^2u^2-2\gamma l^2Mu^3\right].$$

The term in the brackets is the effective potential written in terms of the usual variable $u=1/r$. The first two terms of the potential are the Newtonian gravitational and centrifugal potential; the rest is the relativistic contribution which causes the perihelion rotation. For nearly circular orbits, the perihelion rotation is determined by the second derivative of this relativistic contribution (quadratic approximation). This second derivative, evaluated at the circular orbit $u=M/l^2$ is $(4\beta-8\gamma-8)\,M^4/l^4$. Its general relativity value ($\beta=\gamma=1$) is $-12M^4/l^4$, an attractive potential corresponding to perihelion advance. Hence the correction factor λ_p is

$$\lambda_p=-(4\beta-8\gamma-8)/12=(2+2\gamma-\beta)/3.$$

For the Brans-Dicke theory this factor has the value

$$\lambda_{p\text{B-D}}=(4+3\omega)/(6+3\omega)=0.92.$$

To allow for a possible solar quadrupole moment J_2 one can use the line element (2), or equivalently to lowest order calculate the Newtonian perihelion precession due to such a quadrupole moment, and add its effect to the above λ_p. The result is

$$\lambda_{p,J_2}=(2+2\gamma-\beta)/3+\frac{J_2R^2}{2aM(1-e^2)}.$$

The quadrupole moment term has the value $(3\times10^3)\,J_2$ for Mercury.

Without a knowledge of J_2 we cannot derive what is of interest for testing general relativity, namely the value of λ_p. J_2 can be determined in several ways: (a) Motion of the node of Mercury. (b) Simultaneous fit of λ_p, J_2 and orbital elements to the optical and radar data on planetary motion. (c) Computation from observed rotation of the Sun. (d) Computation from measurement of solar oblateness. None of these gives a unique answer today: (a) gives an upper limit, $J_2<5\times10^{-5}$, (b) has been attempted (Shapiro, 1971) but λ_p and J_2 are not sufficiently independent to be determined separately from the data, (c) leads to $J_2\sim10^{-6}$ if the mean rotation of the Sun is assumed to be the equatorial one, (d) can be done only by means of simplifying assumptions, in particular Von Zeipel's theorem, (see, e.g., Dicke (1970)) which states that in the absence of magnetic and velocity fields, the surfaces of constant P, ϱ, T and Newtonian gravitational potential ϕ all coincide. The observed oblateness of the Sun $\Delta r/r\equiv\varepsilon$ is then the same as that of an equipotential surface, which leads to $J_2=2\varepsilon/3$.

Dicke and Goldenberg (1967) have measured $\varepsilon = (5 \pm 0.7) \times 10^{-5}$, so that J_2 (with correction for the Sun's rotation) becomes 2.6×10^{-5}. This would require a rapidly rotating solar core, with angular velocity up to 20 times as great as the surface angular velocity. Dicke (1971) has supported this view with a solar model whose surface is slowed down by the solar wind torque, and has correlated it with the Lithium depletion in the Sun. Others (see e.g., the references cited in Dicke, 1970, p. 299; also Ingersoll and Spiegel, 1971) have argued that a rapidly rotating core would set up instabilities, and that the observed oblateness is due to temperature variation between solar pole and equator. The experimental data appear to argue against this, but not all the effects involved are entirely understood (Dicke, 1972a). The experiment of Hill mentioned already in connection with light deflection should soon yield additional data on this important question.

In most of the reductions of perihelion motion, J_2 is assumed to be negligible. Although J_2 has a small effect on the other planets, it is fairly accurate to subtract $(3 \times 10^3)\ J_2$ from the quoted λ_p to obtain the value with non-vanishing J_2. The classical optical observations (1750–1970) were reanalyzed in a monumental task by Shapiro, using a computer. He has also analyzed the data on radar reflection from planets. The latter method holds the most promise for the future, since the radar data improve faster than the optical data. The present results are shown in the Table III in Section 8.

3.7. Other Experiments

A number of further general relativity experiments are in the planning or development stage, and I shall not attempt to give a complete picture of them; instead I briefly mention only two which appear particularly instructive.

The Stanford gyroscope experiment (O'Connell, 1972) measures general relativistic effects which go beyond geodesic motion. To derive the equations for the motion of the center of mass and of the angular momentum of a gyroscope one must treat it as an extended body. The motion of the angular momentum is what the experiment attempts to detect. A gyroscope in Earth orbit changes its angular momentum direction with respect to the 'fixed stars' as it is parallely propagated in a metric of type (2). The main contribution to this change occurs even if the Earth's rotation is neglected ($L=0$), and measures the first order (in $1/r$) deviations from Minkowski space of the space and time part of the metric: this effect is about $(1/3) \times (1+2\gamma) \times 7''\ \mathrm{yr}^{-1}$ for a low-altitude orbit. A smaller contribution comes from the 'dragging of inertial frames' by the rotating Earth and amounts to about $\Delta \times 0.05''\ \mathrm{yr}^{-1}$ for a polar orbit (where this second effect is largest). A successful measurement of these precessions would be the first experiment in which the 'magnetic' component of the gravitational field plays a role, and it has been remarked (Davies, 1971) that the significance of the experiment would not be primarily to distinguish general relativity ($\Delta = 1$) from e.g., Brans-Dicke theory $[\Delta = (3+2\omega)/(4+2\omega) = 0.94]$, but to show that this magnetic component is present at all ($\Delta \sim 1$ vs. $\Delta = 0$).

Finally it should be mentioned that there are a number of effects which vanish in general relativity but are present in most of the alternative theories. The most promising one for detection is the 'Nordvedt (1968) effect', a violation of the equivalence principle for bodies for which gravitational interaction energy contributes significantly to the total mass. The bodies must be treated as extended, and other PPN parameters than the ones mentioned here are needed for the computation of this effect. In the Earth-Moon system it would result in a polarization of the orbit that could be detected in the very accurate data made possible by laser ranging of the Moon (see, for example Bender *et al.*, 1971).

3.8. Conclusion

Table III gives a summary of the results of present solar system experiments. From these results one can deduce fairly accurate values for the parameters β and γ, namely $\beta=1.0\pm0.2$, $\gamma=1.0\pm0.1$. They are not accurate enough to exclude with confidence a theory like the Brans-Dicke theory, but a number of other alternative theories (for example, conformally flat theories) are definitely excluded (Thorne *et al.*, 1971), and each improvement in accuracy limits further the range of alternatives. The solar system experiments, the experiments discussed in Chapter 2 and others also place

TABLE III

Some results of solar system experiments

Type of experiment		Measured value/*GR* prediction			Reference
1. Light bending		γ^*			
(a) of visible light at solar eclipse		1.0 ± 0.1			Campbell *et al.* (1923)
(b) of radiowaves from 3C279, using Interferometer					
Frequency (MHz)	baseline (km)				
2400	21	1.04 ± 0.15			Muhlman *et al.* (1970)
9602	1	0.99 ± 0.12			Seielstadt *et al.* (1970)
5000, 2700	0.7 1.4	1.14 ± 0.3 1.07 ± 0.17			K.M.Hill (1971)
8085, 2695	0.7, 1, and 2.7	0.91 ± 0.07 (1970) 0.97 ± 0.08 (1971)			Sramek (1972)
		0.94 ± 0.06 (overall)			
2. Time delay, using		γ^*	± Error		
Reflector	Frequency (Mhz)		Formal	Realistic	
Mercury, Venus	7840	0.9	0.1	0.2	Shapiro (1968)
Mercury, Venus, Mars	7840, 430	1.015	0.02	0.05	Shapiro (1972)
Mariner 6, 7	2200	1.00		0.04	Anderson (1972)
3. Perihelion precession		λ_p, assuming		± Error	
		$J_2=0$	$J_2=2.6\times10^{-5}$	(realistic)	
optical (1750-1970)		1.00	0.92	0.03	Shapiro (1971)
radar		1.006	0.928	0.02	Shapiro (1972)

upper limits on the PPN parameters which measure preferred frame effects and non-conservation of energy (these parameters were not discussed above; see Nordtvedt and Will, 1972).

The possibility of such definite conclusions (contrasted to the less definite arguments from such principles as Mach's) is a very attractive feature of the PPN formalism. It appears best at analyzing theories: one only needs to compute the parameters β, γ, Δ ...predicted by any alternative theory and see whether they lie within the experimentally allowed range. However, one should not overestimate the significance of the PPN formalism in analyzing experiments, since the PPN parameters are based in part on a historical accident, namely what alternative theories happened to have been proposed in the past. If none had conceived of preferred frame theories, Equation (3) might make the gyroscope experiment appear not so significant, since it would not measure a new parameter. Along the same vein, an analysis of experiments in terms of PPN parameters might not attribute enough significance e.g. to the time delay experiments if γ were already very accurately measured by light bending. Finally we must remember that the PPN scheme, like any other theoretical scheme, cannot a priori enumerate all the parameters that are relevant for every actual experiment with real bodies: there is always the possibility of further effects (such as the solar oblateness) which allow possibilities that otherwise would appear excluded. Thus the PPN scheme cannot replace understanding and feeling for the physics involved, but used with such understanding it is an extremely valuable guide in the analysis of the general relativity experiments.

4. Experiments on Gravitational Radiation

4.1. Importance of Waves for Experiment and Theory

If one had to choose one of the first chapter's three views about gravitational experiments to characterize the motivation for gravitational wave experiments, the first view would be most appropriate: Gravitational waves are an interesting consequence of Einstein's theory, and the first task of the experiments is simply to show that they indeed exist in nature. Thus, the original experiments were not designed to measure some basic property of gravity with great accuracy, nor to distinguish very sharply between alternative theories*, although the results of the experiments may have implications for all these questions. Of course, the significance of the experiments goes far beyond that of a test of the particular predictions of general relativity concerning gravitational waves. Gravitational wave detectors open up many new possibilities, such as gravitational wave astronomy, and they may allow us to look into regions of the universe where some of the most important secrets of physics and astrophysics are hidden.

Great experimental ingenuity is necessary to detect gravitational waves experimentally, and even their theoretical discovery (as consequences of Einstein's theory) was not easy. Due to the nonlinearity of Einstein's equations one cannot display the general

* The present PPN scheme cannot treat gravitational waves, but it could be generalized to do so.

wave solution as one can, e.g., in electromagnetism, as an expansion in terms of simple basis functions. But today the existence in general relativity of wave solutions with all physically reasonable properties can be conclusively supported by reasons such as the following: (a) Exact solutions for many particular types of waves have been found (e.g. Ehlers and Kundt, 1962), (b) the existence of whole classes of waves has been shown by consideration of the initial value problem (Brill, 1959), (c) the Einstein gravitational field has two independent degrees of freedom per spacetime point, like other massless fields (Arnowitt *et al.*, 1962), (d) the linearized approximation, where a complete set of solutions exists, can be shown to correspond to exact solutions in physically reasonable situations (Deser and Choquet-Bruhat, 1973; Fischer and Marsden, 1972), (e) total energy can be defined for physically reasonable (e.g., finite) gravitational wavepackets (Brill, 1959; Wheeler, 1964), (f) such waves carry timelike total energy-momentum, with positive total energy, and emission decreases the source's energy (Brill and Deser, 1968; Sachs, 1962), (g) energy density of gravitational waves can be defined in the short-wavelength limit (Brill and Hartle, 1964; Isaacson, 1968).

It is reasonable, then, to expect that, along with the other types of radiation, the universe also contains some gravitational radiation, and we want to consider how the radiation interacts with matter, in particular with gravitational wave antennas.

4.2. Weak Gravitational Waves

Such waves are, roughly speaking, small deviations from Minkowski space (or from some other background, such as that produced by stationary masses or the cosmological background). Hence such waves can be represented by a 'deviation tensor' h from the flat Minkowski metric η, such that the metric of the physical space is

$$g = \eta + h \quad \text{or in components} \quad g_{\mu\nu} = \eta_{\mu\nu} + h_{\mu\nu}. \tag{12}$$

Einstein's equations are demanded to first order in h only, as equations in the Minkowski background: Let a one-parameter family of metrics be defined by

$$g_\lambda = \eta + \lambda h. \tag{13}$$

Then the first order equations are (with G = Einstein tensor)

$$\mathrm{d}G(g_\lambda)/\mathrm{d}\lambda|_{\lambda=0} = 8\pi T. \tag{14}$$

As the condition of 'smallness' we shall take, for any unit vector X in Minkowski space,

$$h(X, X) \ll 1 \quad \text{or in components} \quad |h_{\mu\nu}| \ll 1. \tag{15}$$

The condition is more complicated if one wants to be certain that solutions of (14) correspond to solutions of the full equations, i.e., that there is a power series in λ whose first terms are (13) and which converges to a solution.

The decomposition (12) of the physical metric g is not unique: Let ϕ_λ be a family of mappings of Minkowski space into itself, with ϕ_0 the identity. Another decomposition (defining h_ϕ) is

$$g = \mathrm{d}\phi_1(\eta) + h_\phi. \tag{16}$$

The metric

$$\gamma_\lambda = \eta + \lambda \left(\frac{\mathrm{d}\phi_\lambda(\eta)}{\mathrm{d}\lambda}\bigg|_{\lambda=0} + h_\phi \right) \tag{17}$$

differs from g_λ only to order λ^2 and higher, hence if g_λ satisfies (14), so does γ_λ; and, comparing (13) and (17) we have

$$h_\phi = h - \frac{\mathrm{d}\phi_\lambda(\eta)}{\mathrm{d}\lambda}\bigg|_{\lambda=0}. \tag{18}$$

(In a more general and systematic treatment the last term in (18) is recognized as a Lie derivative). Since this transformation maps solutions of (14) into solutions, and since the physical metric $g = g_{\lambda=1}$ changes only to higher order in λ we call these transformations the gauge group of the weak field approximation. In components, where ϕ_λ is represented by

$$x^\mu \to x^\mu + \lambda \xi^\mu(x)$$

(18) takes the form

$$h_{\phi_{\mu\nu}} = h_{\mu\nu} - \xi_{\mu;\,\nu} - \xi_{\nu;\,\mu}.$$

(The semi-colon indicates covariant differentiation with respect to the background metric).

By a particular choice of gauge one can simplify the sourcefree weak field Equations (14) so that the field equations and gauge conditions take the form

$$\eta^{\alpha\beta} h_{\mu\nu;\,\alpha\beta} = 0 \quad \text{(field equation)} \tag{19}$$
$$\eta^{\alpha\beta} h_{\mu\alpha;\,\beta} = 0, \quad \eta^{\alpha\beta} h_{\alpha\beta} = 0 \quad \text{(transverse and traceless gauge condition)}.$$

As usual the conditions do not fix the gauge uniquely, unless particular boundary conditions are also imposed.

Weak-field theory should be regarded as an approximation, not as self-contained. Thus, test particles are to move on geodesics of the metric g, not according to the Minkowski space conservation law which follows from (14). The reason for this and other inconsistencies regarding the stress-energy tensor is that T is typically of second order in the matter field variables, but the quadratic terms in the gravitational field variables h have been dropped from (14); but these terms describe, in part, the effective stress-energy of the gravitational field. Thus, for questions of energy balance, these terms should be kept and considered together with the matter T as an effective stress-energy T_{eff} of the gravitational waves. This T_{eff} is not gauge-invariant, in accordance with the fact that in general gravitational energy cannot be localized. However, if we restrict attention to solutions h which vary rapidly (and gauge transformations which introduce no long-range variations), an average T_{eff} can be defined over regions large compared to the scale of the variation (Brill and Hartle, 1964; Isaacson, 1968),

$$T_{\mu\nu\,\text{eff}} = \frac{1}{32\pi} \langle h_{\alpha\beta\,;\mu} h^{\alpha\beta}{}_{;\,\nu} \rangle \tag{20}$$

and this term is conserved with the matter tensor over such large regions.

A typical wave solution of Equation (19) has the form

$$h_{\mu\nu} = \mathrm{Re}\left[A_{\mu\nu} e^{ik_\alpha x^\alpha}\right], \quad k_\alpha, A_{\mu\nu} \quad \text{constants} \tag{21}$$

for which Equation (19) implies

$$k_\alpha k^\alpha = 0, \quad A_{\mu\alpha} k^\alpha = 0 \quad \text{and} \quad A^\alpha_\alpha = 0. \tag{22}$$

This is an example of a solution to which no exact solution corresponds: the total mass-energy (computed e.g., by integrating (20)) is infinite, hence the second-order corrections, which respond to this effective energy, cannot remain uniformly small. However, this difficulty does not appear for a wavepacket built out of waves of type (21). We shall always assume that such a wavepacket has been built, and that (21) is valid to good accuracy only in some region near the center of the wavepacket.

The remaining gauge freedom can be used to impose additional conditions on $A_{\mu\nu}$. A gauge transformation generated by

$$\xi_\mu = X_\mu e^{ik_\alpha x^\alpha}, \quad X_\mu = \text{const.} \tag{23}$$

preserves the conditions (22) if

$$X_\alpha k^\alpha = 0.$$

Hence X_μ lies in the 3-dimensional subspace orthogonal to k_μ, and can be fixed by demanding of the transformed amplitude (which will be denoted by the old symbol $A_{\mu\nu}$),

$$A_{\mu\alpha} u^\alpha = 0 \tag{24}$$

for some timelike vector u^α. (Note that these are only 3 conditions, since the contraction with k^μ is automatically satisfied by virtue of (22)). In the restframe of u^α, $A_{\mu\nu}$ therefore has only spacelike components and is transverse to k^μ. Let e_1, e_2 be two transverse vectors ($u \cdot e_i = k \cdot e_i = 0$), then the general amplitude can be decomposed into the two polarization tensors

$$\begin{aligned} &(\text{vertical}) \quad e_+ = e_1 \otimes e_1 - e_2 \otimes e_2 \\ &(\text{diagonal}) \quad e_\times = e_1 \otimes e_2 + e_2 \otimes e_1 \end{aligned} \tag{25}$$

so that we have

$$A = A_+ e_+ + A_\times e_\times \tag{26}$$

as the general solution of (22) and (23). The energy of the wave (20) now becomes

$$T_{00} = \frac{1}{32\pi} k_0^2 (A_+^2 + A_\times^2). \tag{27}$$

Local, invariantly defined effects of this wave, like such effects of any gravitational field, are due to the curvature of the wave geometry; they are measured by the geodesic deviation of pairs of testparticles, or geodesic deviation of a test particle from an inertial (geodesic) observer in a parallely propagated frame. [See the lectures of Ehlers (these proceedings) for exact definition of the frame used to measure the testparticle motion, and note that for a geodesic parallely propagated observer frame, u^i and w^i vanish.] If r^i is the purely spatial separation vector of a nearby particle written in an orthonormal basis we have

$$\frac{d^2 r^i}{dt^2} = R^i{}_{00k} r^k = \frac{1}{2} \frac{\partial^2 h_{ik}}{\partial t^2} r^k. \tag{28}$$

Here we have used the expression for the Riemann tensor valid for weak fields, and ceased to distinguish between upper and lower indices of the spacelike components in the observer's orthonormal frame. We let

$$r^k = a^k + \delta r^k, \quad a^k = \text{const.} \tag{29}$$

and assume δr^k remains small, then to lowest order the solution of (28) is

$$\delta r^j = \tfrac{1}{2} h_{jl} a^l = \tfrac{1}{2} \operatorname{Re}(A_{jl} e^{ik_\alpha x^\alpha}) a^l. \tag{30}$$

A quantity measuring the strain in the a-direction is

$$\frac{a_j \delta r^j}{a_i a^i} = \tfrac{1}{2} \hat{a}^j h_{jl} \hat{a}^l = \frac{1}{(k_0)^2} R(\hat{a}, u, u, \hat{a}). \tag{31}$$

Here $\hat{a}^i$ is a unit vector in the a-direction and we have reintroduced the 4-velocity u of the observer to write the result as an invariant. We see that the strain measures directly the metric deviation, or the Riemann tensor, for this monochromatic wave.

The average amplitude of the oscillation is

$$\begin{aligned} \langle \delta r_i \, \delta r^i \rangle &= \tfrac{1}{8} a^k a^l \langle A_{ik} A^i_l \rangle \\ &= \tfrac{1}{8} (A_+^2 + A_\times^2) \, a^2, \end{aligned} \tag{32}$$

where a^2 is the length of the projection of a in the plane spanned by e_1 and e_2. Thus, particles originally in a circle about the observer in the e_1, e_2 plane oscillate with the same amplitude but in different directions in such a way (see Equation (29)) that the circle of particles is distorted into various ellipses with axes along e_1, e_2 for the vertical polarization, and ellipses with axes at 45° with e_1, e_2 for the diagonal polarization. Observation of such free particle motion in accordance with the directivity pattern implied by Equation (29) could in principle be used to detect gravitational waves.

4.3. Realistic Detectors

To date only the Weber type of detector is working and producing data*. The subsequent discussion is therefore mainly concerned with this type. Here the particles are not free, but subject to nongravitational restoring forces. Fortunately it is not necessary to develop a general relativistic theory of the restoring forces, since one can use a simple extension of the principle of equivalence. In a geodesic, parallely propagated observer frame the Christoffel symbols vanish at the origin. The (strong)

* About a dozen experimental groups are at present working on a repetition and extension of Weber's experiments.

principle of equivalence therefore implies that all forces which do not involve higher than first derivatives must take their special relativity form in this frame. Only on the left side of the equation of motion does a second derivative appear, the acceleration. To lowest order its modification in the presence of gravity must be the same as in the forcefree case treated above. Thus if $F(r)$ is the specific force (i.e., force/mass) on the particle located at $r=a$ calculated in Minkowski space, the equation of motion must be, to lowest order,

$$\frac{d^2 r^i}{dt^2} - R^i{}_{00k} r^k = F^i(r).$$

In the observer's orthonormal frame we may regard $R^i{}_{00k} r^k$ as a simple Newtonian force which is to be added to all the other Newtonian forces acting on the particle to obtain the correct lowest order equation of motion under the combined action of gravity and the other forces. If the observer is not in geodesic motion, additional terms appear on the left side, but they can also be moved to the right side and be identified with the familiar 'inertial' forces.

One version of a Weber detector consists of an aluminum bar with electronic sensors sensitive to vibrations in the lowest mode (Weber, 1961, 1970, 1972). A Fourier analysis of the equation of motion in eigenfrequencies of the bar gives the usual simple harmonic oscillator equation for the lowest mode; the driving force is the 'Fourier' component* of all the applied forces – curvature forces, inertial forces, and matter forces. The experimenter's task is to reduce the Fourier components of all forces except the curvature force to a minimum, so that the Fourier component of the Riemann tensor can be measured via the vibration of the bar which it excites. Inertial forces due to rotation of the Earth (Coriolis and centrifugal forces) have very small amplitude components at the vibration frequency of the bar, but other inertial forces due to any sudden acceleration (e.g., seismic) of the bar must be avoided. Similarly all matter forces, notably electromagnetic forces, with appreciable Fourier component at the vibration frequency should be excluded or accounted for. Among the inevitable forces are: (a) Restoring and damping forces in the bar. These are accounted for and used as part of the detection scheme. (b) Static forces of support at the center (origin of inertial frame). They have no component at the vibration frequency. (c) Forces of the sensors back on the bar. These are accounted for as part of the 'equivalent circuit' of the detector. (d) Random forces due to thermal motion. They are an important limitation on the accuracy. We assume that all other forces can be eliminated or monitored by using sufficient experimental ingenuity.

4.4. The Essential Features of Weber's Apparatus

A typical Weber detector consists of a 1.5 ton bar of aluminum (selected for its solid state properties and for economic reasons), about 150 cm long and 60 cm in diameter. The lowest mode frequency is 1660 cycles, the 'Q' (measure of ringing time) is near 10^5.

* i.e., $\int F^i(r^j, t)\, \delta r_i(r^j, t)\, d^3r\, dt$ where $\delta r_i(r^j, t)$ describes the lowest mode of oscillation of the bar.

The bar is suspended by a sling of piano wire about its center, and hung from acoustical filters in a vacuum. The vibration sensors are ceramic piezoelectric crystals glued to the bar near the center where the strain in the lowest mode is largest. The relaxation time of this electromagnetic degree of freedom is negligibly short ($<10^{-4}$ s). The output of the sensors is amplified and put through a synchronous detector whose signal is proportional to the square of the amplitude of vibration averaged over tens of milliseconds. This output signal is mainly due to the thermal vibration of the bar; it therefore tends to vary only by amounts small compared to its kT value for times small compared to the ringing time (many seconds).

A typical gravitational wave detected by this arrangement is thought to be a short pulse with significant Fourier component near 1660 Hz. The driving force of the wave changes the vibration of the bar in a time short compared to the ringing time, and since this signal amplitude is generally less than the noise amplitude, the change may be an increase or decrease in total measured amplitude, or a change in phase. The present apparatus detects these amplitude changes. Rapid amplitude changes are most likely to represent true signals (rather than thermal fluctuations). In order to allow more positive identification of signals, two detectors are run in coincidence. The detectors are separated by as much as 1000 km (at Argonne National Laboratories and at Maryland). The (rectified) output signal of the Argonne detector is sent over telephone lines in encoded form so as to be unaffected by possible fluctuation in the line, arriving at Maryland 17 ms later. These 'events' (candidates for signals from gravitational wave pulses) are identified by demanding that the output cross an adjustable threshold in an adjustable time. If 'events' occur in the two cylinders within 70 ms of each other, a coincidence is recorded. A typical coincidence rate is currently 5 per day.

A coincidence is not sufficient to identify a true signal, since there are always accidentals, depending on how the thresholds are adjusted. To be able to measure the rate of signal reception with confidence it is essential to have a good understanding of the accidentals. Weber does this roughly as follows (see, for example, Weber, 1970b, for details): After coincidence detection the signal is further averaged in time, so that the peak of the resulting curve is some measure of the energy deposited in each bar. To measure the probability that a peak of height H in detector A is a real signal, Weber (more recently a computer) counts the number N_A of times that the peak amplitude H was exceeded on that day. (High peaks, likely true signals, correspond to low N_A's). From N_A and the corresponding N_B for the other detector the probable number of accidentals is computed. To check that the computation gives the accidental rate correctly, Weber has introduced a time delay in one channel, keeping the rest of the experiment unchanged. Thus, only accidentals should be observed, and agreement with the computed rate of accidentals was found. Various other tests are also made to check that the signals do not correlate with seismic disturbances, magnetic field, electromagnetic radiation or cosmic rays.

The intensity of a detected signal (Weber, 1970c) can be computed by comparing it to the thermal noise signal. Since the thermal vibration amplitude corresponds to an energy kT in the bar, the energy deposited in the bar by the gravitational wave can be computed. This energy in turn is related to the energy carried by the wave in the frequency range where the bar is sensitive (cf. Equations (32) and (27)).

4.5. Results of Gravitational Wave Experiments

4.5.1. *Discovery of Gravitational Radiation (Weber, 1969, 1972)*

In the Maryland-Argonne experiments, many coincidences were found between pulses of such large amplitude that an accidental occurrence of such a coincidence would be extremely unlikely. The computation of the rate of these true coincidences was done both by human and by computer. Numerous tests were made to exclude the possibility of nongravitational generation of these coincidences. On the basis of present evidence it is therefore appropriate to conclude that gravitational waves have been observed.

4.5.2. *Sidereal Anisotropy*

A Weber bar has a definite pattern of sensitivity to waves from different directions and of different polarizations (Weber, 1970c, 1971). This pattern is a simple consequence of the geodesic deviation equation, and is characteristic of the transverse, tensor nature of waves. The pattern is given to good approximation by Equation (31), where $\hat{a}$ is a vector along the axes of the bar. It is seen that the apparatus is most sensitive for 'vertical' polarization (i.e., parallel or perpendicular to the bar) coming from a direction normal to the bar. Equation (13) implies a variation of sensitivity with angle θ away from the normal as $\cos^2\theta$, so the sensitivity pattern is not highly directional; however, by observing for a sufficiently long time it is possible to use this directivity to find out something about the location of the source. If the source is confined in size and located outside the solar system, the intensity of the signal should vary within a period of half a sidereal day (rather than half a solar day). Such variation was found during four independent six-month uninterrupted runs of the coincidence experiment. The phase of the variation is consistent with a (unpolarized) source located in the direction of the center of the Galaxy. It is also consistent with other source locations and degrees of polarization (Tyson and Douglass, 1972) but no particular objects at these other locations have been suggested as the source. The energy flux during the pulse is estimated to be above 10^5 erg/cm^2 s Hz.

4.5.3. *Scalar Component of the Radiation*

In addition to the bar antennas, Weber (1971a) has operated an antenna consisting of a disk about 1 m in radius, resonant at the same frequency as the bars, and with detector crystals mounted between center and edge of the disk and connected in phase. Such an arrangement is sensitive only to scalar waves in the direction normal to the disk ('breathing mode'); in other directions it is also weakly sensitive to tensor waves. Disk-cylinder coincidence experiments showed a siderial anisotropy consistent with a single source located in the direction of the galactic center, emitting tensor waves only*. (Typical numbers are 0 ± 3 true coincidences in the scalar channel, vs. 65 ± 8.5 true coincidences in the tensor channel of a simultaneous two-bar experiment).

4.5.4. *Frequency Bandwidth of the Radiation*

Weber (1972) has recently observed a high rate of coincidence between a 1660 Hz bar and the quadrupole mode of the disk at 1030 Hz. To detect this disk mode the detection crystals were phased so as to be sensitive to quadrupole distortions of the disk. As in the earlier experiments the disk is oriented so that it points toward the center of the Galaxy once a (sidereal) day, and the crystals are so oriented that the mode aligned with the plane of the Galaxy is detected. From the present data one can conclude that the frequency spectrum of the radiation extends over at least 600 Hz, and that there is considerable radiation polarized in the plane of the Galaxy. An experiment in progress with the disk rotated by 45° will allow conclusions about the degree of linear polarization of these signals.

4.6. Conclusions

The Weber experiments have not only shown persuasive evidence of gravitational radiation impinging on the Earth from a distant source outside the solar system; they have also identified important properties of the radiation, with the antennas acting as (as yet crude) gravitational wave telescopes and spectroscopes. If one uses the (probably conservative) estimates of energy flux and bandwidth reported above and assumes that the source is located at the center of the galaxy, and that it is radiating isotropically, a total energy loss of the order of 1000 $m_{\odot}$ per year from the galactic center then comes into view. Such values for the mass loss appear to contradict other astronomical evidence (Sciama *et al.*, 1969). Clearly our theoretical understanding of gravitational radiation will remain incomplete until this energy problem is resolved. To clarify further, in both theory and experiment, the physics of gravitational radiation appears to be the most exciting challenge of present day gravitation theory.

References

Arnowitt, R., Deser, S., and Misner, C. W.: 1962, in L. Witten (ed.), *Gravitation – An Introduction to Current Research*, John Wiley & Sons, New York, Chapter 7.

Anderson, J.: 1972, *Fourth 'Cambridge' Conference on Relativity*.

Anderson, J., Esposito, P. B., Martin, W. L., and Muhlman, D. O.: 1971, in R. W. Davies (ed.), JPL Technical Memorandum 33–449, p. 111.

Bahcall, J. N. and Schmidt, M.: 1967, *Phys. Rev. Letters* **19**, 1294.

Bertotti, B., Brill, D., and Krotkov, R.: 1962, in L. Witten (ed.), *Gravitation – An Introduction to Current Research*, John Wiley & Sons, New York, Chapter 1.

Bender, P., Dicke, R. H., Wilkinson, D. T., Alley, C. O., Currie, D. G., Faller, J. E., Mulholland, J. D., Silverberg, E. C., Plotkin, H. E., Kaula, W. M., and McDonald, G. J. F.: 1971, in R. W. Davies (ed.), JPL Technical Memorandum 33–449, p. 178.

Braginskü, V. B. and Panov, V. I.: 1971, *Zh. Eks. Teor. Fiz.* **61**, 875.

Brault, J. W.: 1962, Thesis, Princeton University.

* This does not necessarily contradict the scalar-tensor theory, since it is quite conceivable that the source does not radiate strongly in the monopole mode (Brill, 1962).

Brans, C. and Dicke, R. H.: 1961, *Phys. Rev.* **124**, 925.

Brill, D.: 1959, *Ann. Phys.* **7**, 466.

Brill, D.: 1962, *Proceedings International School of Physics 'E. Fermi'*, XX course, Academic Press, New York, p. 63.

Brill, D. and Hartle, J.: 1964, *Phys. Rev.* **135**, B271.

Brill, D. and Cohen, J.: 1966, *Phys. Rev.* **143**, 1011.

Brill, D. and Deser, S.: 1968, *Ann. Phys.* **50**, 548.

Campbell, W, and Trumpfer, R.: 1923, *Lick Obs. Bull.* **11**, 41.

Chapman, P.: 1972, *Fourth 'Cambridge' Conference on Relativity.*

Chapman, P. and Hanson, A.: 1971, in R. W. Davies (ed.), JPL Technical Memorandum 33–449, p. 228.

Davies, R. W.: 1971, JPL Technical Memorandum 33–449, p. 225.

Deser, S. and Choquet-Bruhat, Y.: 1973, *Ann. Phys.*, to be published.

Dicke, R. H.: 1962, *Science* **138**, 653.

Dicke, R. H.: 1962b, *Phys. Rev.* **125**, 2163.

Dicke, R. H.: 1964a, in DeWitt, B, and DeWitt, C. (eds.), *Relativity Groups and Topology* (Les Houches, 1963), Gordon and Breach, p. 170.

Dicke, R. H.: 1964b, *loc. cit.*, p. 189.

Dicke, R. H.: 1970, *Ann. Rev. Astron. Astrophys.* **8**, 297.

Dicke, R. H.: 1971, *Phys. Rev. Letters* **27**, 210.

Dicke, R. H.: 1972a, *Trans. New York Acad. Sci. II* **34**, 171.

Dicke, R. H.: 1972b, *Astrophys. J.* **176**, 479.

Dicke, R. H. and Goldenberg, H.: 1967, *Phys. Rev. Letters* **18**, 313.

Drever, R. W. P.: 1961, *Phil. Mag.* **6**, 683.

Dyson, F.: 1967, *Phys. Rev. Letters* **19**, 1291.

Eddington, A. S.: 1934, *The Mathematical Theory of Relativity*, Cambridge University Press.

Ehlers, J. and Kundt, W.: 1962, in Witten, L. (ed.), *Gravitation – An Introduction to Current Research*, John Wiley & Sons, New York.

Everitt, C. W. F.: 1971, in R. W. Davies (ed.), JPL Technical Memorandum 33–449, p. 68.

Fischer, A. and Marsden, J.: 1972, *Comm. Math. Phys.* **28**, 1.

Gilvary, J. and Muller, P.: 1972, *Phys. Rev. Letters* **28**, 1665.

Hughes, V., Robinson, H. G., and Beltran-Lopez, V.: 1960, *Phys. Rev. Letters* **4**, 342.

Hill, H.: 1971, in R. W. Davies (ed.), JPL Technical Memorandum 33–449, p. 89.

Hill, J. M.: 1971, *Monthly Notices Roy. Astron. Soc.* **153**, 7.

Ingersoll, A. P. and Spiegel, E. A.: 1971, *Astrophys. J.* **163**, 375.

Isaacson, R. A.: 1968, *Phys. Rev.* **166**, 1263 and 1272

Klüber, H.: 1960, *Vistas in Astronomy* **3**, 47.

Kreutzer, L. B.: 1968, *Phys. Rev.* **169**, 1007.

Marzke, R. F. and Wheeler, J. A.: 1964, in H. Y. Chiu and W. F. Hoffmann (eds.), *Gravitation and Relativity*, W. A. Benjamin, New York.

Misner, C. W., Thorne, K. S., and Wheeler, J. A.: 1973, *Gravitation*, Freeman Co., San Francisco.

Muhleman, D. O., Ekers, R. D., and Fomalout, E. B.: 1970, *Phys. Rev. Letters* **24**, 1377.

Nordtvedt, K.: 1968, *Phys. Rev.* **169**, 1017.

Nordtvedt, K. and Will, C. M.: 1972, *Astrophys. J.* **177**, 775.

O'Connell, R. F.: 1972, *GRG J.* **3**, 123.

Peebles, P. J. and Dicke, R. H.: 1962, *Phys. Rev.* **128**, 2006.

Peebles, P. J. and Dicke, R. H.: 1962a, *Phys. Rev.* **127**, 629.

Pound, R. V. and Snider, J. L.: 1965, *Phys. Rev.* **140**, B788.

Schiff, L.: 1959, *Proc. Nat. Acad. Sci.* **45**, 69.

Sachs, R.: 1962, *Proc. Roy. Soc.* **A270**, 103.

Sciama, D. W., Field, G. E., and Rees, M. J.: 1969, *Phys. Rev. Letters* **23**, 1514.

Seielstadt, G. A., Sramek, R. A., and Weiler, K. W.: 1970, *Phys. Rev. Letters* **24**, 1373.

Shapiro, I. I. *et al.*: 1968, *Phys. Rev. Letters* **20**, 1265.

Shapiro, I. I. *et al.*: 1971, *Phys. Rev. Letters* **26**, 27 and 1132.

Shapiro, I. I.: 1972, *GRG J.* **3**, 135.

Shapiro, I. I., Pettengill, G. H., Ash, M. E., Ingalls, R. P., Campbell, D. B., and Dyce, R. B.: 1972, *Phys. Rev. Letters* **28**, 1594.

Sramek, R. A.: 1971, *Astrophys. J. Letters* **167**, 55.
Sramek, R. A.: 1972, *Fourth 'Cambridge' Conference on Relativity*.
Thorne, K. S., Will, C. M., and Ni, W. T.: 1971, in R. W. Davies (ed.), JPL Technical Memorandum 33–449, p. 10.
Trautmann, A.: 1965, *Lectures on General Relativity*, Prentice-Hall Inc., Englewood Cliffs, N.J., p. 101.
Trautmann, A.: 1972, *GRG J.* **3**, 171.
Turner, R. and Hill, H.: 1964, *Phys. Rev.* **134B**, 252.
Tyson, J. A. and Douglass, D. H.: 1972, *Phys. Rev. Letters* **28**, 991.
Vessot, R.: 1972, *Fourth 'Cambridge' Conference on Relativity*.
Vessot, R. and Levine, M.: 1971, in R. W. Davies (ed.), JPL Technical Memorandum 33–449, p. 54.
Weber, J.: 1961, *General Relativity and Gravitational Waves*, Interscience, New York.
Weber, J.: 1969, *Phys. Rev. Letters* **22**, 1320.
Weber, J.: 1970a, *Nuovo Cimento* **4**, 653.
Weber, J.: 1970b, *Phys. Rev. Letters* **24**, 276.
Weber, J.: 1970c, *Phys. Rev. Letters* **25**, 180.
Weber, J.: 1971a, *Nuovo Cimento* **4**, 197.
Weber, J.: 1971b, *Phys. Letters* **34A**, 271.
Weber, J.: 1972a, *Pulses of Gravity*, Science Year.
Weber, J.: 1972b, *Popular Science* **200**, 106.
Weber, J.: 1972c, *GRG J.* **3**, 59.
Weber, J.: 1972d, *Nature* **240**, 28.
Weber, J.: 1973, *Proc. 6th Texas Symposium*, Trans. New York Acad. Sci (to appear).
Wheeler, J. A.: 1964, in De Witt, B. and De Witt, C. (eds.), *Relativity Groups and Topology* (Les Houches 1963), Gordon and Breach, Section 6.
Will, C. M. and Nordtvedt, K., Jr.: 1972, *Astrophys. J.* **177**, 757.
Will, C. M.: 1971a, *Astrophys. J.* **163**, 611.
Will, C. M.: 1971b, *Astrophys. J.* **169**, 141.
Witten, L.: 1962, *Gravitation – An Introduction to Current Research*, John Wiley & Sons, New York.

RELATIVISTIC STARS, GRAVITATIONAL COLLAPSE, AND BLACK HOLES

JAMES B. HARTLE*

Dept. of Physics, University of California, Santa Barbara, Calif. 93106, U.S.A.

1. Introduction

The search for explanations of pulsars, quasars, Weber's gravitational wave pulses, X-ray sources and explosions in galactic nuclei has proved the crucial role which relativistic gravity plays in these frontier astrophysical phenomena. Just a decade ago, before these phenomena had been discovered, it was difficult to find a useful application of relativistic gravity to astrophysics outside of cosmology. Pioneers such as Oppenheimer, Landau, Wheeler, Hoyle and others had anticipated, on theoretical grounds, the role which general relativity must play in resolving such issues as the end point of stellar evolution and the importance of gravitational collapse as a source of energy. Their ideas, however, did not prove relevant to observation until much later. A decade ago, therefore, a series of lectures like this on relativistic astrophysics would have been compact and theoretical and yet still could have been comprehensive.

Today after the discovery of the phenomena mentioned above the situation is reversed. It is not possible to treat all of the topics which might be subsumed under the heading of 'relativistic astrophysics' in a series of lectures as short as this. Since this is a summer school in general relativity it seems appropriate to focus on those general aspects of relativistic gravity which workers in the field believe will have application to realistic astrophysical objects, rather than on detailed models for the objects themselves. Further, we will restrict ourselves to purely non-cosmological problems as cosmology is already being covered in other lectures and we will also concentrate only on those ideas which can be elaborated with simple mathematics, leaving the reader to find the more detailed and general calculations and proofs in the literature.

There are two major types of issues which stand out in the subject of relativistic astrophysics and upon which we will concentrate. The first is the theory of the structure of relativistic stars and the second is the theory of the progress and outcome of gravitational collapse. The objects whose properties these theories predict are the white-dwarf and neutron stars, supermassive objects and black holes. Taken together, these objects answer the questions of what are the possible endstates of stars and of collections of stars. We have already seen two of these objects – the white-dwarf and neutron stars. There is every reason to believe that the other objects predicted by relativistic gravity should also exist and be observable. It is the aim of these lectures to develop in an elementary way some aspects of the theory of the basic parameters of these relativistic objects.

* Alfred P. Sloan Research Fellow, supported in part by the National Science Foundation

W. Israel (ed.), Relativity, Astrophysics and Cosmology, 153–195.

Notation

(1) Indices: Greek indices range over space and time, while Latin indices range over space variables only.
(2) Signature: $(-\ +\ +\ +)$.
(3) Units: Geometrized units are used throughout in which $c=$(velocity of light)$=1$ and $G=$(Newton's universal constant of gravitation)$=1$. Masses are then measured in units of cm. Energy densities in units of cm^{-2} and pressures in units of cm^{-2}. Useful conversion factors are:

Mass:	M (cm)	$= 1.47 \times 10^5$ M (solar masses)
Mass:	M (cm)	$= 0.742 \times 10^{-28}$ M (grams)
Density:	E (cm^{-2})	$= 0.742 \times 10^{-28}$ E (grams/cm^3)
Pressure:	P (cm^{-2})	$= 0.826 \times 10^{-49}$ P (dynes/cm^2)

2. The Relativistic Stellar Structure of Spherical Stars

What are the possible equilibrium end states of stellar evolution? What are the properties of an equilibrium configuration of $10^8 M_\odot$? Such questions lead naturally to a theory of the structure of relativistic stars.

Any equilibrium configuration is a balance between pressure forces, gravitational forces and rotational forces. The physics of a Newtonian star and a relativistic star is essentially the same except that the theory of gravity changes from Newton's theory to Einstein's general relativity. In general relativity the description and computation of the gravitational field due to a general equilibrium distribution of matter is often complex. It involves as many as four independent $g_{\mu\nu}$ which have to be computed as solutions to Einstein's equations. It is appropriate therefore, to begin a discussion of relativistic stellar structure by considering the simplifying case when the star has zero angular momentum and a spherical shape. The resulting spherical symmetry greatly simplifies the description of the curved space-time and hence all the dependent calculations of the star's structure.

Fortunately, the assumption of no rotation is actually a good approximation for one of the most interesting classes of relativistic stars – the neutron stars which underly the pulsar phenomenon. These stars are rotating slowly in the sense that the rotational distortion away from spherical symmetry is small. The theory of spherical stars is thus not only simple but important for the most interesting astrophysical case.

We will now outline the theory of the relativistic stellar structure of spherical stars. Basically all we have to do is solve Einstein's equations for a spherically symmetric distribution of matter. Before doing this, however, it is appropriate to discuss how we describe the matter inside a relativistic star.

2.1. Relativistic Perfect Fluids

For most cases of astrophysical interest – neutron stars, supermassive objects – the source of the gravitational field is well approximated by a perfect fluid. The stress-energy in this case has the form

$$T_\mu^\nu = (E + P)\, u_\mu u^\nu + \delta_\mu^\nu P\,. \tag{2.1}$$

Here, E and P are the energy density and pressure as measured by an observer moving with the fluid and u^μ is the fluid's four-velocity. The equations of motions for the fluid are equivalent to the vanishing of the divergence of T_μ^ν.

In addition to satisfying the equations of motion, any fluid motion will also conserve the number of baryons and, therefore, satisfy an equation of continuity. In the language of microscopic physics we say that there is an absolute conservation law – the conservation of baryon number – which any motion of the fluid must satisfy. We can write this law by introducing the number density of baryons $N(x)$. The baryon number current is then Nu^μ and the associated equation of continuity is

$$[Nu^\mu]_{;\mu} = 0\,. \tag{2.2}$$

Corresponding to this local conservation law there is a global conserved quantity – the total number of baryons, A:

$$A = \int \mathrm{d}^3x \sqrt{-g}\, u^t N\,. \tag{2.3}$$

The total rest mass of a star can be found by computing A and multiplying by the rest-mass per baryon μ_B.

The functions E, P and N are not all independent but are related in two ways: (i) by an equation of state and (ii) by the first law of thermodynamics. The equation of state is a relation which gives, for example, the pressure as a function of the energy density and some other thermodynamic variables, typically the temperature or entropy, chemical concentrations, etc. For the two cases of main astrophysical interest – neutron stars and supermassive objects – there are additional constraints which fix these remaining thermodynamic variables. Inside neutron stars the composition is fixed by the requirement that the matter be at the endpoint of thermonuclear evolution (see Section 2.3). In addition the matter is highly degenerate,

$$kT \ll (\text{Fermi energies of constituents})\,, \tag{2.4}$$

and so it is a good approximation to take the temperature $T=0$. For supermassive objects the energy transport mechanisms usually lead to adiabatic or almost adiabatic temperature gradients and so the entropy is nearly constant throughout the star and fixed for a given configuration. Therefore, in all the following discussions, we will assume that the matter is described by a *one parameter* equation of state $P=P(E)$. Equally well the one parameter equation of state could be given in the form $E=E(N)$, $P=P(N)$ and a variety of other combinations.

The second constraint on E, P, and N is the first law of thermodynamics which takes a simple form for these isentropic equations of state

$$\begin{aligned}&\mathrm{d}(\text{energy in volume occupied by } A \text{ baryons}) = \\ &-(\text{pressure}) \times \mathrm{d}(\text{volume occupied by } A \text{ baryons}).\end{aligned} \tag{2.5}$$

Since the volume occupied by A baryons is A/N we have

$$\mathrm{d}\left[E\left(\frac{A}{N}\right)\right] = -P\,\mathrm{d}\left(\frac{A}{N}\right). \tag{2.6}$$

or

$$\frac{\mathrm{d}E}{\mathrm{d}N} = \frac{E+P}{N}. \tag{2.7}$$

Thus, there is essentially only one functional relation between E, P, and N.

We will further assume that the equation of state satisfies several additional restrictions: (i) Positive energy density E. (ii) Microscopic stability – Consider a volume with a fixed number of particles. If we contract the volume, the energy should rise as otherwise it would be energetically favorable for the small volume to collapse. This is the requirement $P \geqslant 0$. Further, as the volume contracts the pressure itself should increase as otherwise the system would be unstable to small volume fluctuations. This is the requirement that $\mathrm{d}P/\mathrm{d}(1/N) \leqslant 0$ or, using Equation (2.7), that $\mathrm{d}P/\mathrm{d}E \geqslant 0$. (iii) Velocity of sound less than the velocity of light – this seems a reasonable requirement from the point of view of causality and it will be assumed here. [See, however, Bludman and Ruderman (1968, 1970) and Ruderman (1968) for detailed discussion of this question.] Since at long wavelengths the velocity of sound is $(\mathrm{d}P/\mathrm{d}E)^{1/2}$ we have (using (ii)) the general restriction

$$0 \leqslant \frac{\mathrm{d}P}{\mathrm{d}E} \leqslant 1. \tag{2.8}$$

2.2. Equations of Structure

The geometry outside a spherical star of total mass M is given in the familiar Schwarzschild coordinates by the line element

$$\mathrm{d}s^2 = -\left(1 - \frac{2M}{r}\right)\mathrm{d}t^2 + \left(1 - \frac{2M}{r}\right)^{-1}\mathrm{d}r^2 + r^2\,\mathrm{d}\Omega^2. \tag{2.9}$$

Here $\mathrm{d}\Omega^2$ is an abbreviation for the line element on a sphere, $\mathrm{d}\theta^2 + \sin^2\theta\,\mathrm{d}\varphi^2$. Inside the star it takes two functions of r to describe the geometry

$$\mathrm{d}s^2 = -e^{2\Phi(r)}\,\mathrm{d}t^2 + \left(1 - \frac{2M(r)}{r}\right)^{-1}\mathrm{d}r^2 + r^2\,\mathrm{d}\Omega^2. \tag{2.10}$$

In the Newtonian limit the function $\Phi(r)$ becomes the Newtonian gravitational potential and $M(r)$, as we will see, can be thought of as the mass* interior to the radius r. To determine the two functions $\Phi(r)$ and $M(r)$, Einstein's equations

$$G_\mu^\nu = R_\mu^\nu - \tfrac{1}{2}\delta_\mu^\nu R = 8\pi T_\mu^\nu, \tag{2.11}$$

must be solved inside the star for a T_μ^ν given by Equation (2.1) and the solutions

* For notational convenience we have not distinguished between $M(r)$ and the total mass M.

matched to the exterior geometry in Equation (2.9). In our coordinate system the fluid is at rest so ($u^i=0$) and as a consequence of $u_\mu u^\mu=-1$ one has

$$u^t = \exp[-\Phi(r)]. \tag{2.12}$$

It is then not difficult to write out the two Einstein equations $G^t_t=8\pi T^t_t$ and $G^r_r=8\pi T^r_r$ and the divergence equation $T^{\nu}_{r;\nu}=0$ to find the three equations of relativistic spherical stellar structure

$$\frac{dM}{dr} = 4\pi r^2 E. \tag{2.13a}$$

$$\frac{d\Phi}{dr} = \frac{M + 4\pi r^3 P}{r(r-2M)}, \tag{2.13b}$$

$$-\frac{dP}{dr} = (E+P)\frac{d\Phi}{dr}. \tag{2.13c}$$

To solve these three equations for the three unknowns M, P and Φ an equation of state relating E and P must also be specified.

To integrate Equations (2.13) and obtain a relativistic stellar model it is convenient to eliminate $d\Phi/dr$ between Equations (2.13b) and (2.13c) and use the two equations

$$\frac{dM}{dr} = 4\pi r^2 E, \tag{2.14a}$$

$$-\frac{dP}{dr} = (E+P)\left(\frac{M+4\pi r^3 P}{r(r-2M)}\right). \tag{2.14b}$$

This is a system of two first order differential equations for the unknowns $P(r)$ and $M(r)$. To solve them begin at the center, $r=0$, specifying a value for the central energy density E_c (and hence a value of the central pressure P_c through the equation of state) and the boundary value $M(0)=0$ which follows from the requirement that space be locally flat at the center. Integrate outward from the center following the decreasing pressure. When the pressure drops to zero the surface of the star has been reached giving the star's radius $R=R(E_c)$ and total mass $M=M(R)\equiv M(E_c)$. The exterior geometry is then given for $r>R(E_c)$ by the Schwarzschild geometry [Equation (2.9)] with this value of M. The rest of the interior solution can then be obtained by integrating Equation (2.12b) for $\Phi(r)$ subject to the boundary condition $\Phi(\infty)=0$ so that the metric becomes asymptotic to the Minkowski metric.

Thus, with a given equation of state there is one unique stellar model with a mass $M(E_c)$ and a radius $R(E_c)$ for each value of the central density E_c. Each equation of state leads to a one-parameter family of spherical stellar models.

The equations of structure [Equations (2.13)] have simple physical interpretations obtained by looking at their Newtonian limits. In that limit $P \ll E$ and $M(r)/r \ll 1$. Equation (2.13a) defines the mass interior to a given radius. Equation (2.13b) becomes

$$\frac{d\Phi}{dr} = \frac{M}{r^2} \tag{2.15}$$

which is the Newtonian expression for the gravitational potential. Equation (2.13c) then expresses the balance between gravitational and pressure forces on an individual fluid element. In general relativity the basic interpretation of these equations summarized in Table I, is unchanged. The differences in their form, from the Newtonian equations, however, have important consequences for stellar models.

(a) The pressure appears on the right hand side of the equation for the effective gravitational force [Equation (2.13b)] and in the effective mass of a fluid element in the force balance equation [Equation (2.13c)]. This means that as the pressure rises the effective gravitational compression becomes even stronger than would have been expected on the basis of Newtonian theory. Essentially this effect is responsible for the maximum mass theorems for spherical stars. If we imagine increasing the mass of an equilibrium configuration then, no matter how stiff the equation of state, the gravitational compression always rises faster than the pressure forces. Beyond some maximum mass gravity always wins and no equilibrium is possible.

TABLE I

Equations of structure for spherical stars

Equation	Physical interpretation
$\frac{dM}{dr} = 4\pi r^2 E$	defines $M(r)$ – the mass interior to radius r
$-\frac{dP}{dr} = (E+P)\left(\frac{M+4\pi r^3 P}{r(r-2M)}\right)$	(pressure force on a fluid element) = (gravitational force on a fluid element)
$-\frac{dP}{dr} = (E+P)\frac{d\Phi}{dr}$	(pressure force on a fluid element) = (mass of fluid element) × (derivative of gravitational potential)

(b) The curvature of space-time modifies the inverse square law of the Newtonian gravitational force to an inverse $r(r-2M(r))$ law. This has the important consequence that no region of the star can be inside its gravitational radius. That is, for all r

$$2M(r)/r < 1\,. \tag{2.16}$$

To see this imagine integrating the equation of structure outward and suppose at some point $r = r_0$, $2M(r_0) = r_0$. If P and E do not both vanish at r_0 then P must become logarithmically infinite. However, P and E cannot *both* vanish at r_0 for near such a point

$$\frac{dP}{E+P} = \text{const.}\,(r - r_0)^{-1}\,dr \tag{2.16a}$$

and so

$$\log\left(\frac{E+P}{N}\right) = \text{const.}\,\log(r - r_0) \tag{2.16b}$$

and the quantity $(E+P)/N$ would have to vanish. For physical equations of state,

$(E+P)/N$ does not vanish when P does but approaches the rest mass per baryon. Thus, P must have a logarithmic infinity at r_0 and, since $(\mathrm{d}P/\mathrm{d}r)<0$, this means that P must have crossed zero and the surface been reached at a radius $r<r_0$. Thus, $2M/r$ is always less than 1 inside the star.

This type of analysis has been done more carefully by Bondi (1964) and Buchdahl (1959) to give upper limits to the value of $2M/R$. The maximum value Bondi obtains under the assumptions on the equation of state used here ($E>0$, $P>0$, $0<\mathrm{d}P/\mathrm{d}E<1$) is

$$2M/R < 0.78\,. \tag{2.17}$$

This is a useful result because it gives an upper limit to the redshift coming from the surface of any relativistic object. If λ_e is the wavelength of radiation v emitted by an observer on the surface of the star and λ_r the wavelength at which it is received by an observer at infinity, the redshift, z, is defined by

$$z = \frac{\lambda_r - \lambda_e}{\lambda_e}\,. \tag{2.18}$$

The redshift can be computed by noting that (i) in this static Schwarzschild geometry the component p_t of the photon's energy-momentum vector is conserved along its trajectory and (ii) the relation between a photon's frequency and the energy as measured by an observer whose four-velocity is u^μ is

$$\hbar\omega = -\,p_\mu u^\mu\,. \tag{2.19}$$

Immediately one has

$$\omega_r = e^{\Phi(R)}\omega_e\,. \tag{2.20}$$

and so

$$z = \left(1 - \frac{2M}{R}\right)^{-1/2} - 1\,. \tag{2.21}$$

The maximum redshift from the surface of any spherical star is thus

$$z < 1.1\,. \tag{2.22}$$

(c) The equation [Equation (2.13a)] which defines $M(r)$ is unchanged from that in Newtonian physics. At first this seems surprising because, as a consequence of the curvature of space, an element of volume is no longer $4\pi r^2\mathrm{d}r$ but rather is $(1-2M(r)/r)^{-1/2}\,4\pi r^2\mathrm{d}r$. One might have expected that $M(r)$ should have been E integrated over this correct volume factor. What would be left out of such a calculation is the contribution of the energy in the gravitational field. Remarkably this gravitational contribution exactly compensates the factor of $(1-2M(r)/r)^{-1/2}$.

2.3. The Endstate of Stellar Evolution

The evolution of a star is the history of the interplay of gravitational collapse and nuclear burning. In broad outline, the life history of a star begins with the collapse and the consequent compressional heating of a cloud of interstellar gas. When a sufficiently high central temperature is reached the hydrogen which makes up most of the gas ignites and burns to make helium. The star rapidly arrives on the main sequence where it remains for a large portion of its life, supporting itself against the everpresent forces of gravitational collapse by the thermal pressure of hot gases. Eventually in the core a substantial fraction of the hydrogen is exhausted and, having no further source of pressure, it must begin to collapse. As it collapses it heats up until the temperature rises to the point where the thermonuclear reactions which burn helium to make carbon are ignited. The core once again stabilizes. Eventually, however, a substantial part of the helium is exhausted and collapse must proceed again.

The question we want to attack here is: How does this process end? What is the endstate of stellar evolution? This is a question clearly relevant to observations as we may expect to find in nature the endstates of massive stars which proceed through their evolution rapidly.

Classifying the *possible* endstates of stellar evolution is a far easier task than tracing the evolution of a single star. The endstate must either be an equilibrium star or a state of perpetual gravitational collapse. We will look first at the equilibrium stars and consider only non-rotating, spherical configurations whose theory of structure we have just developed.

The basic input needed to find the possible equilibrium endstates of stellar evolution is the equation of state of matter in its ground state; that is matter for which all thermonuclear burning reactions which can take place have taken place and which has cooled to zero temperature. A convenient way of parameterizing the equation of state is in terms of the adiabatic index

$$\gamma = \left(\frac{E + P}{P}\right)\frac{\mathrm{d}P}{\mathrm{d}E} \tag{2.24}$$

whose magnitude roughly indicates the stiffness of the material.

We will now outline in rough terms the important sources of pressure in the various density regimes and give a qualitative indication of the corresponding behavior of γ (Figure 1). The essential features are shown in Table II. The equation of state of dense matter is a subject which involves essentially every area of physics *except* general relativity (it is a purely local problem) and we can treat it only briefly and crudely here. For more details the reader is referred to Harrison *et al.* (1965), Ruderman (1972), Salpeter (1961), Baym *et al.* (1971a, b), Pandharipande (1971, 1972), Sawyer (1972), Scalapino (1972), Sawyer and Scalapino (1972), and the many further references cited therein.

(1) $E \lesssim 10^4$ gm/cc. – The nucleus with the highest binding energy per nucleon is Fe^{56}. Thus while other nuclei may be burned to yield Fe^{56} there is no way to burn Fe^{56} to get another nucleus. In the low density regime, the equation of state of matter in its ground state is therefore simply the equation of state of pure Fe^{56}. The dominant source of the pressure is the electromagnetic interactions of the electrons and iron

TABLE II

Equation of state of matter in its ground state

Density range (gm/cc)	Constituents	Dominant source of pressure
$E \lesssim 10^4$	Lattice of Fe^{56} nuclei + electrons	Electromagnetic forces of electrons and nuclei
$10^4 \lesssim E \lesssim 4 \times 10^{11}$	Lattice of increasingly neutron rich nuclei + electrons	Degenerate electrons
$4 \times 10^{11} \lesssim E \lesssim 2 \times 10^{14}$	Lattice of increasingly neutron rich nuclei + electrons + free neutron fluid	Degenerate neutrons, electrons, and nuclear forces
$2 \times 10^{14} \lesssim E \lesssim 5 \times 10^{14}$	Neutron fluid + small amounts of electrons and protons	Degenerate neutrons and nuclear forces
$E \gtrsim 5 \times 10^{14}$	Neutrons, protons, electrons, mesons, hyperons	Strong interactions and degenerate particles

nuclei. The adiabatic index γ approaches ∞ because, as the pressure goes to zero, iron remains a solid with finite sound speed. This density regime is not of great importance for neutron stars. The density scale height at the surface of a typical neutron star is at most a few meters. Moving in from the surface one very rapidly reaches the next regime.

(2) 10^4 gm/cc$\lesssim E \lesssim 4 \times 10^{11}$ gm/cc. – Imagine that the density of iron is increased from its value at zero pressure. The dimension, λ, of the typical volume occupied by an electron will correspondingly decrease. The Fermi energy of the electrons will rise

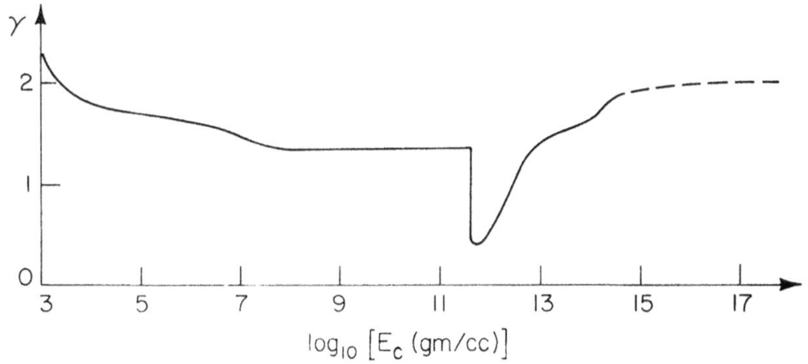

Fig. 1. The qualitative features of the equation of state for cold matter at the endpoint of thermonuclear evolution are portrayed here by a plot of $\gamma = [(E+P)/P]\,(\mathrm{d}P/\mathrm{d}E)$ vs density, E. The size of γ is a measure of the stiffness of the equation of state. The behavior of this curve is discussed in the text and in Table II. The particular data plotted here below nuclear densities have been taken from Baym *et al.* (1971a, b) and are typical in their qualitative features of most other calculations. Above nuclear densities the physics of dense matter is much less well understood than in the nuclear and subnuclear regimes and a crude possible behavior has been indicated by the dotted line.

faster than the Coulomb interactions ($p_F^2/2m \sim \hbar^2/2m\lambda^2$ vs $Z^{2/3}e^2/\lambda$). At a density of about 10^4 gm/cm^3 the degeneracy pressure of the electrons dominates that arising from Coulomb forces. At first γ is close to 5/3, the value appropriate to a non-relativistic Fermi gas. As the density is increased above $\sim 10^7$ gm/cc, however, the electrons become relativistic and γ drops in a smooth way to the value 4/3 appropriate for a relativistic Fermi gas.

(3) 4×10^{11} gm/cc$\lesssim E \lesssim 2 \times 10^{14}$ gm/cc. – At a density of about 8×10^6 gm/cc the Fermi energy of the electrons has risen to where electrons can now be absorbed on protons to make neutrons

$$e^- + p \rightarrow n + \nu . \tag{2.25}$$

At first more and more neutron rich nuclei are created but the dominant source of the pressure remains the degenerate electrons. At a density of about 4×10^{11} gm/cc the extra

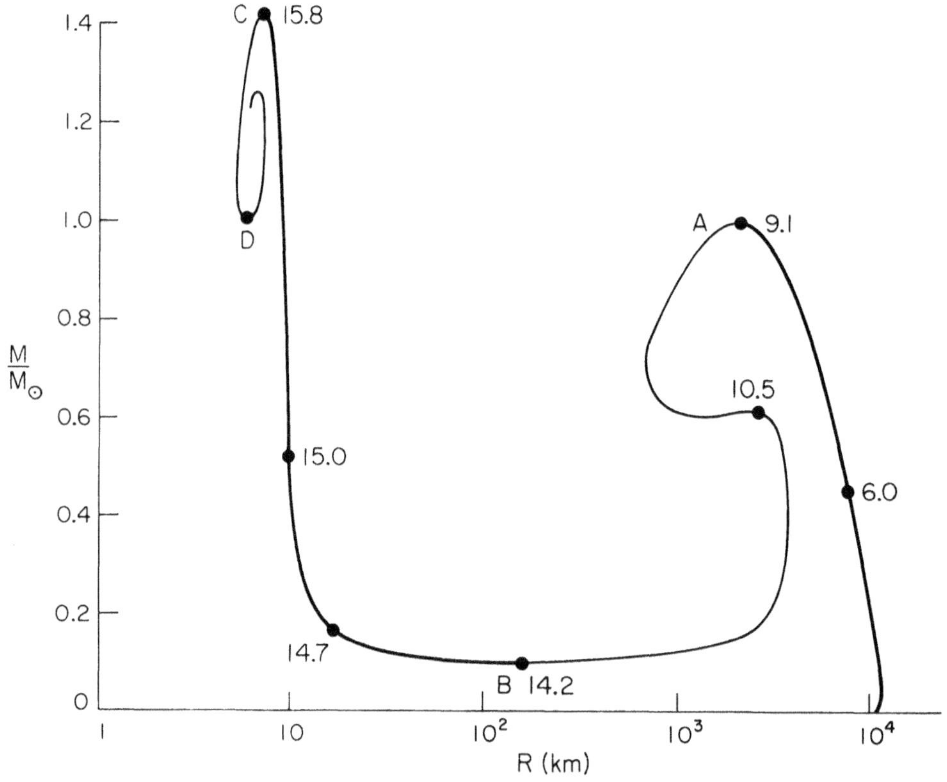

Fig. 2. The total mass, M, vs radius R for spherical stars made from matter obeying the equation of state of Figure 1. (Taken from Baym *et al.*, 1971 below about 6×10^{15} gm/cc and approximately sketched above this.) Each point on this curve represents one star corresponding to a unique value of the central density. Selected values of the central density in gms/cc. have been indicated along the curve, i.e. 6 corresponds to 10^6 gm/cc. The stable equilibrium configurations according to the static stability analysis are indicated by a heavier curve. The low density stable region is the white dwarf stars, while the high density region is the neutron and hyperon stars. Above a mass of about $1.4 M_\odot$ no equilibrium configurations are possible.

neutrons are not all bound to the nuclei and so there is an additional free neutron component. The extra neutrons add little to the pressure but do contribute to the mass and so γ drops sharply. Gradually, as the density increases, the Fermi pressure of the free neutrons plus the nuclear forces act to increase the pressure so that γ rises again.

(4) 2×10^{14} gm/cc$\lesssim E\lesssim 5\times10^{14}$ gm/cc. – At 2×10^{14} gm/cc there is a phase transition in which the nuclei dissolve. The resulting liquid consists mostly of neutrons with a small ($\lesssim 5\%$) number of electrons and protons. The effects of the nuclear forces must be included to obtain the pressure and γ rises steadily.

(5) $E\gtrsim 5\times10^{14}$ gm/cc. – Above nuclear densities, calculations of the equation of state are very uncertain. Other particles besides electrons and nucleons are produced and there is little detailed knowledge of the forces between them. Crude calculations show increasing repulsion with γ steadily rising. Sawyer and Scalapino (1972), however, have suggested that the conventional picture of the ground state of matter at these densities may be radically wrong and that γ may dip down as a condensed pion phase is produced. We should bear these uncertainties in mind.

A typical curve of mass vs radius parameterized by central density is shown in Figure 2. The most important feature evident from a curve like this is the existence of a maximum mass. Beyond about $1.4M_\odot$ no equilibrium configurations are possible. Gravity always wins and a collapse must ensue.

2.4. The Static Stability Analysis

It is not enough to calculate the possible equilibrium configurations from an equation of state for cold matter to know which of these equilibrium configurations can exist in nature. One must also test them for stability.

To test a star for stability one sets the star into a mode of small oscillation. If the amplitude of oscillation grows with time then that mode is unstable. If it dies away or is periodic in time then the mode of oscillation is stable. For the whole star to be stable all the modes must be stable.

The modes which govern whether a spherical star is stable or unstable are the purely radial modes of oscillation. The non-radial modes turn out to be damped by gravitational radiation and do not contribute to the star's instability [see Thorne (1969)]. By restricting attention to the purely radial modes, the analysis is greatly simplified.

A radial mode is described by an amplitude $\xi(r, t)$ giving the radial displacement as a function of time of a fluid element originally located at a coordinate radius r. A general displacement can be decomposed into a discrete set of normal modes which have the form

$$\xi_n(r, t) = (Ae^{i\omega_n t} + Be^{-i\omega_n t})\,\xi_n(r)\,. \tag{2.26}$$

Here, ω_n are the discrete set of eigenfrequencies and A and B are arbitrary constants. Since for purely radial modes there is no dissipation by gravitational radiation, these frequencies will be either purely real corresponding to stable modes, or purely imaginary corresponding to unstable ones. To test a star for stability one simply needs to calculate the ω_n^2. Any one $\omega_n^2<0$ will mean an unstable star.

A careful calculation of the ω_n^2 would involve solving the dynamical equations governing the star's pulsations. It is interesting that with one simple approximation and a few plausible assumptions a great deal can be learned about the regions of stability of a sequence of equilibrium configurations just from the curve of mass versus radius (Figure 9). Appropriately, since no dynamical calculations are involved, the method is called the static stability analysis [Oppenheimer and Volkoff (1939) and Harrison *et al.* (1965)].

The basic approximation which is made in the static stability analysis is that the pressure forces which arise from a pulsation-induced change in density at frequency ω are the same as would arise for the same change in density made slowly and adiabatically. That is, the change in pressure ΔP produced by a change in density ΔE can be calculated directly from the equation of state

$$\Delta P = \left(\frac{\mathrm{d}P}{\mathrm{d}E}\right)_{\text{eqn. of state}} \Delta E. \tag{2.27}$$

Realistically this will not be precisely true. When the matter is compressed slowly certain nuclear reactions take place to bring the matter to the new ground state at higher density. If it is compressed rapidly, these reactions might not have time to take place and Equation (2.27) would incorrectly represent the change in pressure. Fortunately the most interesting range of frequencies for a study of stability are the frequencies near zero where a mode might be changing stability. For these, the pulsation time is long.

The plausible assumptions necessary for the static stability analysis are, first, that the low density configurations – balls of pure iron – are stable. Second, there are three assumptions concerning how the frequency of a given normal mode changes as we move from one equilibrium configuration to another, i.e., the functions $\omega_n(E_c)$. These assumptions are: (a) The different normal frequencies $\omega_0^2(E_c)<\omega_1^2(E_c)<\omega_2^2(E_c)<\cdots$ are all non-degenerate. This means that no curves of $\omega_n(E_c)$ ever intersect. (b) Each normal mode can be characterized as even or odd according to whether $\xi_n(r)$ has an even or odd number of nodes. The modes alternate being even or odd with the lowest mode having no nodes. (c) The curves $\omega_n(E_c)$ are never tangent to the line $\omega=0$. All these assumptions are familiar properties of typical eigenvalue problems and all can be proved from the dynamical pulsation equation.

With these assumptions we can analyse the stability of a sequence of equilibrium configurations from the M vs R curve. We start at small central densities where the configurations are assumed stable; $\omega_n^2>0$ for all n. As we increase the central density we ask, "When can a mode change stability"? A mode changes stability at the central density where ω^2 passes through zero and becomes negative. In other words, changes in stability take place where there is a zero frequency mode. A zero frequency mode however, is an infinitesimal *time independent* displacement. That is, it is a displacement which carries one equilibrium configuration into another. As the pulsations being described here generally conserve the total mass, it does this without changing the mass of the configuration. A mode changes stability, therefore, where we can make a small spherical deformation of one equilibrium configuration into another without changing the total mass. In other words, modes change stability at the extrema of the M versus R curve.

At a given extremum we can determine whether it is an even or odd mode which is changing stability. Consider the zero frequency mode $\xi(r)$, at an extremum which carries us along the curve of M vs. R in the direction of increasing central density. (The argument can equally well be made if we move in the direction of decreasing central density.) At the center $\xi(0)$ must vanish since the center of the star is not

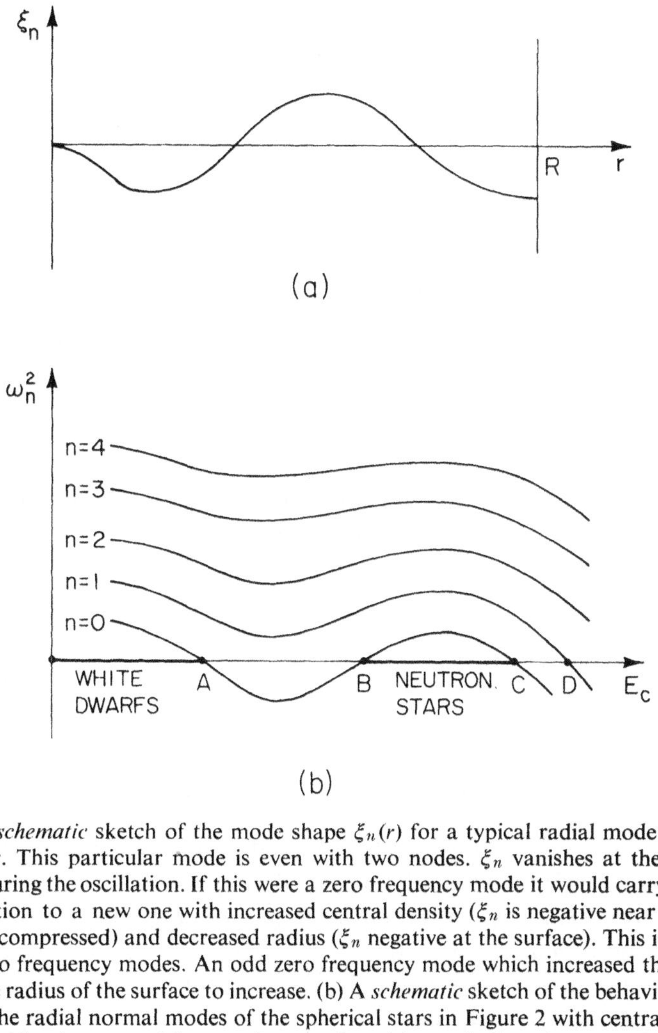

Fig. 3. (a) A *schematic* sketch of the mode shape $\xi_n(r)$ for a typical radial mode of oscillation of a spherical star. This particular mode is even with two nodes. ξ_n vanishes at the center which is not displaced during the oscillation. If this were a zero frequency mode it would carry the old equilibrium configuration to a new one with increased central density (ξ_n is negative near the center so the matter is being compressed) and decreased radius (ξ_n negative at the surface). This is a general property of even zero frequency modes. An odd zero frequency mode which increased the central density would cause the radius of the surface to increase. (b) A *schematic* sketch of the behavior of the squared frequencies of the radial normal modes of the spherical stars in Figure 2 with central density E_c. For low central densities the squared frequencies are all positive and all modes are stable. At the central density corresponding to the first maximum. (A) of the M vs R curve the lowest (even) mode becomes unstable. It becomes stable again at the first minimum (B) and finally unstable at the second maximum (C). The first odd mode becomes unstable at the second minimum (D). The two stable regimes where all the squared frequencies are positive lie between zero density and A and between B and C.

displaced by the pulsation. Near the center, ξ must be negative (compression of central region) so that the displacement carries us to a higher central density. At the surface a positive value of ξ means that the radius is increased during the displacement, while a negative value means that the radius decreases. It is then not difficult to convince oneself (see Figure 3a) of the following: If R increases as one moves through an extremum of the M vs R curve in the direction of increasing central density then it is an *odd* mode which is changing stability. If R decreases then it is an even mode which is changing stability.

With these tools we can now read off the regions of stability of any sequence of spherical equilibrium configurations. Let us take Figure 2 as an example. The *schematic* behavior of the frequencies as a function of central density for this example is shown in Figure 3b.

At low densities the stars have been assumed to be stable and all the ω_n^2 are positive. At the first (moving with increasing E_c) maximum (A) the lowest mode must change stability since the frequencies cannot cross. This mode is even and R is decreasing as required by the above analysis. At the first minimum (B) the radius is decreasing and so it must be an even mode which is changing stability. Since the modes do not cross it must be the lowest mode which is again becoming stable. At the next maximum (C) it again becomes unstable. Thereafter if the curve behaves as shown it is not difficult to show that there are an increasing number of unstable modes (see Figure 3b). For this family of equilibrium configurations there are then only two stable regimes of central density: The lower density stable regime where degenerate electrons provide the dominant source of pressure are the white dwarf stars. The higher density stable regime where degenerate neutrons and strong interaction forces provide the dominant source of pressure are the neutron and hyperon stars.

Thus by (1) computing the equation of state for matter in its ground state, (2) solving the equations of hydrostatic equilibrium for the possible equilibrium configurations composed of this matter, and (3) testing these equilibrium configurations for stability we have found the possible equilibrium end states of stellar evolution in the case of spherical symmetry. They are the white dwarf and neutron stars.

3. Spherically Symmetric Gravitational Collapse

If the endstate of a star's evolution is not an equilibrium configuration then it can only be a state of perpetual gravitational collapse. Because of the maximum non-rotating neutron-white dwarf star mass, it would seem that systems with low angular momentum which are more massive than this limit, might, if they were unable to eject a sufficient amount of their mass in the course of their evolution, wind up in such a state. The properties of this third possible endstate of stellar evolution are the subject of this section and the next.

To begin our study of gravitational collapse we will consider the case of a spherically symmetric collapsing object. This has the great advantage that the problem actually can be solved analytically for some interesting cases and the important features of the collapse can be brought out explicitly. The great simplicity of the spherically symmetric collapse is due to the symmetry which guarantees that there will be no gravitational radiation and that the geometry outside the collapsing star will be the familiar Schwarzschild geometry characterized by the constant total mass of the system. While spherically symmetric gravitational collapse is clearly a very idealized physical situation, many relativists would hope that its essential features are common to more realistic non-spherical gravitational collapse.

3.1. A REVIEW OF KRUSKAL COORDINATES

Outside any spherically symmetric collapsing star of total mass M the metric in terms of familiar Schwarzschild coordinates is

$$\mathrm{d}s^2 = -\left(1 - \frac{2M}{r}\right)\mathrm{d}t^2 + \left(1 - \frac{2M}{r}\right)^{-1}\mathrm{d}r^2 + r^2\,\mathrm{d}\Omega^2. \tag{3.1}$$

This will be valid for $r > R(\tau)$ where $R(\tau)$ is the world line of the surface of the star. From the point of view of relativistic gravity the most interesting era of the collapse is when R is near to and smaller than $2M$. At $r=2M$ the Schwarzschild coordinates are singular and it is appropriate to look at the collapse in a non-singular coordinate system. The Kruskal coordinates are the most useful for this purpose. They are related to Schwarzschild coordinates by the transformation

$$\left.\begin{aligned} u &= (r/2M - 1)^{1/2}\, e^{r/4M} \cosh(t/4M) \quad &(3.2\text{a})\\ v &= (r/2M - 1)^{1/2}\, e^{r/4M} \sinh(t/4M) \quad &(3.2\text{b})\end{aligned}\right\} \quad \text{for} \quad r > 2M$$

and

$$\left.\begin{aligned} u &= (1 - r/2M)^{1/2}\, e^{r/4M} \sinh(t/4M) \quad &(3.3\text{a})\\ v &= (1 - r/2M)^{1/2}\, e^{r/4M} \cosh(t/4M) \quad &(3.3\text{b})\end{aligned}\right\} \quad \text{for} \quad r < 2M.$$

In Kruskal coordinates the metric takes the form

$$\mathrm{d}s^2 = (32M^3 e^{-r/2M}/r)\,(-\,\mathrm{d}v^2 + \mathrm{d}u^2) + r^2\,\mathrm{d}\Omega^2, \tag{3.4}$$

where r is to be regarded as a function of u and v determined by Equations (3.2) and (3.3). This metric is manifestly non-singular at $r=2M$. There is still a singularity at $r=0$ but this is a real singularity where the tidal gravitational forces (Riemann tensor) are infinite.

Aside from not being singular at $r=2M$, Kruskal coordinates have two other important advantages: Unlike t, v is always a time and radial light rays move on lines $u = \pm v + \text{const}$. These two properties make it easy to study the properties of radially moving test particles and light rays.

In Figure 4 the lines of constant Schwarzschild coordinates r and t are plotted in terms of Kruskal coordinates u and v. Curves of constant r are the hyperbolae

$$u^2 - v^2 = (r/2M - 1)\, e^{r/2M}, \tag{3.5}$$

while curves of constant t are the radial straight lines

$$\begin{aligned} v/u &= \tanh(t/4M), \quad r > 2M,\\ u/v &= \tanh(t/4M), \quad r < 2M \end{aligned} \tag{3.6}$$

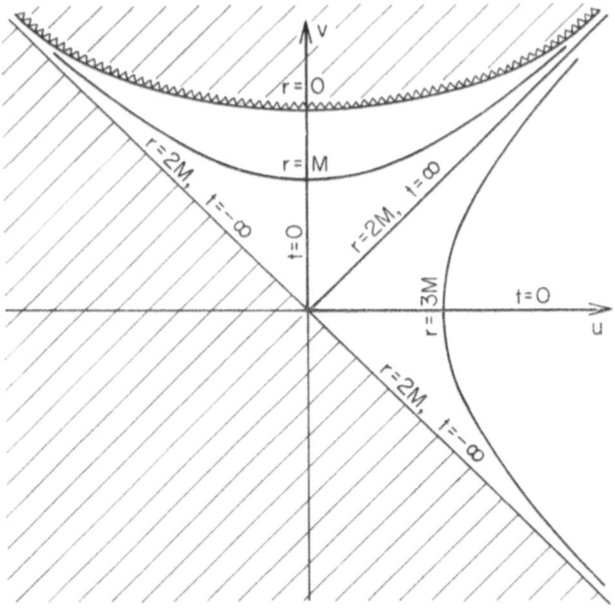

Fig. 4. Curves of constant Schwarzschild coordinates r and t plotted in a Kruskal diagram as a function of u and v. The region covered by the Schwarzschild coordinates $0<r<\infty$, $-\infty<t<\infty$ lies above the line $u = -v$ and is unshaded.

In Kruskal coordinates, the region covered by the original Schwarzschild coordinates lies to the upper-right of the line $v=-u$, corresponding to $r=2M$, $t=-\infty$. There is no physical reason why this line should be a boundary of space-time. The geometry is regular on the line. Furthermore, as can be seen directly from Equation (3.4) the proper time measured by an observer who travels from a point on this line to any finite time is finite.

A space like this, whose geodesics cannot be extended to infinite values of their affine parameters without reaching a non-singular boundary, is called geodesically incomplete. The Schwarzschild geometry can be made into a geodesically complete space simply by adding the reflection of Figure 4 through the origin. This gives Figure 5. To the lower left of $u=-v$ the metric is given by Equation (3.4) with $r(u, v)$ still defined by Equation (3.5). Clearly, any radial geodesic when extended will either hit infinity or the singularity at $r=0$.

This completion sometimes is called the maximal analytic extension of the Schwarzschild geometry. While it is not very important for physics – the added half of the space never arises in a realistic collapse – it is useful as an example that allowable coordinate ranges are to be determined by physical and geometrical requirements and not by arbitrary restrictions such as $u>-v$.

3.2. The Exterior Geometry of a Collapsing Star

The world line of the surface of a collapsing spherical star is a curve in the Kruskal

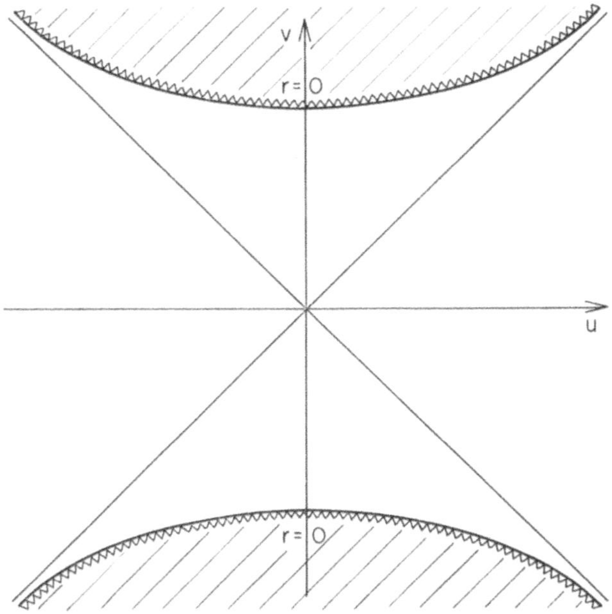

Fig. 5. The maximal analytic extension of the Schwarzschild geometry is the unshaded region which lies between the two $r = 0$ hyperbolae in the above Kruskal diagram.

diagram which starts at a large value of r at $t = -\infty$ and proceeds to $r = 0$, always with slope greater than 1. (Particles on the surface must have timelike world lines.) A typical example is shown in Figure 6. Outside the star the geometry is the Schwarzschild geometry characterized by the constant total mass of the star M.

Let us consider the history of two observers: One who rides down on the surface of the collapsing star and another who remains at a large fixed radius r_0. Let us suppose the falling observer communicates with the fixed observer by sending out radial light rays at fixed intervals according to a clock he carries with him. How does the distant observer receive these signals? The following properties are easily read off the Kruskal diagram in Figure 6.

(1) The light signals which are emitted at equal proper time intervals by the falling observer will be received at increasingly longer proper time intervals by the fixed observer. The signal emitted as the surface crosses $r = 2M$ will never reach the distant observer until $t = +\infty$. Since the fixed observer also sees the frequency of arrival of the crests of light waves to decrease, he will see the frequency of the light as decreasing. The light from the star will be redshifted and the redshift of the light emitted as the radius $r = 2M$ is crossed will be infinite. Since each photon he receives is losing energy and, since he is receiving photons less and less frequently, the luminosity of radiation in these radial signals will go to zero as $t \to \infty$.

More quantitatively, the redshift of photons emitted at radius R can be found by an analysis similar to that used in deriving Equation (2.21). Suppose that u^μ is the four-velocity of a point on the star's surface. If we denote the proper radial velocity

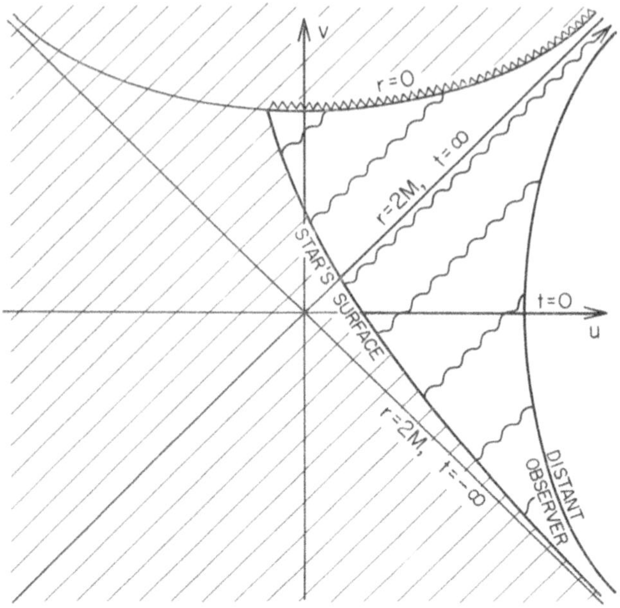

Fig. 6. The Kruskal diagram for the exterior geometry of a spherically symmetric collapsing star. The geometry outside the star is the Schwarzschild geometry shown here. Inside the geometry is not the Schwarzschild geometry and therefore only the unshaded portion of this Kruskal diagram is of physical significance. The world line of the surface of the collapsing star is a timelike line (slope greater than unity) which starts at $t=-\infty$ at some large value of r, proceeds through $r=2M$ and ends in a singularity of zero radius, infinite density and infinite tidal gravitational forces. The world line of a distant observer who remains at fixed r is also shown. Light signals emitted at equal intervals by an observer on the surface of the collapsing star appear as wavy 45° lines. Light signals emitted after the surface has crossed $r=2M$ fall into the singularity at $r=0$ and never reach the distant observer. Light signals emitted before $r=2M$ do reach the distant observer but at increasingly longer intervals i.e. they are red-shifted and the red-shift is infinite for the light emitted at $r=2M$. Once the star has crossed the radius $r=2M$ there is no escape from collapse to a singularity since there is no timelike line for the surface to follow which does not lead to $r=0$.

of the surface u^r by $\dot{R}$ then the redshift of the light emitted radially at radius R is

$$z+1=\left[\left(1-\frac{2M}{R}\right)^{-1}+\dot{R}^2\left(1-\frac{2M}{R}\right)^{-2}\right]^{1/2}-\dot{R}\left(1-\frac{2M}{R}\right)^{-1}. \tag{3.7}$$

These outgoing radial light rays move along curves $t(r)$ which satisfy $ds^2=0$ or in the Schwarzschild geometry

$$-\left(1-\frac{2M}{r}\right)\left(\frac{dt}{dr}\right)^2+\left(1-\frac{2M}{r}\right)^{-1}=0. \tag{3.8}$$

The light ray which is emitted at radius R at time T then satisfies

$$t-T=r-R+2M\log\left(\frac{r-2M}{R-2M}\right). \tag{3.9}$$

The time T is the time when the trajectory of the surface of the star crosses the radius R. If R_1 is the radius of the surface at time T_1, then T is fixed by

$$T - T_1 = \int_{R_1}^{R} \mathrm{d}r \left(\frac{u^t}{\dot{R}}\right). \tag{3.10}$$

The light rays which arrive at the distant observer at time t are emitted at values of R close to $2M$ and as a consequence of Equation (3.10) at late times T. Putting these two behaviors together the time t at which the ray emitted at an R close to $2M$ arrives at the distant observer behaves as

$$t/4M \sim -\log(R - 2M). \tag{3.11}$$

The redshift as seen by a distant observer [Equation (3.7)] then becomes infinite as

$$z \propto \exp(t/4M). \tag{3.12}$$

It is more difficult to calculate the behavior of the total luminosity of the collapsing star. At late times the bulk of the luminosity comes not from the radial light rays but from light rays which were emitted nearly tangentially at $r \approx 3M$ and have spiraled slowly out from the star. Ames and Thorne (1968) have calculated the time dependence of the luminosity L at late times and find

$$L \propto \exp\left(-\frac{2}{3\sqrt{3}}\frac{t}{2M}\right). \tag{3.13}$$

The important point of each of these results is that the time scale for the extinction of the luminosity and the approach to infinite redshift is of order of magnitude M. In usual units

$$(\text{Characteristic time for decay of luminosity}) \sim 10^{-5}\left(\frac{M}{M_\odot}\right)\text{s}. \tag{3.14}$$

Thus, from the point of view of any exterior observer, a star in the late stage of collapse very rapidly approaches a totally black object.

(2) Another property of spherical collapse which can be read off the Kruskal diagram is that an exterior observer never sees the surface of the star cross the Schwarzschild radius $r = 2M$. He would only receive the signals emitted at $r = 2M$ at $t = +\infty$. Signals which are emitted after the radius $r = 2M$ is crossed never reach the observer.

Thus, in a time scale given approximately by Equation (3.14), the distant observer sees the collapsing object approach a black quasi-equilibrium object whose geometry for $r > 2M$ is indistinguishable from the exact static vacuum Schwarzschild solution. The collapsing star is said to have become a black hole. This third endstate of stellar evolution – in the case of spherical symmetry – can be regarded for all practical purposes as another equilibrium object and its observable properties can be deduced from the properties of the Schwarzschild solution.

(3) What happens to the observer who follows the surface of the collapsing star

down? In Kruskal coordinates there is a finite coordinate separation between the observer's position at the start of the collapse and when the radius $r=2M$ is crossed. Since the metric is non-singular in Kruskal coordinates, the surface $r=2M$ must be crossed in a finite amount of *proper* time after the collapse starts.

Once across $r=2M$, the light signals sent out by the observer on the star will never reach infinity. The gravitational field is so strong that eventually they will fall back into the singularity at $r=0$. Another way of putting this is to say that the spheres of radius $r<2M$ are trapped surfaces. A trapped surface is any closed space-like two-surface such that if the surface emits a pulse of light in the inward and outward directions, then the area of the wavefronts of *both* of these pulses decreases. The wavefront of the inward going light rays would be expected to decrease in area, and it is clear from the Kruskal diagram that if $r<2M$ the spherical wavefront of the outward moving light rays also proceeds to smaller and smaller radii and hence to smaller and smaller area. There is no way in which the falling observer, once inside $r=2M$, can ever communicate with anyone outside this radius. Any light ray or test particle sent out will end up in the singularity.

But it is not only the signals which cannot get out; the star itself can never re-expand through the radius $r=2M$. It cannot do so because its surface must follow a time-like world line in the Schwarzschild geometry and no time-like world lines starting from $r<2M$ ever cross back over this radius. They inevitably lead to the singularity at $r=0$. Furthermore, they arrive at $r=0$ in a finite amount of proper time since

$$\tau=-\int_r^0 \mathrm{d}r\left\{\frac{32M^3}{r}e^{-r/2M}\left[\left(\frac{\mathrm{d}u}{\mathrm{d}r}\right)^2-\left(\frac{\mathrm{d}v}{\mathrm{d}r}\right)^2\right]\right\}^{1/2} \tag{3.15}$$

is always integrable and finite.

Thus, once the radius $r=2M$ is crossed, the star inevitably collapses to zero radius and infinite density, and a singularity is formed in the geometry. This is an example of the inevitability of gravitational collapse.

The surface $r=2M$ which plays such an important role in the collapse is an example of an *event horizon*. It is the boundary of all events which can be connected by light rays to future infinity. It separates the space into two parts, one part which can communicate with infinity and another part which cannot. The distinguishing feature of a black hole is that, as here, all singularities lie inside an event horizon.

3.3. The Interior Geometry of a Collapsing Ball of Dust

To complete the picture of spherically symmetric gravitational collapse the evolution of the geometry in the interior of a star must be discussed. There is no unique interior geometry. It depends on the properties of the matter and the physical processes which take place inside the collapsing star. A simple exactly solvable example, however, will give us a good deal of insight into the general case. The example is the collapse of an initially momentarily static uniform density sphere of pressureless dust.

To solve the problem of collapsing dust we will first write down the formalism for spherically symmetric gravitational collapse, including pressure, following Misner and Sharp (1964). Their formulation of the equations makes their interpretation much more transparent.

It is convenient to follow the collapse in a co-moving coordinate system in which the spatial velocities u^i are zero. The spatial coordinates of a given particle thus remain the same throughout the collapse. One way of doing this is to use for a radial coordinate the number of baryons interior to a sphere of that coordinate radius, a. If we ride on a given particle, this number never changes due to the baryon number conservation law.

In co-moving coordinates, the spherically symmetric metric can be written

$$ds^2 = -e^{2\Phi(a,t)}\,dt^2 + e^{\lambda(a,t)}\,da^2 + r^2(a,t)\,d\Omega^2 . \tag{3.16}$$

The stress energy for a perfect fluid is

$$T_t^t = -E, \quad T_t^i = 0, \quad T_j^i = P\delta_j^i . \tag{3.17}$$

All one has then to do is write down Einstein's equations and solve them.

Writing out the equation $T^{\nu}_{i;\nu}=0$, one finds it can be immediately integrated to give

$$\Phi(a,t) = -\log\left(\frac{E+P}{N}\right) + \text{(arbitrary function of time)}. \tag{3.18}$$

The arbitrary function can always be eliminated by an appropriate redefinition of the time coordinate.

Misner and Sharp find that it is convenient to introduce a function $M(a,t)$ *defined* by

$$e^{\lambda(a,t)} = \left[1 + \dot{r}^2 - 2M/r\right]^{-1} (\partial r/\partial a)^2 , \tag{3.19}$$

where here and in the following for any function f

$$\begin{aligned} \dot{f} &= e^{-\Phi}(\partial f/\partial t)_{a=\text{const.}} \\ &= \begin{pmatrix}\text{derivative along world line of the particle at} \\ a \text{ with respect to its proper time}\end{pmatrix}. \end{aligned} \tag{3.20}$$

The equations of spherically symmetric gravitational collapse and their physical interpretations are given in Table III. In all of these equations $\partial/\partial r$ is defined by

$$\frac{\partial f}{\partial r} = \left(\frac{\partial r}{\partial a}\right)_t^{-1}\left(\frac{\partial f}{\partial a}\right)_t . \tag{3.21}$$

The interpretations given in Table III result essentially from the Newtonian limit of these equations and are for purposes of intuition only. The use of words such as 'pressure force', and 'mass interior to a given radius' does not mean that these concepts can be precisely defined.

Equations I and II in Table III are dynamical equations. Given M, r, E and P at one instant of time, they show how to find M and r at the next instant of time. From

TABLE III

Equations of spherically symmetric collapse

	Equation	Physical interpretation
I	$\dot{M} = -4\pi r^2 P \dot{r}$	$\begin{pmatrix}\text{rate of change of}\\ \text{mass inside a}\\ \text{sphere of radius } r\end{pmatrix} = \begin{pmatrix}\text{rate at which work}\\ \text{is done on sphere}\\ \text{by pressure}\end{pmatrix}$
II	$\ddot{r} = -\left[\dfrac{1+\dot{r}^2-2M/r}{E+P}\right]\left(\dfrac{\partial P}{\partial r}\right) - \dfrac{M+4\pi r^3 P}{r^2}$	$(\text{acceleration}) = \begin{pmatrix}\text{pressure force}\\ \text{per unit mass}\end{pmatrix} + \begin{pmatrix}\text{gravitational force}\\ \text{per unit mass}\end{pmatrix}$
III	$\dfrac{\partial M}{\partial r} = 4\pi r^2 E$	$\begin{pmatrix}\text{mass in a shell}\\ \text{of radius } r\\ \text{thickness } \partial r\end{pmatrix} = \begin{pmatrix}\text{energy}\\ \text{density}\end{pmatrix} \times \begin{pmatrix}\text{volume}\\ \text{factor}\end{pmatrix}$

this, Equation III and from the equation of state, E and P at this later time can be determined and the integration continued.

In the case of dust ($P = 0$), these equations reduce to

$$\dot{M} = 0 \tag{3.22a}$$
$$\ddot{r} = -M/r^2 \tag{3.22b}$$
$$\partial M/\partial r = 4\pi r^2 E. \tag{3.22c}$$

These equations are to be solved for the evolution of a ball of dust which initially is momentarily static and has uniform density E_0 and total mass M. The solution can be obtained by making the guess that the initially uniform density remains uniform for all times: $E = E(t)$. The solution is then conveniently expressed in terms of a length R_0 which sets the initial density

$$E_0 = 3/8\pi R_0^2, \tag{3.23a}$$

a scale parameter $R(t)$ which controls how it evolves

$$E(t) = E_0 (R_0/R(t))^3 = 3R_0/8\pi R^3, \tag{3.23b}$$

and an arbitrary increasing function $r_0(a)$ which relates the initial value of the coordinate $r(a, t)$ to the comoving coordinate a through

$$r(a, 0) = r_0(a)\, R_0. \tag{3.23c}$$

(The function $r_0(a)$ could be fixed, for example, by requiring that a be the number of baryons interior to a sphere of that coordinate radius.) It is then not too difficult to verify that the solution to Equation (3.22) corresponding to these initial conditions is (3.23) and

$$M(a, t) = (4/3)\, \pi r_0^3(a)\, R_0^3 E_0, \tag{3.24a}$$
$$r(a, t) = r_0(a)\, R(t). \tag{3.24b}$$
$$\Phi(a, t) = 0. \tag{3.24c}$$

where $R(t)$ is determined parametrically by

$$R(\eta) = (R_0/2)(1 + \cos\eta), \tag{3.25a}$$
$$t(\eta) = (R_0/2)(\eta + \sin\eta). \tag{3.25b}$$

With this solution the metric inside becomes

$$\mathrm{d}s^2 = -\,\mathrm{d}t^2 + R^2(t)\left[\frac{\mathrm{d}r_0^2}{1 - r_0^2} + r_0^2\,\mathrm{d}\Omega^2\right]. \tag{3.26}$$

Equation (3.26) is the same geometry as inside a closed, homogeneous and isotropic universe. Equation (3.25) is just the equation for the evolution of the scale factor in such a universe. This coincidence is a consequence of the fact that the pressure is zero and of the analog of the Newtonian theorem that the gravitational field in a spherically symmetric situation depends only on the matter located inside the radius under consideration.

The geometry inside the dust cloud is part of the surface of a sphere in four dimensions like that of a closed Friedmann cosmology. Writing $r_0 = \sin\chi$, the metric becomes

$$\mathrm{d}s^2 = -\,\mathrm{d}t^2 + R^2(t)\left[\mathrm{d}\chi^2 + \sin^2\chi\,\mathrm{d}\Omega^2\right], \tag{3.27a}$$
$$= R^2(\eta)\left[-\,\mathrm{d}\eta^2 + \mathrm{d}\chi^2 + \sin^2\chi\,\mathrm{d}\Omega^2\right]. \tag{3.27b}$$

Of course, the range of χ is not 0 to π as in the cosmological case but 0 to χ_s where χ_s is the coordinate of the surface of the collapsing ball. It has to be checked that the interior solution obtained here can be matched smoothly onto the exterior Schwarzschild geometry. This is not completely trivial since we have expressed them in different coordinate systems. The solutions, however, can be matched correctly (see e.g. Misner *et al.*, 1972) and the equation of the surface in Schwarzschild coordinates is then

$$r_s(t) = R(t)\sin\chi_s. \tag{3.28}$$

The total mass is $M = (4/3)\pi \sin^3\chi_s R_0^3 E_0$, and it is easily verified that Equation (3.28) is the equation of a test particle geodesic in the Schwarzschild geometry of this mass as one would expect in this pressureless case. Equation (3.28) shows that there are actually two interior solutions for a given value of initial radius and density; one corresponding to $0 < \chi_s < \pi/2$ and the other to $\pi/2 < \chi_s < \pi$. The one corresponding to $\chi_s > \pi/2$ is, however, singular (see Leibovitz and Israel, 1970) and we will consider only $\chi_s < \pi/2$.

A diagram analogous to the Kruskal diagram can be constructed to represent the geometry in the interior of a collapsing ball by plotting the radial coordinate χ and the time coordinate η. This is shown in Figure 7. Also plotted there is the standard Kruskal geometry for the exterior solution. The solutions join along the line representing the surface $SA_2H_2B_2F$. The singularity of infinite density and zero radius occurs at $\eta = \pi$, a finite proper time from the start of the collapse for an observer who rides with the star.

In both diagrams radial light rays move along 45° lines. Three typical light rays which start from the center of the collapsing star have been drawn. Ray A eventually gets out to infinity. Ray B gets out of the star, but only after the surface has crossed

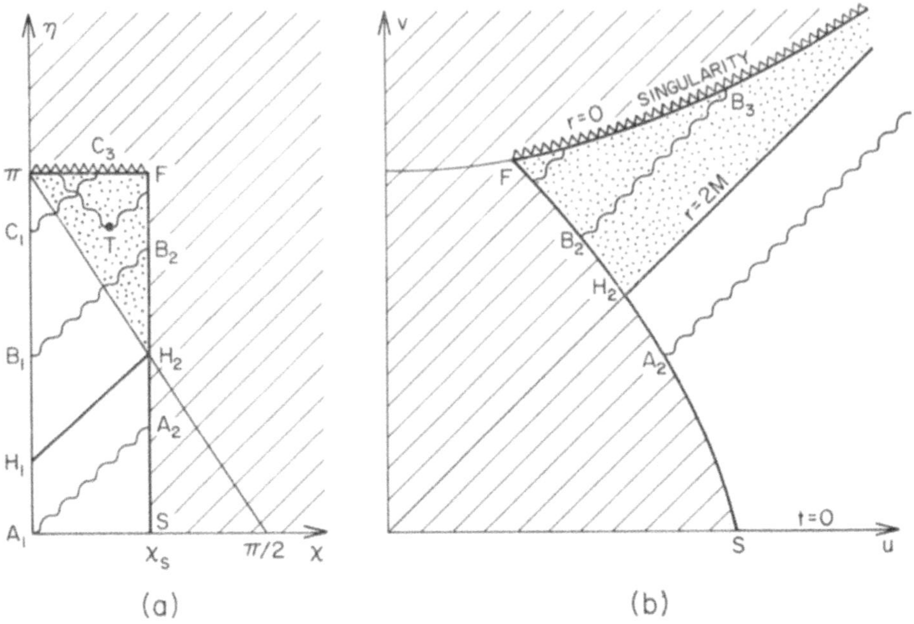

Fig. 7. Interior and exterior geometry of a collapsing sphere of dust. A diagram showing the interior geometry of a collapsing sphere of dust which initially ($t=\eta=0$) was momentarily static and had uniform density is plotted in (a). Also shown is the Kruskal diagram for the exterior geometry in (b). Only the unshaded region of both diagrams has any physical significance. The interior and exterior geometries join along the line $SA_2H_2B_2F$ which represents the surface of the collapsing star. In both diagrams points correspond to spheres and radial light rays move along 45° lines. Three typical light rays A, B, and C starting from the center of the star are indicated. The ray A gets out of the star and out to infinity. B gets out of the star, but only after it has crossed the radius $r=2M$, and so eventually falls into the singularity at $r=0$. C never gets out of the star before the singularity is formed. The horizon (H) is the boundary of those events which can be connected with future infinity by a light ray. It is formed in the interior of the star at H_1, expands outward until it crosses the surface of the star as the surface is crossing the radius $r=2M$, and thereafter remains at $r=2M$. The stippled region indicates the domain of trapped spheres. If a typical trapped sphere (T) emits a light pulse in both the inward and outward directions then the area of *both* wavefronts decreases.

$r=2M$, so that it eventually falls back into the singularity at $r=0$. Ray C never gets out of the star before it has collapsed to a singularity around it.

The event horizon is the null surface generated by those radial light rays which neither get to infinity nor fall back into the singularity. Outside the star it is the surface $r=2M$. Inside, it is the surface formed by those radial light rays which just reach the surface as the surface crosses the radius $r=2M$. This occurs when

$$\tfrac{1}{2} R_0 \sin\chi_s (1+\cos\eta) = 2M = R_0 \sin^3\chi_s, \tag{3.29}$$

or when

$$\eta = \pi - 2\chi_s. \tag{3.30}$$

The equation of the horizon inside the star is then

$$\eta = \chi + \pi - 3\chi_s, \tag{3.31}$$

and the horizon starts to form at a coordinate time

$$\eta = \pi - 3\chi_s, \tag{3.32}$$

from the start of the collapse at $\eta = 0$. The area of the two-surface formed by the intersection of the event horizon and an $\eta = \text{const.}$ surface is [from Equation (3.31)]

$$\begin{aligned} A &= 4\pi R^2(\eta)\sin^2\chi(\eta) \\ &= \pi R_0^2[(1 + \cos\eta)\sin(\eta + 3\chi_s)]^2 . \end{aligned}$$

This area monotonically increases from when the horizon is formed at $\eta = \pi - 3\chi_s$ to when it passes outside the star at $\eta = \pi - 2\chi_s$. Once outside, the area of the horizon is a constant

$$A = 16\pi M^2 .$$

Later, we will see that event horizons in general have non-decreasing areas.

The trapped spheres on the inside of the star are those spheres from which the outgoing light rays start to converge, i.e. those for which

$$\frac{\mathrm{d}}{\mathrm{d}\lambda}[R(\eta + \lambda)\sin(\chi + \lambda)]^2|_{\lambda=0} < 0 . \tag{3.33}$$

It is not too difficult to verify that the inner boundary of these trapped spheres is the line

$$\eta = \pi - 2\chi . \tag{3.34}$$

The region in which the trapped surfaces lie is stippled in Figure 7. This shows that the region interior to the event horizon is not coextensive with the region of trapped surfaces but is larger. A trapped surface is determined by the local geometry, by whether light rays converge or diverge when they are emitted from a closed surface. The event horizon is a global property, however, since it is determined by which light rays can get out to infinity.

3.4. Summary of Spherical Collapse

The basic features of spherically-symmetric gravitational collapse can now be summarized. [Another type of diagram which displays these features is shown in Figure 8.]

(1) Initiation of the collapse when the object is well outside its Schwarzschild radius.

(2) The formation of an event horizon inside the star which expands outward with increasing area until it just contains the star at the point when the star's surface crosses the radius $r = 2M$ after which it remains stationary at $r = 2M$.

(3) The formation of a domain of trapped surfaces after the star has collapsed through radius $r = 2M$.

(4) The inevitable collapse to a singularity of infinite density and infinite curvature in a finite proper time as measured by an observer who travels with the star.

(5) The rapid approach of the geometry outside $r = 2M$ to the static, spherically-

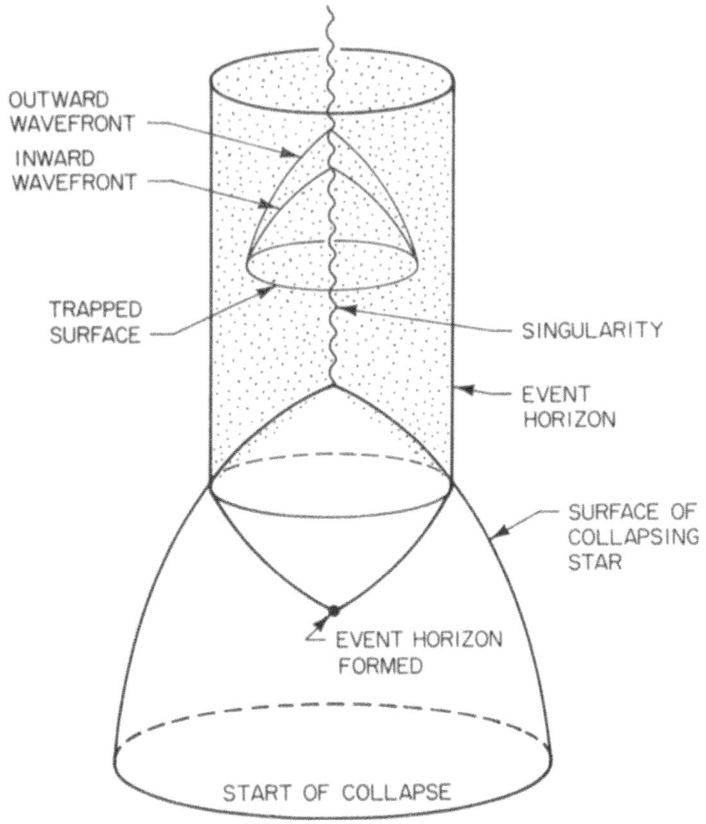

Fig. 8. The essential features of spherically symmetric gravitational collapse. In this schematic space-time diagram a time direction runs vertically and two space directions are portrayed horizontally in perspective. The surface of the collapsing star is shown proceeding to smaller and smaller area and eventually to zero area where a singularity is formed. The event horizon is formed in the interior of the star and expands outward to a radius $r = 2M$ after which it remains stationary. After the stars' surface has passed through $r = 2M$ a domain of trapped spheres (shown stippled) is formed. A typical trapped sphere is shown together with its inward and outward directed radial light rays.

symmetric vacuum solution of Einstein's equations, as a consequence of the formation of an event horizon; that is the formation of a black hole.

4. Non-Spherical Gravitational Collapse and Black Holes

The outstanding issue today in the theory of gravitational collapse is whether the distinctive features of spherical gravitational collapse are common to more general non-spherical gravitational collapse. In particular, the crucial question is whether every gravitational collapse leads to the formation of an event horizon which hides the eventually formed singularity, or whether a naked singularity unshielded by an event horizon sometimes can evolve. No counter-example is known to the conjecture that an event horizon forms in a general gravitational collapse in three space dimenisons, but neither is there a demonstration that an event horizon is inevitable. (For two-dimensional collapse, however, counter-examples are known, see Thorne, 1972.)

The issue is important for astrophysical applications because if an event-horizon is formed, then gravitational collapse always leads either to a degenerate star or to a black hole. In the case of a black hole it leads to a geometry which, from the point of view of exterior observers (us), is indistinguishable from a *stationary* solution of the vacuum Einstein's equations with event horizon. Thus if an event horizon is always formed, gravitational collapse would always end in an object which for all practical purposes is stationary. Furthermore, if a black hole is formed, nature is very restrictive about the type of black hole it can be. It essentially can be only a Kerr black hole characterized by two parameters, the total mass and the total angular momentum. If the conjecture of the inevitability of an event horizon is correct, we are led from a complicated dynamical gravitational collapse requiring the specification of an infinite number of parameters to a simple quasi-equilibrium object with a small number of parameters. We may hope that this simple final state yields a simple observable signature.

In this section we will present some of the evidence for the inevitability of the event horizon and will discuss the types of black holes which can arise if this conjecture is correct. The treatment will necessarily be incomplete because the basic problem is still unsolved and because the theorems on the uniqueness of black holes require more time to prove than we have here.

4.1. Non-spherical scalar field perturbations of spherically symmetric gravitational collapse

The most obvious way to test the conjecture about the inevitability of the formation of the event-horizon is to study small perturbations away from spherical collapse. (By small perturbations we mean those whose energy is too small to significantly perturb the gravitational field.) Such perturbations then can be considered as test fields propagating in the background geometry of a spherically-symmetric gravitational collapse. The goal of such an analysis is to determine whether the property that an event horizon always forms in spherical collapse remains stable under small perturbations away from spherical symmetry. If it does, this is one small piece of evidence in favor of the conjecture that event horizons are inevitable in a general collapse.

The problem of the linear perturbations away from spherically symmetric gravitational collapse essentially has been solved by Price (1972a, b). He solved the problem for perturbing fields of any integer spin; the most interesting case being the spin-2 gravitational perturbations corresponding to small lumps and bumps on the collapsing star. The treatment of this case, however, does not differ in any essential way from the treatment of *scalar* spin-0 perturbations. Since the scalar perturbations are somewhat simpler kinematically, we will treat these first in order to convey the essential ideas and will comment later on the extensions to the gravitational and other spin cases.

(i) *Equations for Scalar Perturbations*

Let us suppose that there is a scalar field Φ coupled to a scalar charge density J associated with matter. The field equation which Φ satisfies we take to be*

$$\Box^2\Phi = \Phi_{;\alpha}{}^{;\alpha} = 4\pi J\,. \tag{4.1a}$$

or

$$(-g)^{-1/2}\left[(-g)^{1/2}g^{\alpha\beta}\Phi_{,\beta}\right]_{,\alpha} = 4\pi J\,. \tag{4.1b}$$

The $g^{\alpha\beta}$ are just the metric components of the background Schwarzschild geometry. As a consequence of the background geometry's spherical symmetry the scalar field can be expanded in spherical harmonics

$$\Phi = \sum_{l=0}^{\infty}\sum_{m=-l}^{+l} r^{-1}\Phi_l(r,t)\,Y_l^m(\theta,\varphi), \tag{4.2}$$

and the linear wave equation [Equation (4.1)] will separate into a radial equation for each l. In writing these equations out, it is not convenient to use the usual Schwarzschild radius r but rather a new coordinate r^* first introduced by Regge and Wheeler (1957) and defined by

$$r^* = r + 2M\log(r/2M - 1)\,. \tag{4.3}$$

In terms of this coordinate the Schwarzschild geometry metric takes the form

$$\mathrm{d}s^2 = \left(1 - \frac{2M}{r}\right)(-\mathrm{d}t^2 + \mathrm{d}r^{*2}) + r^2\,\mathrm{d}\Omega^2\,. \tag{4.4}$$

where $r = r(r^*)$ from Equation (4.3). These coordinates have the advantages that radial light rays travel on 45° lines and that they are regular at spatial infinity (unlike the Kruskal coordinates). The horizon is located at $r^* = -\infty$ in these coordinates which are singular there.

In the Regge-Wheeler coordinates, the wave equation for $\Phi_l(r)$ outside the star becomes

$$\frac{\partial^2\Phi_l}{\partial t^2} - \frac{\partial^2\Phi_l}{\partial r^{*2}} + V_l(r^*)\,\Phi_l(r^*) = 0\,. \tag{4.5}$$

$$V_l(r^*) = \left(1 - \frac{2M}{r}\right)\left(\frac{2M}{r^3} + \frac{l(l+1)}{r^2}\right). \tag{4.6}$$

We now will analyse some of the solutions of these equations.

(ii) *No Black Holes with Weak Scalar Fields*

Imagine a spherical distribution of matter endowed with a weak scalar charge density

* It is sometimes argued that the field equation for a massless spin-zero particle should be conformally invariant in which case, for a scalar field, it would be

$$\Box^2\Phi - \tfrac{1}{6}R\Phi = 4\pi J\,.$$

This will give the same results as Equation (4.1.) for our analysis.

J which is not necessarily spherically symmetrical. Collapse ensues. If a horizon is formed, the solution exterior to the event horizon will approach asymptotically a stationary black hole solution of Einstein's equations with perturbing scalar field. What kind of black hole can it be? To answer this question we simply have to look for suitably regular *stationary* solutions to Equation (4.5). At large values of r^* solutions for Φ behave like

$$r^{-1}\Phi_l \sim r^{*l}, \quad \text{or} \quad r^{*-l-1}, \quad r^* \to \infty . \tag{4.7}$$

Near the event horizon they behave like

$$r^{-1}\Phi_l \sim r^*, \quad \text{or} \quad \sim \text{const.} \quad r^* \to -\infty . \tag{4.8}$$

At large r^* we clearly must have only the decaying solution for Φ if the source of the scalar field is the black hole. The solution on the horizon then *must* be a linear combination of the two possible behaviors in Equation (4.8). That is, Φ must diverge on the horizon. Since Φ is a scalar with direct physical significance this divergence is impossible if we are to have a non-singular event horizon. The only way out is that Φ must vanish identically outside the horizon.

This result is a simple example of a black hole uniqueness theorem. The only black hole with a non-singular spherical event horizon and a weak scalar field is the Schwarzschild black hole with *no* scalar field.

This conclusion was reached on the basis of perturbation theory and therefore applies only to weak scalar fields. Chase (1970), however, has given a general proof, including the energy density in the field, that the only static black hole with a non-singular event horizon and a scalar field is the Schwarzschild black hole with no scalar field outside.

(iii) *The Evolution of the External Scalar Field of a Collapsing Star*

Imagine a spherically symmetric star endowed with some distribution of scalar charge J and, as a consequence, possessing a long range scalar field outside the star. Let the star begin to collapse at $t=0$ from a previously static configuration and follow the evolution of the scalar field in the geometry of the collapsing star. Viewed in any non-singular coordinates there is nothing special about the local geometry as the star crosses the event horizon. The equations for the scalar field and therefore the scalar field itself remain finite and well behaved. The weak scalar field does not impede the formation of the event horizon.

If this picture is to be consistent with our previous result that there are no black holes with weak external scalar field, the scalar field must all be radiated away as the exterior geometry settles down to a static black hole. This includes the scalar monopole. Unlike the case of electromagnetism there is no dynamically conserved flux integral to prevent the monopole from being radiated away.

Richard Price (1972a, b) first clarified the mechanism by which the scalar field is radiated away and calculated the asymptotic time dependence of the decaying field. In the following we will present a brief account of the basic ideas of Price's work employing in part, some methods of other of his expositors (Thorne, 1972; Fackerell, 1971).

In Kruskal coordinates the trajectory of the star's surface as it crosses the event horizon is

$$\begin{aligned} v &= \alpha s + \beta \\ u &= -\gamma s + \beta \end{aligned}, \quad 0 < \gamma/\alpha < 1, \tag{4.9}$$

where s ranges over a small interval about zero. The transformation from Kruskal to Regge-Wheeler coordinates near the horizon is

$$r^* \approx 2M \log[(u^2 - v^2)/4M^2], \tag{4.10a}$$

$$t \approx -2M \log(1 - v/u). \tag{4.10b}$$

The trajectory of the star's surface near the horizon in Regge-Wheeler coordinates is then

$$r^* \approx 2M \log[(\alpha + \gamma) s], \tag{4.11a}$$

$$t \approx -2M \log[(\alpha + \gamma) s]. \tag{4.11b}$$

Thus, in Regge-Wheeler coordinates the trajectory of the star's surface approaches the null line $r^* = -t$ as t approaches infinity (see Figure 9). Of course, the surface does not actually become null; it is only the pathology of the coordinate system which makes it appear so.

We are interested in the behavior of the scalar field at large times and fixed r^* in the region exterior to the star. The field in this region may be taken to be completely determined by the value of the scalar field on the two characteristic null surfaces $\mathscr{C}_1$ and $\mathscr{C}_2$ defined in Figure 9. The field on $\mathscr{C}_2$ is just the static field of the pre-collapse star. The field on $\mathscr{C}_1$ (the ingoing radial light ray which hits the star's surface as it crosses the horizon) is always finite. In fact, on $\mathscr{C}_1$ the field very rapidly approaches its value Φ_0 on the horizon. Using $V = t - r^* + \text{const.}$ as a coordinate along $\mathscr{C}_1$ and transforming from a reasonable behavior in Kruskal coordinates, one has in Regge-Wheeler coordinates

$$(\Phi)_{\text{on } \mathscr{C}_1} = \Phi_0 + 0[\exp(-V/4M)]. \tag{4.12}$$

In the region external to the star, the scalar field can always be written as a superposition of free scalar waves traveling to the left and to the right

$$\Phi_l(r^*, t) = \int_0^\infty d\omega e^{-i\omega t} [A^{(+)}(\omega)\, \Phi_\omega^{(+)}(r^*) + A^{(-)}(\omega)\, \Phi_\omega^{(-)}(r^*)]. \tag{4.13}$$

Here the $\Phi_\omega^{(\pm)}(r^*)$ solve

$$\frac{d^2\Phi_\omega^{(\pm)}}{dr^{*2}} + [\omega^2 - V_l(r^*)]\, \Phi_\omega^{(\pm)}(r^*) = 0 \tag{4.14}$$

for the particular value of l in question. The function $\Phi_\omega^{(+)}$ is defined as the solution corresponding to a unit amount of scalar wave incoming from the left while $\Phi_\omega^{(-)}$

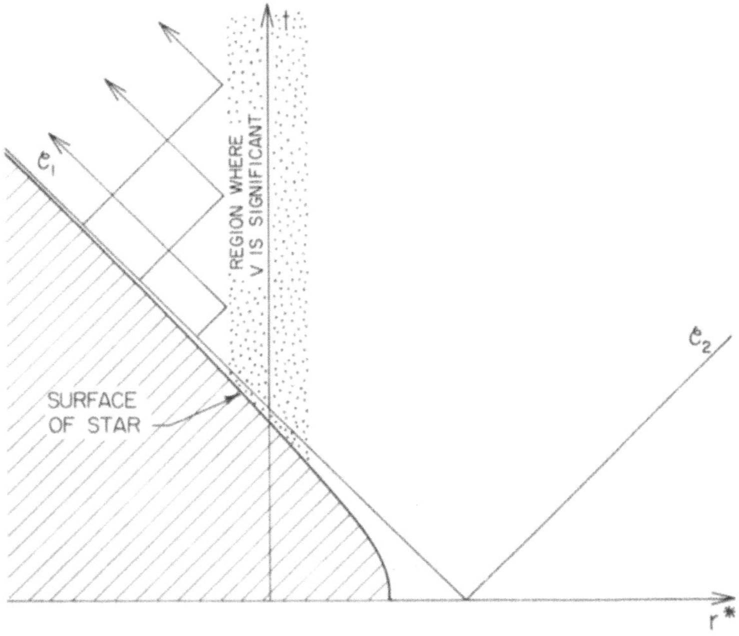

Fig. 9. Spherical collapse with scalar field in Regge-Wheeler coordinates. In these coordinates radial light rays move on 45° lines and the horizon is located at $r^* = -\infty$. The surface of the star appears as a line which becomes tangent to a 45° line as it approaches the horizon. The characteristic surface $\mathscr{C}_1$ is defined as this 45° line and the surface $\mathscr{C}_2$ as the oppositely directed null surface which joins $\mathscr{C}_1$ at $t=0$. The region where the effective potential is important is stippled. At late times and fixed r^* the most important contributions to the scalar field arise from the low frequency data on $\mathscr{C}_1$. This is almost totally reflected by the potential barrier and changed in phase. For values of r^* to the right of the barrier the large t field decays because the low frequency components do not penetrate while for r^* to the left of the barrier it decays because the incident and reflected low frequency components interfere destructively.

corresponds to a unit amount from the right. The coefficients $A^{(\pm)}(\omega)$ are the amplitudes of the waves traveling to the left and right respectively.

The potential, V_l, in which these waves propagate is shown in Figure 10. It dies off at large positive and negative r^*. As a consequence, at large values of $|r^*|$, the $\Phi_\omega^{(\pm)}(r^*)$ have the behaviors

$$\Phi_\omega^{(+)} \to e^{i\omega r^*} + R^{(+)} e^{-i\omega r^*}, \quad r^* \to -\infty \tag{4.15a}$$

$$\Phi_\omega^{(+)} \to T^{(+)} e^{i\omega r^*}, \quad r^* \to +\infty \tag{4.15b}$$

$$\Phi_\omega^{(-)} \to T^{(-)} e^{-i\omega r^*}, \quad r^* \to -\infty \tag{4.15c}$$

$$\Phi_\omega^{(-)} \to e^{-i\omega r^*} + R^{(-)} e^{i\omega r^*}, \quad r^* \to +\infty\,. \tag{4.15d}$$

The quantities $R^{(+)}$, $T^{(+)}$ have the interpretation of reflection and transmission coefficients for a wave incident from the left, and $R^{(-)}$ and $T^{(-)}$ a similar interpretation for a wave incident from the right.

In principle, the functions $A^{(\pm)}(\omega)$ are completely determined by the value of the

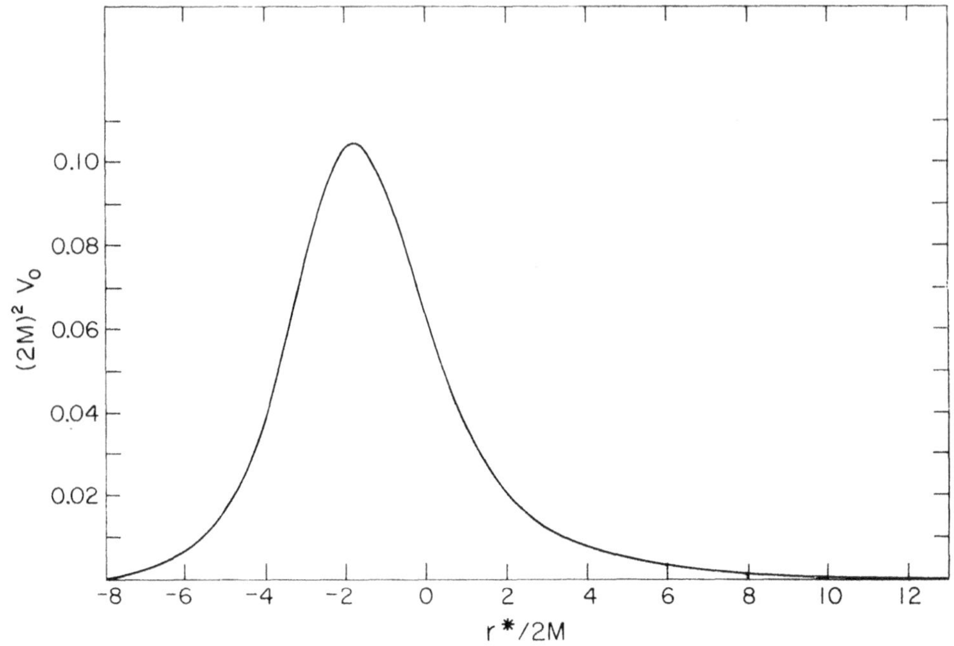

Fig. 10. The $l = 0$ effective potential $V_l(r^*)$ in which the scalar field propagates outside a Schwarzschild black hole. (Taken from Price (1971a).)

scalar field on the characteristic surfaces $\mathscr{C}_1$ and $\mathscr{C}_2$. In practice to determine them one would have to continue the solution back to $t=0$ and use the orthogonality of $\Phi_\omega^{(\pm)}(r^*)$. One important result of such an analysis is that $A^{(\pm)}(\omega)$ must at least diverge no faster than ω^{-1} for small ω. This is a consequence of the fact that the scalar field on $\mathscr{C}_1$ and $\mathscr{C}_2$ is at most constant as we move to large positive and negative values of r^*, and it is the small ω behavior of $A^{(\pm)}$ which controls this large $|r^*|$ behavior. Typically, if $a(\omega) \sim \omega^p$ for small ω, then integrals of the form $\int a(\omega) \exp(i\omega x)\, d\omega$ behave like $x^{-(p+1)}$ for large x. In actual fact, the behavior of $A^{(\pm)}$ is much better than this for small ω but rather more work is required to prove it.

At fixed r^* the large t behavior of Φ_l is determined through Equation (4.13) by the small ω behavior of both $A^{(\pm)}$ and $\Phi_\omega^{(\pm)}(r^*)$. In regions where the potential is negligible this is in turn connected to the small ω behavior of $R^{(\pm)}$ and $T^{(\pm)}$. Calculation of these quantities, however, is familiar from barrier penetration problems in quantum mechanics. For small ω the barrier completely reflects any incident wave and, furthermore, changes its phase so that the reflected wave interferes destructively with the incident one. Thus,

$$R^{(\pm)} = -1 + 0\,(\omega \text{ or higher power}), \tag{4.16a}$$

$$T^{(\pm)} = 0\,(\omega \text{ or higher power}). \tag{4.16b}$$

As a consequence of this in the regions where the potential is negligible, $A^{(\pm)}\Phi_\omega^{(\pm)}(r^*)$

behaves at most like a constant as $\omega \to 0$ for fixed r^*. By simple extension of these arguments this also holds where the potential is non-negligible. Thus, $\Phi_l(r^*, t)$ must decay at least as fast as $1/t$. In actual fact the fields decay much faster. The more thorough analysis of Price shows that for any multipole initially present in the pre-collapse scalar field

$$\Phi_l(r^*, t) \sim t^{-(2l+2)}. \tag{4.17}$$

If the multipole is not initially present, the field decays one power of t faster. The exterior field represented by the initial scalar field on $\mathscr{C}_1$ and $\mathscr{C}_2$ is thus completely radiated away in the course of collapse.

In physical language what is happening is that the scalar field on $\mathscr{C}_1$ is acting as a source of outgoing gravitational waves while that on $\mathscr{C}_2$ is acting as a source of ingoing gravitational waves. These waves scatter off the geometry as represented by the effective potential V_l (see Figure 9). At late times the source which is most important for the field is that at large values of $|r^*|$ on $\mathscr{C}_1$ and $\mathscr{C}_2$. This either vanishes (on $\mathscr{C}_2$) or is of vanishing frequency (on $\mathscr{C}_1$). The low frequency waves are completely reflected by the potential barrier and, upon reflection interfere destructively with the incident waves. The result is that the field decays at late times. The space curvature represented by the potential V_l is at once reponsible for the non-existence of stationary black holes with weak scalar fields and, at the same time, for the decay of any pre-existing scalar field during gravitational collapse.

(iv) *Gravity and Higher Spin Fields*

The arguments above have been extended by Price to higher spin fields including spin-two gravitational perturbations. He finds that any multipole which is not preserved by a dynamically conserved surface integral (such as the charge monopole in the electro-magnetic case) is radiated away in the process of collapse. As he puts it, "anything which can be radiated will be radiated." What is always left is a black hole with a non-singular event horizon. Thus, at least in the case of small perturbations from spherical collapse the formation of an event horizon seems inevitable.

4.2. The Kerr Black Hole

If stars or black holes are always formed in the process of gravitational collapse then the enumeration and classification of the possible black hole solutions to Einstein's equations becomes an important question. By black hole solution we mean a stationary solution of the source-free Einstein's equations with an asymptotically flat and Euclidean infinity which has a single, non-singular event horizon outside of which there are no singularities. (For simplicity we are excluding from most of the rest of the discussion black holes with external electromagnetic or other spin fields.) Today there is only one example of such a solution. This is the two-parameter family of Kerr black holes and it is appropriate to begin our discussion of non-spherical holes by a consideration of some of their properties.

The Kerr metric in the coordinates introduced by Boyer and Lindquist (1967) is

$$\mathrm{d}s^2 = -\frac{R^2\Delta}{A}\,\mathrm{d}t^2 + \sin^2\theta\,\frac{A}{R^2}\left(\mathrm{d}\varphi - \frac{2Mar}{A}\,\mathrm{d}t\right)^2 + \frac{R^2}{\Delta}\,\mathrm{d}r^2 + R^2\,\mathrm{d}\theta^2 \tag{4.18a}$$

where

$$\Delta = r^2 - 2Mr + a^2\,, \tag{4.18b}$$

$$R^2 = r^2 + a^2\cos^2\theta\,, \tag{4.18c}$$

and

$$A = (r^2 + a^2)^2 - a^2\Delta\sin^2\theta\,. \tag{4.18d}$$

For large values of r the solution is

$$\mathrm{d}s^2 = -\left(1 - \frac{2M}{r}\right)\mathrm{d}t^2 + \left(1 + \frac{2M}{r}\right)\mathrm{d}r^2 + r^2\,\mathrm{d}\Omega^2 - \\ - \frac{4Ma}{r}\sin^2\theta\,\mathrm{d}\varphi\,\mathrm{d}t + 0\left(\frac{1}{r^2}\right). \tag{4.19}$$

The quantity M can thus be identified as the total mass of the black hole and a with its angular momentum per unit mass. The inverse metric is given by

$$\left(\frac{\partial}{\partial s}\right)^2 = -\frac{A}{R^2\Delta}\left(\frac{\partial}{\partial t}\right)^2 - \frac{4Mar}{R^2\Delta}\frac{\partial}{\partial\varphi}\frac{\partial}{\partial t} + \frac{\Delta}{R^2}\left(\frac{\partial}{\partial r}\right)^2 + \\ + \frac{1}{R^2}\left(\frac{\partial}{\partial\theta}\right)^2 + \frac{\Delta - a^2\sin^2\theta}{R^2\Delta\sin^2\theta}\left(\frac{\partial}{\partial\varphi}\right)^2. \tag{4.20a}$$

The Kerr geometry is clearly stationary and axisymmetric. In Boyer-Lindquist coordinates, the corresponding Killing vectors are $\xi_{(t)} = \partial/\partial t$ and $\xi_{(\varphi)} = \partial/\partial\varphi$.

To see that the Kerr geometry is a black hole solution we must locate the event horizon. In a stationary solution like this one the event horizon will be a closed stationary surface generated by those light rays which never collapse toward the singularity nor escape to infinity. That is, it will be the largest stationary null surface. The equation of a stationary surface has the form

$$F(x^i) = 0, \quad x^i = (r, \theta, \varphi)\,. \tag{4.21}$$

and the normal to a null surface must be null

$$g^{ij}\frac{\partial F}{\partial x^i}\frac{\partial F}{\partial x^j} = 0, \quad \text{on} \quad F(x^i) = 0\,. \tag{4.22}$$

In the case of the Kerr geometry the horizon will also be clearly axisymmetric and so Equation (4.22) reduces to

$$\Delta\left(\frac{\partial F}{\partial r}\right)^2 + \left(\frac{\partial F}{\partial\theta}\right)^2 = 0\,. \tag{4.23}$$

The largest solution of this is the surface $r=r_+$ where r_+ is the largest root of the equation $\Delta=0$,

$$r_+ = M + (M^2 - a^2)^{1/2}. \tag{4.24}$$

This surface is the event horizon. To actually establish this we should continue the metric in Equation (4.20) beyond r_+ where the Boyer-Lindquist are singular and show that $r=r_+$ is indeed the boundary between those events which can be reached by observers starting from infinity and ending at infinity, and those events which cannot be so reached. This demonstration can be found in the works of Boyer and Lindquist (1967) and Carter (1968). In order for the horizon to exist the value of a must not exceed M. Kerr geometries with $a>M$ have naked singularities and are not black holes.

A second important surface in the Kerr geometry is the surface $r=r_e(\theta)$ where

$$r_e^2 - 2Mr_e + a^2\cos^2\theta = 0 \tag{4.25}$$

on which g_{tt} vanishes. This surface lies outside the horizon except where it touches it at $\theta=0$ and $\theta=\pi$. The region between the horizon and this surface has been dubbed the ergosphere by Ruffini and Wheeler. The importance of this surface arises because the Killing vector which is timelike at infinity becomes spacelike in the ergosphere,

$$\xi_{(t)}\cdot\xi_{(t)} = g_{tt} > 0, \quad r_+ < r < r_e(\theta). \tag{4.26}$$

As a consequence there can be no stationary observers (velocity along $\xi_{(t)}$) inside the ergosphere. The t-direction has become spacelike. An observer who remains at fixed r and θ must move in the φ-direction. His angular velocity Ω must be such that

$$(\xi_{(t)} + \Omega\xi_{(\varphi)})^2 < 0. \tag{4.27}$$

Since both $\xi_{(t)}$ and $\xi_{(\varphi)}$ are spacelike this implies that Ω must be greater than a certain minimum value but not greater than a maximum value

$$\Omega_{\min} < \Omega < \Omega_{\max},$$

$$\Omega_{\substack{\max\\ \min}} = -\frac{g_{\varphi t}}{g_{\varphi\varphi}} \pm \left[\left(\frac{g_{\varphi t}}{g_{tt}}\right)^2 - \left(\frac{g_{tt}}{g_{\varphi\varphi}}\right)\right]^{1/2}. \tag{4.28}$$

4.3. The Uniqueness of the Kerr Geometry

It is a remarkable fact that the Kerr family of black holes are essentially the only single black hole solutions of the source-free Einstein equations. We have already seen a hint of this type of uniqueness theorem from our discussion of Price's results on perturbations of spherically symmetric gravitational collapse. The unique spherically symmetric black hole with weakly coupled scalar field is the Schwarzschild geometry with no scalar field. We owe the proof that the Kerr family of black holes are the unique black holes in Einstein's source-free general relativity to a series of powerful theorems by Carter, Hawking and Israel (see also the work of Wald, 1971).

(1) Hawking: [See, Hawking (1971), Hawking and Ellis (1973)]. A stationary

black hole must be either static or axially symmetric. The horizon of a black hole must have spherical topology.

(2) Israel: [See, Israel (1967, 1968) and Müller Zum-Hagen *et al.* (1973)]. The Schwarzschild geometry is the unique *static* black hole for which the exterior is empty.

(3) Carter: [See, Carter (1971, 1973)]. All *stationary*, axisymmetric black holes whose exteriors are empty and whose horizons have spherical topology fall into disjoint families depending on only two parameters: the mass M and the angular momentum J. The Kerr family of black holes are the unique family which have a zero angular momentum limit (the Schwarzschild geometry).

Taken together these theorems almost prove that the Kerr solution is the unique black hole in general relativity. They fall short in two respects. First, there are a number of technical assumptions which are not of direct physical significance and would be desirable to dispense with. Second, Carter's theorem does not rule out the possibility that there are other two parameter black hole families distinct from the Kerr family and not possessing a zero angular momentum limit. Most relativists would hope that these caveats exist only for technical reasons and that the Kerr geometry will turn out to be the unique family of black holes in general relativity.

The uniqueness of the Kerr geometry, coupled with the validity of the conjecture that a black hole is always formed in the process of gravitational collapse, would yield exceptionally stringent conditions on the third endstate of stellar evolution. Starting from a pre-collapse geometry which would require an infinite number of parameters to specify (all the gravitational multipole moments for example), gravitational collapse would rapidly lead to an essentially equilibrium configuration needing only two parameters. The black-hole endstate is thus essentially very simple and we may hope that this simplicity might yield some unique observable signature.

5. Black Hole Dynamics

If we are ever to observe black holes in nature it will not be through a study of black holes in isolation but only through a study of black holes interacting with realistic astrophysical environments. The questions which are relevant to observations are those dealing with the interaction of black holes with exterior matter and, possibly, with other black holes. How does the matter left over from gravitational collapse affect a black hole and how does the black hole affect the matter? What are the observable consequences of a black hole member of a binary star system? In this section we will analyse a few of the issues which are relevant to questions such as these. Our selection of topics necessarily will be limited to those which can be covered briefly and which involve little additional physics besides relativity.

5.1. The Area of the Horizon Must Increase

A powerful general result on the interaction of black holes with matter and with other black holes is that of Hawking (1971, 1972) that the area of a black hole must either remain constant or must increase. In the following we will give a heuristic demonstration of this result. We begin with some definitions. The event horizon is the surface generated by those light rays which neither escape to infinity nor collapse into the singularity. Let us parametrize each of the generating light rays by an affine parameter v in a way which is smooth and continuous between neighboring light rays. The parameter v may be viewed as a kind of time along the horizon. (See Figure 11a). Surfaces of constant v are closed two surfaces (Figure 11b). These two surfaces (also sometimes called 'the horizon') sweep out the complete three dimensional horizon as v progresses. Each point on one of these two-surfaces thus represents the position of a light ray at time v. Consider at one instant v a small circle of light rays of radius r. A time $v + \mathrm{d}v$ later, this circle will have changed in area and sheared into an ellipse. Two numbers are needed to specify the new principal axes of the ellipse (Figure 11c).

$$(\text{new semi-major axis}) = r\left[1 - (\varrho - |\sigma|)\,\mathrm{d}v\right], \tag{5.1a}$$

$$(\text{new semi-minor axis}) = r\left[1 - (\varrho + |\sigma|)\,\mathrm{d}v\right]. \tag{5.1b}$$

The quantity ϱ is called the *convergence* of the null geodesic generators. It governs the rate of change of area. If ΔA was the original area of the circle, then the area of

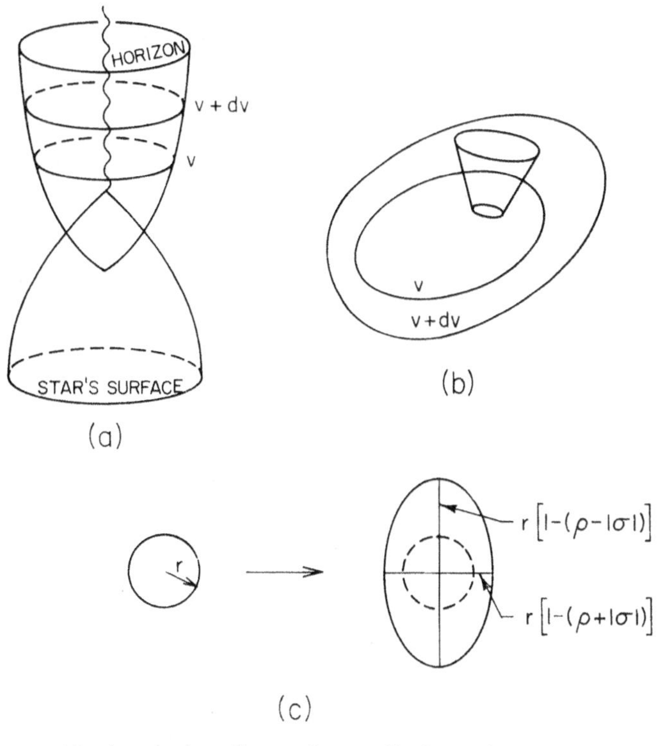

Fig. 11. (a) The event horizon in the collapse of a star. Surfaces of constant v are closed two-surfaces shown in (b). A small circle of light rays on the surface at v changes into an ellipse at $v + \mathrm{d}v$. The parameters necessary to describe this change are shown in (c).

the ellipse will have increased by an amount

$$\mathrm{d}(\Delta A) = -\,2\varrho \Delta A\,\mathrm{d}v\,. \tag{5.2}$$

Integrating over the whole two surface, we obtain an expression for the rate of change of the area of the horizon A

$$\frac{\mathrm{d}A}{\mathrm{d}v} = -\,2\int \varrho\,\mathrm{d}A\,. \tag{5.3}$$

Thus, if ϱ is positive over the whole horizon, the area is decreasing; if it is negative, the area is increasing.

The quantity σ is called the *shear* and governs the evolution of the ellipticity of the circle of light rays.

The quantities ϱ and σ basically describe how the distance between neighboring light rays changes with v. Equations for the change of ϱ and σ themselves can thus be derived from the equation of geodesic deviation. The derivation of these equations is given with slightly different notation in Prof. Sachs' lectures in this volume. Here we will quote only the result for ϱ:

$$\frac{\mathrm{d}\varrho}{\mathrm{d}v} = \varrho^2 + |\sigma|^2 + \Phi_{00}\,. \tag{5.4}$$

where, if $l^\mu = \mathrm{d}x^\mu/\mathrm{d}v$ is the tangent vector to the null geodesic generators,

$$\Phi_{00} = \tfrac{1}{2}R_{\mu\nu}l^\mu l^\nu = 4\pi T_{\mu\nu}l^\mu l^\nu\,. \tag{5.5}$$

If energy is positive, $T_{\mu\nu}t^\mu t^\nu$ should be positive for any time-like vector t^μ. By continuity we would expect the same result to hold for any null vector l^μ and the familiar energy momentum tensors bear this out. Thus, Φ_{00} and the entire right hand side of Equation (5.4) is positive. This has the consequence that if ϱ is positive at some value v_0, then it must increase and become infinite at a finite value of v not more than $[\varrho(v_0)]^{-1}$ later. An infinite value of ϱ means that the area of the circle of light rays previously discussed must go to zero. In effect, the rays then have ceased to generate a surface. Putting it another way, if ϱ ever became positive some of the generators of the horizon would have a future end point at a finite value of their affine parameters. From the very definition of a horizon, however, the generators cannot have any endpoints before infinite values of their affine parameters. Hawking therefore concludes that ϱ must be negative or zero over the whole horizon. An immediate consequence of Equation (5.3) is that the area of the horizon must always increase or must remain stationary.

It is important to understand the origin of the apparent time asymmetry of this result. The horizon is defined as the surface of light rays which never collapse and never get out to *future* infinity. They, therefore, cannot have endpoints in the future but only can have endpoints in the past as is the case, for example, at the point where the horizon forms in spherically symmetric collapse. Since there can be no endpoints in the future, the area of the horizon must increase in that direction.

5.2. Gravitational Radiation in Black Hole Collisions

A simple application of Hawking's result (suggested by him in the same paper) is to set upper limits on the energy which can be emitted as gravitational radiation if two black holes coalesce.

Suppose we have two black holes with masses M_1 and M_2 and specific angular momenta a_1, and a_2. Suppose that they collide to form a third black hole with parameters M_3, a_3. The area of a single Kerr black hole is given by

$$A = 8\pi M\left[M + (M^2 - a^2)^{1/2}\right]. \tag{5.6}$$

Since the area of the event horizon must always increase, one has

$$\begin{aligned} M_3(M_3 + (M_3^2 - a_3^2)^{1/2}) \geqslant M_1(M_1 + (M_1^2 - a_1^2)^{1/2}) + \\ + M_2(M_2 + (M_2^2 - a_2^2)^{1/2}). \end{aligned} \tag{5.7}$$

The energy radiated is $M_1 + M_2 - M_3$ and the fraction f of energy radiated is

$$f = 1 - \frac{M_3}{M_1 + M_2}. \tag{5.8}$$

It is not too difficult to see that the inequality in Equation (5.7) implies

$$f < \tfrac{1}{2}. \tag{5.9}$$

Thus, at most 50% of the initial energy can be emitted as gravitational radiation if two black holes coalesce.

5.3. Extracting the Rotational Energy of a Black Hole

A rotating black hole possesses rotational energy much like a rotating star. Several methods have been suggested for extracting the rotational energy.

(i) *General Limitations*

Whatever the energy extracted from a black hole, its area must increase during the process. This has led Christodoulou (1970) to define the 'irreducible mass' of a black hole by the relation

$$A = 16\pi M_{ir}^2. \tag{5.10}$$

Using Equation (5.6) for the area one can then express M as

$$M^2 = M_{ir}^2 + \frac{J^2}{4M_{ir}^2}, \tag{5.11}$$

where J is the black hole's angular momentum. The second term represents the rotational contribution to the black hole's mass. The maximum rotational energy which can be extracted from the black hole is $M - M_{ir}$.

(ii) *The Penrose Process*

Penrose (1969) has suggested an interesting way of extracting the rotational energy of a black hole which utilizes its ergosphere. Suppose a particle falls into the ergosphere of a rotating black hole and splits into two pieces. One piece falls through the horizon and the other is ejected back out to infinity. (See Figure 12.) Let $\mathbf{p}^{(0)}$, $\mathbf{p}^{(1)}$ and $\mathbf{p}^{(2)}$ be respectively the energy-momentum four-vectors of the incident particle, the ejected particle and the particle which falls through the horizon. The conserved components $-p_t$ and p_φ represent the energy and angular momentum of the particles when they are at infinity. The difference $(-p_t^{(1)})-(-p_t^{(0)})$, if positive, is the energy which is extracted from the black hole in this process.

In the decay process energy momentum is conserved,

$$\mathbf{p}^{(0)} = \mathbf{p}^{(1)} + \mathbf{p}^{(2)}. \tag{5.12}$$

The energy change of the black hole is thus

$$\Delta w = p_t^{(1)} - p_t^{(0)} = - p_t^{(2)} = - \xi_{(t)} \cdot \mathbf{p}^{(2)}. \tag{5.13}$$

Outside the ergosphere, $\xi_{(t)} = \partial/\partial t$ is a timelike vector so that $-p_t^{(2)}$ represents the energy measured by some local observer and must always be positive. If the decay takes place outside the ergosphere, therefore, the black hole always will gain energy and no energy will be extracted. Inside the ergosphere, however, $\xi_{(t)}$ is spacelike and there is no reason why $-p_t^{(2)}$ cannot be arranged to be negative. Energy would then be extracted from the black hole.

If rotational energy is being extracted from the black hole by this process, then we

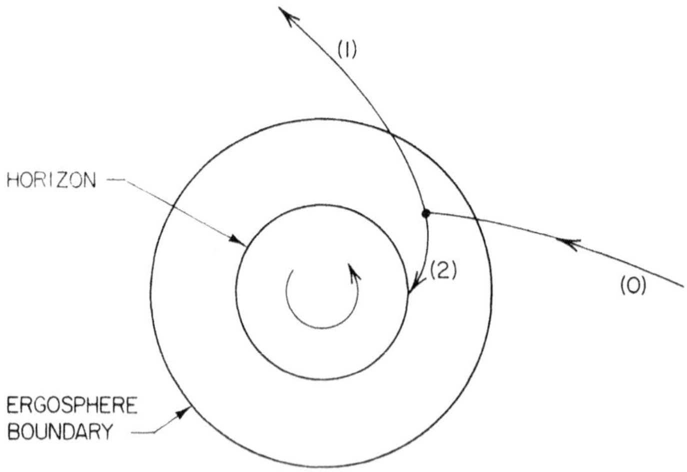

Fig. 12. The Penrose process. Figure shows a black hole viewed along the pole of its rotation axis. Particle (0) enters the ergosphere and splits into particle (1) and (2). Particle (2) falls into the black hole in a way which decreases its angular momentum. Particle (1) escapes with an energy greater than particle (0).

would expect it also to lose angular momentum. It is easily seen that this is the case. The four-velocities of observers in the ergosphere at fixed r, θ will be proportional to $\xi_{(t)}+\Omega\xi_{(\varphi)}$ where $\Omega_{\min}<\Omega<\Omega_{\max}$ [cf. Equations (4.27), (4.28)]. The energy of the particle which falls into the black hole as seen by these local observers must always be positive

$$-\mathbf{p}\cdot(\xi_{(t)}+\Omega\xi_{(\varphi)})\geqslant 0\,. \tag{5.14}$$

Thus if $-p_t^{(2)}$ is negative, $p_\varphi^{(2)}$ is also negative. The angular momentum lost by the black hole is $p_\varphi^{(1)}-p_\varphi^{(0)}=-p_\varphi^{(2)}$ and is thus always positive. Christodoulou (1970) has shown that it is possible to extract all of the rotational energy of a black hole by a sequence of suitably chosen Penrose processes.

(iii) *Tidal Friction*

The Penrose process, while an interesting example of the unusual physics which occurs near black holes, does not seem likely to be of great importance in realistic astrophysical situations because of the special way in which the orbits and breakup must be arranged. A process which may occur in more general situations is the transfer of angular momentum between a black hole and exterior matter via gravitational tidal friction (Hawking, 1971; Press, 1972; Hawking and Hartle, 1972; Hartle, 1973).

If a chunk of matter is rotating with respect to infinity and if it is not symmetric about its axis of rotation, then it will radiate gravitational radiation to infinity. This gravitational radiation has the effect of dissipating the relative rotation between the object and infinity, i.e. the object slows down.

In many respects the event horizon of a black hole is analogous to infinity. Both infinity and the event horizon, for example, are one-way surfaces through which if radiation passes it can never return. One might expect that a relative rotation between a black hole and a non-axially symmetric distribution of matter would lead to gravitational dissipation which would decrease the relative rotation between the black hole and the exterior matter. A stationary exterior Moon, for example, should cause the black hole to slow down.

The existence of such an effect is clear from Hawking's theorem that a stationary black hole must be either static or axially symmetric. A non-axially symmetric distribution of matter will make the solution non-stationary. Either the black hole must tend to a static (non-rotating) black hole or the matter must become axially symmetric.

This slowing down mechanism is closely related to the slowing down of a planet by dissipation in a viscous sea due to the tide raised by a moon. The presence of a stationary external moon outside a black hole will distort the event horizon. As the horizon rotates under this distortion, gravitational dissipation will convert the rotational energy of the black hole into irreducible mass and the black hole will slow down. The Moon as a consequence will speed up.

A general prescription for calculating this process has been given by Hawking and Hartle (1972). Application of this prescription to the case of a slowly rotating black hole ($a\ll M$) being slowed down by a mass μ located at Schwarzschild coordinates

R, Θ gives for the rate of change of black hole angular momentum

$$\frac{dJ}{dt} = -\frac{4}{5}\frac{\mu^2 J M^3}{R^6} \sin^2 \Theta, \tag{5.15}$$

if $R \gg M$ and

$$\frac{dJ}{dt} = -\frac{1}{64}\frac{\mu^2 J}{M^3}\frac{\log(R/M - 2)}{1-(2M/R)} \sin^2 \Theta \tag{5.16}$$

if $(1-2M/R) \ll 1$. This rate is only significant when the matter is very close to the black hole. Far away, the rate of angular momentum gained by the Moon in an orbiting binary system is more than balanced by its gravitational radiation loss to infinity.

5.4. Accretion and Gravitational and Electromagnetic Radiation

We have only mentioned a few of the mechanisms by which black holes can interact with a realistic astrophysical environment. A very important mechanism which we have not discussed is the accretion of matter into black holes. In a realistic collapse one might expect the black hole and the residual matter to be rotating in the same direction. Accretion will tend to increase the rotational energy in this case, opposing the rotational energy extraction mechanisms mentioned earlier. In fact, neglecting all other mechanisms, Bardeen (1970) has shown that accretion will tend to keep the angular momentum of a black hole at the maximum value $a = m$. A more important consequence of accretion is the electromagnetic and gravitational radiation of the infalling particles. There is the exciting possibility that accretion onto a black hole can explain the recently discovered pulsating X-ray cources and also Weber's gravitational events.

The dynamical interaction of a black hole and realistic astrophysical environment is clearly a complex process. It involves the interplay of the dynamics of accreting particles, the emission of electromagnetic and gravitational energy both into the black hole and out to infinity, and the various energy extraction mechanisms. We may hope, however, that this very complexity will yield a rich harvest of observational predictions and, perhaps someday soon, the positive identification of a black hole.

Acknowledgements

These lectures were written while the author was at the Aspen Center for Physics. Conversations with colleagues and students too numerous to mention were useful in their preparation.

References

(No attempt at bibliographic completeness has been made in citing these references. They are simply the ones which seemed most directly relevant to the material in the text.)

Ames, W. and Thorne, K. S.: 1968, *Astrophys. J.* **151**, 659.
Bardeen, J. M.: 1970, *Nature* **226**, 64.

Baym, G., Bethe, H., and Pethick, C. J.: 1971a, *Nucl. Phys.* **A175**, 225.
Baym, G., Pethick, C. J., and Sutherland, P.: 1971b, *Astrophys. J.* **170**, 299.
Bludman, S. A. and Ruderman, M. A.: 1968, *Phys. Rev.* **170**, 1176.
Bludman, S. A. and Ruderman, M. A.: 1970, *Phys. Rev.* **D1**, 3243.
Bondi, H.: 1964, *Proc. Roy. Soc.* **A282**, 303.
Boyer, R. H. and Lindquist, R. W.: 1967, *J. Math. Phys.* **8**, 265.
Buchdahl, H.: 1959, *Phys. Rev.* **116**, 1027.
Carter, B.: 1966, *Phys. Rev.* **141**, 1242.
Carter, B.: 1968, *Phys. Rev.* **174**, 1559.
Carter, B.: 1971, *Phys. Rev. Letters* **26**, 331.
Carter, B.: 1973, de Witt (ed.), *Black Holes*, Gordon and Breach, New York.
Chase, J.: 1970, *Comm. Math. Phys.* **19**, 276.
Christodoulou, D.: 1970, *Phys. Rev. Letters* **25**, 1596.
Fackerell, E.: 1971, *Astrophys. J.* **166**, 197.
Harrison, B. K., Thorne, K. S., Wakano, M., and Wheeler, J. A.: 1965, *Gravitation Theory and Gravitational Collapse*, University of Chicago Press, Chicago.
Hartle, J. B.: 1973, *Phys. Rev.* **D8**, 1010.
Hawking, S. W.: 1972, *Comm. Math. Phys.* **25**, 152.
Hawking, S. W.: 1971, *Phys. Rev. Letters* **26**, 1344.
Hawking, S. W. and Ellis, G. F. R.: 1973, *Large Scale Structure of Space Time*, Cambridge University Press.
Hawking, S. W. and Hartle, J. B.: 1972, *Comm. Math. Phys.* **27**, 283.
Israel, W.: 1967, *Phys. Rev.* **164**, 1776.
Isreal, W.: 1968, *Comm. Math. Phys.* **8**, 245.
Leibovitz, C. and Israel, W.: 1970, *Phys. Rev.* **D1**, 3226.
Misner, C. and Sharp, D.: 1964, *Phys. Rev.* **136**, B571.
Misner, C., Thorne, K., and Wheeler, J. A.: 1973, *Gravitation*, W. H. Freeman, Co., San Francisco.
Muller Zum-Hagen, H., Robinson, D. C., and Seifert, H. J.: 1973, *GRG J.* **4**, 53.
Oppenheimer, J. R. and Volkoff, G.: 1939, *Phys. Rev.* **55**, 374.
Pandharipande, V.: 1971, *Nucl. Phys.* **A178**, 123.
Pandharipande, V.: 1972, *Nucl. Phys.* **A181**, 33.
Penrose, R.: 1969, *Revista del Nuovo Cimento* **1**, 252.
Press, W.: 1972, *Astrophys. J.* **175**, 243.
Price, R.: 1972a, *Phys. Rev.* **D5**, 2419.
Price, R.: 1972b, *Phys. Rev.* **D5**, 2439.
Regge, T. and Wheeler, J. A.: 1957, *Phys. Rev.* **108**, 1063.
Ruderman, M. A.: 1968, *Phys. Rev.* **172**, 1286.
Ruderman, M. A.: 1972, *Ann. Rev. Astron. Astrophys.* **10**, 427.
Salpeter, E.: 1961, *Astrophys. J.* **134**, 669.
Sawyer, R.: 1972, *Phys. Rev. Letters* **29**, 382.
Sawyer, R. and Scalapino, D.: 1972, to be published.
Scalapino, D.: 1972, *Phys. Rev. Letters* **29**, 386.
Thorne, K. S.: 1969, *Astrophys. J.* **158**, 1.
Thorne, K. S.: 1972, in J. Klauder (ed.), *Magic Without Magic*, W. H. Freeman, San Francisco.
Wald, R.: 1971, *Phys. Rev. Letters* **26**, 1653.

COSMOLOGY

R. K. SACHS

Departments of Mathematics and Physics, University of California at Berkeley, Calif., U.S.A.

1. Introduction

These lectures are intended mainly for relativists with a geometrical background. I hope they will provide access to the astrophysical literature on cosmology. Five main references are used: Ehlers' (1973) lectures on general relativity; Ellis' (1971) lectures on theoretical cosmology; the books by Peebles (1971) and Weinberg (1972) for astrophysical cosmology; and Longair's (1971) article on observational cosmology. Together with the forthcoming volume 2 of Novikov and Zeldovich (1973?), these cover almost all the points which a nonspecialist needs. An attempt has here been made to keep the topics covered and further references to a minimum. Unless explicitly indicated otherwise I will use Ehlers' (1973) conventions, including the units $8\pi G = c = h = k = 1$. A subscript 'o' always denotes 'here and now' and/or 'observed'; 'e' means 'emitted' and/or 'earlier'.

The basic observational data are discussed in Peebles (1971), Weinberg (1972) and Longair (1971). The present value H of the Hubble constant is approximately $H = (14 \cdot 10^9 \text{ yr})^{-1} \cong 70 \text{ km s}^{-1} \text{ Mpc}^{-1}$. Typical ages of stars and rocks are of order H^{-1}. Define the so-called 'critical' energy density (or, equivalently in our units, mass density) ϱ_c by

$$\varrho_c = 3H^2 \cong 10^{-29} \text{ g cm}^{-3}. \tag{1.1.1}$$

The present energy density ϱ_o from all kinds of matter is probably within an order of magnitude of ϱ_c. Probably ϱ_o is dominated by the contribution from the rest mass energy of protons and neutrons, and we shall always assume such to be the case. Then (1.1) corresponds to a Baryon number density b_o given by

$$b_o \cong 0.6 \cdot 10^{-5} \text{ cm}^{-3}. \tag{1.1.2}$$

There seems to be an admixture of primordial helium with mass-abundance of 25–30%.

Cosmological backgrounds of photons and high energy protons have been investigated in detail during the last ten years. For photons, values or upper limits are known over much of the range from wavelengths of $3 \cdot 10^4$ cm to those of 10^{-12} cm*, with hopes of extending the results to higher energies in the forseeable future. In particular, the microwave background is probably Planck at a temperature of 2.7 K. As discussed in 2.3 this corresponds to an energy density close to

$$\varrho_{\gamma o} \cong 5 \cdot 10^{-34} \text{ g cm}^{-3} \cong 5 \cdot 10^{-5} \varrho_c, \tag{1.1.3}$$

* Rough but convenient conversion factors are: $\lambda = 1$ cm $\to h\nu = 10^{-4}$ eV and $kT = 10^{-4}$ eV $\to T = 1$ K.

W. Israel (ed.), Relativity, Astrophysics and Cosmology, 197–236. All Rights Reserved and, using (1.2), a photon number density $n_{\gamma o}$ of order

$$n_{\gamma o} \sim 10^8 b_o . \tag{1.1.4}$$

The microwave background is isotropic to astonishing accuracy, better than 1%. The values or upper limits for the photon background at other frequencies are roughly consistent with the numbers obtained by assuming present discrete sources radiating for one Hubble time, though that in the radio suggests there has probably been some evolution of sources. All the observed backgrounds appear to be isotropic to within 5%. There is no convincing evidence for magnetic fields on scales comparable to $H^{-1} = cH^{-1} \sim 10^{10}$ light years, (compare Mitton and Reinhardt, 1972).

It will here be assumed that quasars are cosmological. Optical and radio observations of distant discrete sources also indicate that significant evolution of sources has probably occurred. The classical tests – red shift vs. luminosity, apparent sizes, number vs. flux, and number vs. red shift – are at present still consistent with a wide variety of models, including most of the big-bang models.

The observed lumps in the universe, from stars to galactic clusters or superclusters, seem to occur with a maximum scale of perhaps 10^8 light years and a maximum mass of perhaps 10^{18} solar masses.

The most appropriate way to analyze this data might be by dimensional arguments. Dimensional analysis would respect the uncertainties, the ambiguities, and the enormous number of viable alternatives which are the curse of theoretical cosmology. But dimensional analysis can't go very deep. I shall instead proceed rather deductively, writing down some of the mathematics that is actually used and relying on the main references for the necessary grain (or ton) of salt.

The following assumptions will always be made. Spacetime is described and interpreted by the standard rules of general relativity. Thus the essential spacetime structure is the metric tensor g_{ij}.* It is assumed that the macroscopic electromagnetic field is negligible. The Einstein field equations

$$G^{ij} = T^{ij} \tag{1.1.5}$$

are postulated. All matter, including the photons we observe, will be described by general relativistic one-particle kinetic theory (Ehlers, 1973). The matter variables are the distribution functions f_A ($A = 1, \ldots,$ # of species considered) for the various species; the matter equations are Boltzmann's equation

$$Lf_A = c_A \tag{1.1.6}$$

for each species. Here L is the Liouville operator (geodesic spray) and c_A is the

* In the language of current differential geometry (Schmidt, 1973; Bishop and Goldberg, 1968) we shall use a real, Hausdorf, 4-dimensional, oriented, time-oriented, C^∞ Lorentzian manifold (M, g). Given a coordinate neighbourhood $U \subseteq M$ we write $g|_U = \Sigma_{i,j}\, g_{ij}\, \mathrm{d}x^i \otimes \mathrm{d}x^j \equiv g_{ij}\, \mathrm{d}x^i\, \mathrm{d}x^j$. Thus $g_{ij} = g(\partial/\partial x^i, \partial/\partial x^j)$. Except in Sections 5 & 6 this kind of notation will be relegated to the footnotes. When the notation is used ϕ_* will denote the differential of a map $\phi: M \to M$ and ϕ^* will denote the pullback. For example an isometry ϕ of M onto M is a diffeomorphism ϕ such that $\phi^* g = g$.

collision integral (Ehlers, 1973). Thus at each point T^{ij} is

$$T^{ij} = \sum_A \int f_A p^i p^j \pi_A \tag{1.1.7}$$

where the integral is over the appropriate submanifold in the tangent space to the point (Ehlers, 1973).

Equations (1.5)–(1.7) determine the time evolution of the metric and the distribution functions uniquely from appropriate initial values*. In principle they are the basic equations we shall use; they conceal a multitude of sins.

The assumptions mentioned exclude many current cosmological models. But they give a systematic framework for discussing the essential effects: the non-linear effects of gravity on gravity as built into G^{ij}; the effects of spacetime on matter (and the properties of geodesics) as summarized in the Liouville operator L; the very complex effects of matter on matter *via* the collision terms; and the observable effects of photons, determined by the photon distribution function here and now. Section 2 following develops the general formalism required to apply (1.1.4)–(1.1.6) to cosmology.

Perhaps the best way to learn theoretical cosmology is to choose one reasonable spacetime and try to analyze all the effects of interest using it. Which particular spacetime one chooses is not so important; but it seems to be essential to analyze one in depth, trying to pull things together, rather than superficially analyzing models based on many different spacetimes. Sections 2 and 3 are therefore devoted to the hot Einstein de Sitter model, which is on almost everyone's short-list of favorite candidates. The spacetime is (spatially) homogeneous and (spatially) isotropic with zero spatial curvature. 'Hot' refers to the fact that one assumes thermal equilibrium at some very early time. Together with the assumptions already mentioned and the final values for ϱ_o and $n_{\gamma o}$ these restrictions yield an essentially unique model.

The following open questions are perhaps the most important and puzzling. First, what accounts for the observed properties of the lumps? It seems that no arguments restricted to the present epoch of the universe can furnish an answer; one is driven back, willy-nilly, to very early times and conceivably even to the big-bang in a search for the explanation. Apart from its intrinsic interest, one needs an answer to the question in order to make those corrections for the time evolution of discrete sources which are apparently indispensible for interpreting the classical observational tests. Second is the converse problem: why is the universe as isotropic and homogeneous in the large as the observations indicate? Simple models assert that we are now rather directly observing parts of the universe which had not had time to communicate when they emitted the signals we see. How did they know to send signals which appear isotropic to us? Third, one may ask why the number in (1.4) is 10^8 rather than any other number; estimates indicate that the ratio n_γ/b

* Here and throughout no systematic attempt is made to distinguish between local results or definitions and global ones. For example, a local group of local isometries will be called an isometry group, etc.

remains roughly constant when we extrapolate backwards in time by factors of 10^{16} or more. Explanations for such puzzles have been sought by analyzing inhomogeneous models and homogeneous anisotropic ones. Sections 4 and 5 give a very brief introduction to some of the current work.

2. General Methods

This section summarizes methods used for almost all cosmological models. 2.1 discusses matter. 2.2 and 2.3 discuss matter symmetries and thermal equilibrium. 2.4 concerns cosmological spacetimes and their classification. 2.5 and 2.6 discuss how photons carry information from distant sources or initial conditions to us.

2.1. Kinetic Theory

Some matter influences the geometry by contributing significantly to T^{ij} in (1.1.5), some matter (mainly photons) carries information. In any case we use kinetic theory and the calculations ultimately center on integrating (1.1.6) in some manageable approximation. The geometry-influencing matter is treated in a smoothed-out way, so that galaxies are replaced by a gas of appropriate density. We shall need the following results of kinetic theory (Ehlers, 1973).

Consider the Ath species of particles; suppose each such particle has rest-mass m_A baryon number B_A, etc. Here and throughout 'etc.' refers to the other absolutely conserved quantities: electric charge, electronic lepton number, and muonic lepton number. We shall always average over spin states. The particles are described by the distribution function $f_A(x^i, p_\mu)$*. f_A is particle number per 3-volume per mode. More specifically, let π_A be the natural volume element in momentum space. Thu (Ehlers, 1973)

$$\pi_A = \sqrt{-g}\,\mathrm{d}p^1\,\mathrm{d}p^2\,\mathrm{d}p^3\,(-p_4)^{-1} = \mathrm{d}p_1\,\mathrm{d}p_2\,\mathrm{d}p_3\,(p^4\sqrt{-g})^{-1}. \quad (2.1.1)$$

Here $p^4(x^i, p_\mu)$ and $p_4(x^i, p_\mu)$ are determined by $p^ip_i = -m^2$ and the fact that p^i is future-pointing. The defining equation for f_A is that the number of particles $\mathrm{d}N$ in momentum range π_A crossing an infinitesimal 3-space (in spacetime) with 3-area element σ_i is given by (Ehlers, 1973)

$$\mathrm{d}N = f_A p^i \sigma_i \pi_A. \quad (2.1.2)$$

From (2.1.2) one argues (Ehlers, 1973) that the contribution of f_A to the stress energy tensor, the entropy flux density, to the particle 4-current density for this

* As the notation suggests we shall here consider f_A as a function whose domain is the mass shell in the cotangent bundle T^*M rather than in the tangent bundle TM. There is no essential difference from Ehlers' treatment since the Lorentzian structure g provides a canonical bundle isomorphism $\tilde{g}: T^*M \to TM$. Working on T^*M is a little more convenient in detailed computations. It also exhibits the relation to Hamiltonian mechanics a little more clearly. For example, let $\omega: T^*M \to T^*T^*M$ be the canonical 1-form on T^*M (Bishop and Goldberg, 1968). Then, apart from a numerical factor, Ehlers' volume form on TM is just $\tilde{g}^*(\mathrm{d}\omega \wedge \mathrm{d}\omega \wedge \mathrm{d}\omega \wedge \mathrm{d}\omega)$.

kind of particle, to the baryon 4-current density, etc. are respectively given by

$$T^{ij}_A = \int f_A p^i p^j \pi_A ; \tag{2.1.3}$$

$$S^i_A = - \int [f_A \ln f_A \mp (1 \pm \quad _A) \ln (1 \pm f_A)] \, p^i \pi_A ; \tag{2.1.4}$$

$$N^i_A = \int f_A p^i \pi_A ; \tag{2.1.5}$$

$$B^i_A = B_A N^i_A, \text{ etc.} \tag{2.1.6}$$

In (2.1.4) and throughout, the top sign refers to bosons, the bottom one to fermions. The total stress energy tensor, entropy flux density, baryon 4-current density, etc. are additive:

$$T^{ij} = \sum_A T^{ij}_A ; \quad S^i = \sum_A S^i_A ; \quad B^i = \sum_A B^i_A, \quad \text{etc.} \tag{2.1.7}$$

Each f_A obeys Boltzmann's equation (Ehlers, 1973)

$$L f_A \equiv p^i \nabla_i f_A = c_A . \tag{2.1.8}$$

Here ∇_i means 'take $\partial/\partial x^i$ holding p_μ parallel displaced'* and c_A is some very messy collision integral. (1.1.8) implies (Ehlers, 1973)

$$T^{ij}{}_{;j} = 0; \quad S^i{}_{;i} \geqq 0; \quad B^i{}_{;i} = 0, \text{ etc}. \tag{2.1.9}$$

If $c_A = 0$ we also have

$$T^{ij}_{A\,;j} = 0 = S^i_{A;i} = N^i_{A;i} = B^i_{A;i}, \text{ etc}. \tag{2.1.10}$$

We shall often wish to consider the case where spacetime has certain symmetries (isometries) and a distribution function f has the 'same' symmetries. For example**, let ξ^i be a Killing vector field on spacetime, the generator of a 1-parameter symmetry group. Thus (Schmidt, 1973; MacCallum, 1973) $\xi_{i;j} = -\xi_{j;i}$. Then f is invariant under the isometry group iff

$$\xi^i \nabla_i f + \xi^i{}_{;j} p_i (\partial/\partial p_j) f = 0 . \tag{2.1.11}$$

2.2. Isotropy

The only feature common to all the many spacetimes used in cosmology is the assumed existence of a preferred reference frame. A reference frame f^i is a unit, timelike, future-pointing vector field f^i on spacetime. Thus $f^i f_i = -1$. In discussing recent epochs f^i is identified with the average galactic 4-velocity. For example, the world line of the center of mass of our local group of galaxies is taken to have f^i as tangent.

* Thus ∇_1 is that horizontal vector field on the cotangent bundle which is projection-related to $\partial/\partial x^1$. The restriction of ∇_1 to the Ath mass shell is the object of interest.

** The general criterion is the following. Let $\phi: M \to M$ be an isometry, $\psi = (\phi, \phi^{-1*}): T^*M \to T^*M$ be its natural lift. f is called invariant under ϕ iff $f \circ \psi = f$.

Formally, f^i is usually introduced by using T^{ij} as in (2.2.1) below and/or by some kind of isotropy argument, as we shall now discuss. However (though people don't like to say so) the most compelling reason for introducing f^i is really that using it simplifies life. In this subsection and 2.3 we analyze how the geometry-influencing matter is usually related to f^i; subsection 2.4 discusses some of the ways f^i is related to classifications of cosmological spacetimes.

Except in some weird cases, not of interest here, (2.1.3) and (2.1.7) imply algebraically that there is precisely one reference frame f^i such that

$$T^i_j f^j \propto f^i . \tag{2.2.1}$$

We will henceforth always take (2.2.1) as the basic definition of the preferred reference frame f^i. Very often one assumes, motivated by the apparent isotropy of our observations, that any observer at a point with world-velocity f^i measures isotropy. Roughly, this means that no direction within his rest space, the set of vectors orthogonal to f^i, is preferred. It is made precise in various ways, some strong, some weaker.

A very strong form of the isotropy assumption is the following. We assume that for any point p in spacetime and any pair of unit vectors u, $\tilde{u}$ in the rest space at p there is a symmetry (isometry) of spacetime which leaves p fixed, carries u into $\tilde{u}$, and leaves all the distribution functions invariant.* This assumption leads to the Robertson-Walker models, as discussed in Section 2.

Of the various weaker forms, the following is the one of immediate interest to us. For a particular species described by f_A one assumes that none of the quantities in (1.1.3)–(1.1.6) algebraically determines a preferred direction in any rest-space $(f^i)^\perp$. The only isotropic vector in Euclidean 3-space is the zero vector; the only isotropic 2-tensor is $\delta_{\mu\nu}$. The latter can be identified with the 3-metric or projection operator h_{ij} defined by (Ehlers, 1973; Ellis, 1971)

$$h_{ij} = g_{ij} + f_i f_j . \tag{2.2.2}$$

In fact, $h_{44} = h_{4\mu} = 0$ and $h_{\mu\nu} = \delta_{\mu\nu}$ in a local Lorentz frame for which $f^i = \delta^i_4$. The weak assumption on f_A is thus equivalent to the assumption that for some partial energy density $\varrho_A(x^i)$, partial pressure p_A, partial entropy density s_A, number density n_A, partial baryon number density $b_A = B_A n_A$, etc. we must have the algebraic forms

$$T^{ij}_A = \varrho_A f^i f^j + p_A h^{ij} ; \tag{2.2.3}$$

$$S^i_A = S_A f^i ; \tag{2.2.4}$$

$$N^i_A = n_A f^i ; \tag{2.2.5}$$

$$B^i_A = b_A f^i, \text{ etc} . \tag{2.2.6}$$

Moreover, (2.1.3) implies algebraically that

$$0 \leqslant p_A \leqslant \tfrac{1}{3} \varrho_A . \tag{2.2.7}$$

* Thus if $\phi : M \to M$ is the isometry we have $\phi p = p$, $\phi^* g = g$, $\phi_* u = \tilde{u}$, $f \circ \psi = f$.

If each geometry-influencing species is similarly isotropic then we get

$$T^{ij} = \varrho f^i f^j + p h^{ij}, \quad 0 \leqslant p \leqslant \tfrac{1}{3}\varrho \tag{2.2.8}$$

with $\varrho = \sum_A \varrho_A$, $p = \sum_A p_A$. The analogous equations for S^i, B^i, etc. are of course

$$S^i = s f^i, \quad s = \sum_A s_A ; \tag{2.2.9}$$

$$B^i = b f^i, \quad b = \sum_A b_A ; \text{etc.} \tag{2.2.10}$$

In a perfect-fluid approach one often starts directly with (2.2.3)–(2.2.6) or (2.2.8)–(2.2.10), with or without the inequality on the total pressure p. One postulates $T^{ij}{}_{;j}=0$, $S^i{}_{;i} \geqslant 0$, $B^i{}_{;i}=0$, etc. and interrelates the energy density ϱ, pressure p, entropy density s, baryon number density b, and the various other unknowns such as ϱ_A by enough equations of state to get a unique time development. One sometimes generalizes to imperfect fluids (Ehlers, 1973). Though often very convenient, the fluid approach is rather obscure because it is hard to see where the various extra conditions required to get a unique time development come from. In cosmology, though not in general, it is legitimate and easier to take the Boltzmann Equations (2.1.8) as in principle the only basic matter equations. On the other hand, it must be admitted that no one ever really integrates (2.1.8) with an actual collision integral taken from microphysics. Rather one uses (2.1.8) to guess at appropriate approximations in a reasonably systematic way.

2.3. Complete Local Thermal Equilibrium

When all collisions are very rapid compared to typical times associated with the space-time curvature or with the derivatives of f^i, as in most models of the very early universe, it is appropriate to assume, as an approximation, complete local thermal equilibrium (Ehlers, 1971; Weinberg, 1972). The assumption can be obtained by specializing 2.2 as follows. Define $E = -p_i f^i$, the energy measured by an observer with world-velocity along the preferred reference frame. Let r_A be the number of spin states; thus $r_A = 2$ for photons, for electrons or for positrons; $r_A = 1$ for e-neutrinos or e-antineutrinos; and so on. One postulates that each distribution function has the form

$$f_A = r_A [\exp(-\alpha_A + T^{-1} E) \mp 1]^{-1} . \tag{2.3.1}$$

Here $T(x^i)$ is the local temperature. $T\alpha_A(x^i)$ is the relativistic chemical potential. For each A, α_A can be determined in terms of four basic chemical potentials α_B, etc. for the four conserved quantities baryon number, etc. The rule is

$$\alpha_A = B_A \alpha B + \cdots . \tag{2.3.2}$$

Thus to determine the distribution functions completely one must specify five quantities: the temperature T, the baryon chemical potential α_B, etc. One can specify the total baryon number density b instead of α_B, etc. (2.3.1) implies various thermodynamic relations. For example, integrating (2.1.4) by parts implies

$$T(s_A + \alpha_A n_A) = \varrho_A + p_A . \tag{2.3.3}$$

The limits $m_A \ll T$ and $m_A \gg T$ are of particular interest. Suppose $m_A \ll T$ and $\alpha_A = 0$. Then (2.3.1) with $m_A = 0$ gives in (2.1.3)

$$\varrho_A = 3p_A; \quad \varrho_A = (\tfrac{1}{2})\, r_A a T^4 \,(\text{bosons}); \quad \varrho_A = (\tfrac{7}{16})\, r_A a T^4 \,(\text{fermions}), \tag{2.3.4}$$

where $a \cong 7.6 \cdot 10^{-15}$ erg cm^{-3} (K)$^{-4} \cong 8.4 \cdot 10^{-36}$ g cm^{-3} (K)$^{-4}$ is the blackbody constant. The entropy density, by (2.3.3), is

$$s_A = (\tfrac{4}{3})\, \varrho_A T. \tag{2.3.5}$$

(2.1.5) gives $n_A \propto s_A$ where the proportionality constant, different for fermions and bosons, is of order magnitude unity in our units. For the limit $m_A \gg T$ we similarly get to lowest approximation

$$p_A = 0, \quad \varrho_A = m_A n_A. \tag{2.3.6}$$

Normally (2.3.1) is not consistent with the Boltzmann equations and using it requires replacing them by approximations as in section 4 (Stewart, 1971).

2.4. Cosmological Spacetimes

The spacetimes used in cosmology are classified by either of two related methods: specifying their isometry group or specifying the geometric properties of the preferred reference frame f^i. MacCallum (1973) discusses the former method in detail and we shall later give examples. We now outline the latter, which is somewhat more powerful. Here and in some later sections we shall need a few properties of vector fields. Let v^i be any non-zero vector field. Then it determines integral curves $x^i(u)$, *via* the equation

$$\mathrm{d}x^i/\mathrm{d}u = v^i(x(u)). \tag{2.4.1}$$

u is the parameter; one integral curve goes through each spacetime point. For example, if v^i is a reference frame each integral curve is the world line of an observer in that reference frame and $u = s$ is the proper time of the observer. A 'connecting' vector $n^i(u)$ along an integral curve of v^i is defined by the equation of Lie transport (Ehlers, 1973; Schmidt, 1973):

$$Dn^i/\mathrm{d}u \equiv n^i{}_{;j} v^j = v^i{}_{;j} n^j. \tag{2.4.2}$$

Any such n^i has the following intuitive interpretation: if $n^i(0)$ connects $u=0$ on the given integral curve c with a point on an infinitesimally neighbouring integral curve $\bar{c}$ then $n^i(u)$ connects the point $x^i(u)$ on c with that point on $\bar{c}$ obtained by moving u parameter units along $\bar{c}$.

Suppose in particular v^i is geodesic, $v^i{}_{;j} v^j = 0$, and has constant norm, $v^i v_i =$ constant. Then we have $d(n^i v_i)/\mathrm{d}u = D(n^i v_i)/\mathrm{d}u = n^i{}_{;j} v^j v_i + n^i v_{i;j} v^j = v^i{}_{;j} n^j v_i + 0 = (\tfrac{1}{2}) \times \times (v_i v^i)_{,j} n^j = 0$. In particular

$$n^i v_i \big|_o = 0 \rightarrow n^i v_i = 0 \quad \text{for all} \quad u \tag{2.4.3}$$

in this case. Finally, given v^i there exist (many) comoving coordinate systems, defined by the property that

$$v^i = \delta^i_4 \tag{2.4.4}$$

in such a coordinate system. For example, if v^i is a reference frame any comoving coordinate system can be obtained as follows. Choose any hypersurface $\sum$ which cuts each observer world line once; $\sum$ need not, and in general cannot, be chosen orthogonal to v^i. Label the points of $\sum$, and thus the observers, by spatial coordinates x^μ; each observer is to retain his given values of x^μ permanently (hence the name comoving). Now label each point of spacetime by (x^μ, s) where x^μ is the observer who goes through the point and s is the distance of the point from $\sum$ along that observer's world line. Then $dx^i/ds = (dx^\mu/ds, ds/ds) = \delta^i_4$ so the coordinate system is comoving.

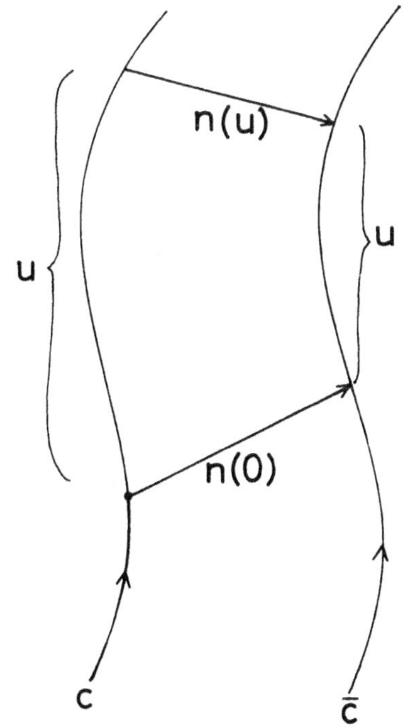

Fig. 1.

The classification of cosmological spacetimes involves analyzing the derivatives of f^i. One first asks whether the f^i observers are accelerating; they are freely falling iff $f^i{}_{;j} f^j = 0$. The general scheme (Ellis, 1971) simplifies considerably if we assume no acceleration; we now restrict attention to this subcase to indicate the ideas without the full machinery.

Now consider a connecting vector n^i from a given observer A to a neighboring one. By (2.4.3) we can suppose that n^i always lies in A's rest space, i.e.

$$Dn^i/ds = f^i{}_{;j} n^j ; \quad n^i f_i = 0 . \tag{2.4.5}$$

$\sqrt{n^i n_i} = R(s)$ is the spatial distance to the neighbor and n^i/R is the spatial direction to the neighbor. Their rate of change is determined by $f_{i;j}$; for example $\mathrm{d}(R^2)/\mathrm{d}s = 2f_{i;j}n^i n^j$; the directional rate of change is with respect to gyroscopes carried by A since in the geodesic case any gyroscope axis g^i obeys $Dg^i/\mathrm{d}s = 0$, which is to be compared to (2.4.5). $f_{i;j}$ obeys $f_{i;j}f^i = 0 = f_{i;j}f^j$ so it can be regarded as a tensor in each rest space. To get more details we split $f_{i;j}$ into 'irreducible' bits, essentially as in the Wigner-Eckart theorem*. We here get a volume expansion rate Θ, a vorticity ω_{ij}, and a shear rate σ_{ij}:

$$f_{i;j} = \tfrac{1}{3}\Theta h_{ij} + \omega_{ij} + \sigma_{ij}; \quad \sigma_{ij} = \sigma_{ji}; \quad \sigma^i_i = 0; \quad \omega_{ij} = -\omega_{ji}. \tag{2.4.6}$$

Since everything takes place in a rest space, the three pieces have their standard interpretation from Newtonian hydrodynamics (Ehlers, 1973; Ellis, 1971). For example $\Theta = (\mathrm{d}V/\mathrm{d}s)/V$, where V is the intrinsic 3-volume, in the local rest space, of a small lump of comoving observers centered on A. By algebra (2.4.6) is equivalent to

$$\Theta = f^i{}_{;i}, \quad \omega_{ij} = \tfrac{1}{2}(f_{i;j} - f_{j;i}), \quad \sigma_{ij} = \tfrac{1}{2}(f_{i;j} + f_{j;i}) - \tfrac{1}{3}\Theta h_{ij}. \tag{2.4.7}$$

In discussing collapse theorems (Hawking and Ellis, 1973), classifying cosmological spacetimes, and drawing analogies to lightlike geodesics (Subsection 2.6 below) one often needs the proper time derivatives of Θ, ω_{ij} and σ_{ij}. To get these here, note that

$$f_{i;j;k}f^k = f_{i;k;j}f^k + R_{mijk}f^m f^k = -f_{i;k}f^k{}_{;j} - R_{imjk}f^m f^k. \tag{2.4.8}$$

(2.4.7) and (2.4.8) give

$$D\omega_{ij}/\mathrm{d}s = -\sigma_{ik}\omega^k{}_{.j} + \sigma_{jk}\omega^k{}_{.i} - \tfrac{2}{3}\Theta\omega_{ij}. \tag{2.4.9}$$

In particular $\omega_{ij}|_{x(s)} = 0$ iff $\omega_{ij}|_{x(o)} = 0$. Similarly one can obtain $D\sigma_{ij}/\mathrm{d}s$ and $\mathrm{d}\Theta/\mathrm{d}s$. We shall here use these equations only when two additional restrictions hold: $\omega_{ij} = 0$ and $G^{ij} = T^{ij}$ for a T^{ij} of the perfect-fluid form (2.2.8). Algebra gives

$$D\sigma_{ij}/\mathrm{d}s = -\sigma_{ik}\sigma^k_j + \tfrac{1}{3}h_{ij}\sigma_{km}\sigma^{km} - \tfrac{2}{3}\Theta\sigma_{ij} - E_{ij}; \tag{2.4.10}$$

$$\mathrm{d}\Theta/\mathrm{d}s = -\tfrac{1}{3}\Theta^2 - \sigma_{ij}\sigma^{ij} - R_{ij}f^i f^j = -\tfrac{1}{3}\Theta^2 - \sigma_{ij}\sigma^{ij} - \tfrac{1}{2}(\varrho + 3p). \tag{2.4.11}$$

Here $E_{ij} = G_{imjk}f^m f^k$ with G_{imjk} the conformal tensor (compare Ellis, 1971).

An intuitive interpretation of (2.4.10)–(2.4.11) is the following. The only term in (2.4.10) which would not be present in a flat space is the conformal tensor term; thus the anisotropic part of the gravitational field drives the shear rate. In (2.4.11) the gravitational term is $(-\tfrac{1}{2})(\varrho + 3p)$; it corresponds to an active mass driving the expansion rate toward more negative or less positive values; any extra shear generated by E_{ij} similarly causes pulling together in this sense.

Most models that have been investigated in detail have $\omega_{ij} = 0$; if we demand weak

* Let ψ_s be an element of the flow of f, P_s denote parallel transport a distance s along an integral curve, R be the rest space at a point on that curve, and $h = g|_R \equiv (g + f\otimes f)|_R$. Then $P_s(\psi_s)_*$ is a linear transformation: $R \to R$; so is $F = \mathrm{d}[P_{-s}(\psi_s)_*]|\mathrm{d}s$. As always, one can decompose F with respect to h into a trace Θ, a skew adjoint part ω, and a trace free self-adjoint part σ.

isotropy in the sense that ω_{ij} does not determine a preferred rest space direction then we must have $\omega_{ij}=0$, since otherwise $\omega_{ij}n^j=0=n^i f_i$ defines a unique direction. By (2.4.7) $\omega_{ij}=0$ in our geodesic case iff there is (locally) a function t such that

$$f_i = - t_{,i}. \tag{2.4.12}$$

When such a function t exists it is determined up to an additive constant by (2.4.12) and is called the cosmological time. More detailed examples of the classification by Θ, ω_{ij} and σ_{ij} will be given later.

2.5. Photons

To complete our general discussion we will give two subsections on information-carrying matter, using unpolarized photons as the example. The only equations needed are (2.1.2) and (2.1.8) for the photon distribution function f.

All the photon quantities of interest in observational cosmology – magnitude, surface brightness, antenna temperature, etc. – are built from $f_o=f(x^i_0, p_\mu)$, the distribution function here and now. (2.1.2) indicates the relation; we give the two main examples. Use (2.1.2) in a local Lorentz frame of some observer with the infinitesimal 3-space the rest space. Let $E=\omega=2\pi\nu$ be the energy he measures, $\boldsymbol{P}$ be the physical components of the 3-momentum. Then $\sigma_i=\delta_i^4\,\mathrm{d}^3x$ for the observer, $p^i\sigma_i=E\,\mathrm{d}^3x$, and by (2.1.1) $\pi=\mathrm{d}^3P/E$. Thus

$$\mathrm{d}N = f\,\mathrm{d}^3x \times \mathrm{d}^3P. \tag{2.5.1}$$

This equation indentifies f as the photon number per mode, which is often taken as the basic quantity in astrophysics (Peebles, 1971). Alternately, suppose the infinitesimal 3-space $\sum$ is chosen timelike and orthogonal to the photon direction in the rest space of the observer. Let $\mathrm{d}\Omega_P$ be the measured angle in 3 dimensional momentum space, $\mathrm{d}A$ the measured 2-area of $\sum$, $\mathrm{d}t$ its measured duration. Then some algebra in (2.1.2) gives for the energy contribution

$$\omega\,\mathrm{d}N = \omega^3 f\,\mathrm{d}\nu\,\mathrm{d}\Omega_P\,\mathrm{d}A\,\mathrm{d}t. \tag{2.5.2}$$

This equation identifies

$$\omega^3 f = i_\nu \tag{2.5.3}$$

as the radiation brightness per unit frequency interval i_ν (erg cm^{-2} s^{-1} ster^{-1} Hz^{-1}; Peebles, 1971; i_ν is sometimes called 'specific intensity'). The observations are often discussed in terms of i_ν and those quantities obtained from it by integrating (in any combination!) over one of the differentials on the right side of (2.5.2) (Peebles, 1971). For example if a distant discrete source cannot be resolved, one measures the energy flux per unit frequency interval,

$$\omega^3 \int f\,\mathrm{d}\Omega_P, \tag{2.5.4}$$

where the integral extends over the directions subtended by the source. And so on;

what is important for us is that the single quantity f is the basic object of interest. If we can find f and can handle $d\Omega_p$ and $d\nu$ in (2.5.2) geometrically we have full information. Handling $d\nu$ amounts to knowing the red shift, as shown in later examples; $d\Omega_p$ will be discussed in the next subsection. We now make a few comments about integrating (2.1.8) to find f.

The key approximation almost always made is to replace the collision term for photons by the simpler form

$$c \to \alpha(x^i, p_\mu) - \beta(x^i, p_\mu) f \tag{2.5.5}$$

where α and β are independent of f. Since the collision term is the number of photons emitted per mode per unit invariant spacetime volume (Ehlers, 1973) α corresponds to spontaneous emission of photons, say by a gas. For example, if a gas is at rest in the preferred frame f^i, has number density $n(x^i)$ in that frame and each gas particle isotropically emits $e(E)$ photons per mode per unit time at energy E then $\alpha = ne(p^i f_i)$ (with p^4 determined from $p^i p_i = 0$ as always). β similarly corresponds to absorption, stimulated emission, or scattering out of a narrow beam. An example of a situation where (2.5.5) is *not* applicable is inelastic scattering within a large region. In inelastic scattering the number of photons created at one momentum depends on the number present at different momenta so that a more sophisticated approximation or the whole collision integral is required.

If we can integrate for the lightlike geodesics in any spacetime, and α and β are known, then (2.5.5) can be integrated directly. We shall always parametrize timelike and lightlike geodesics in such a way that their tangent is a 4-momentum; thus the equations for a geodesic are

$$dx^i/du = p^i, \quad Dp^i/du = 0, \quad p^i p_i = -m^2, \quad p^i \text{ future-pointing}. \tag{2.5.6}$$

Consider a geodesic on our past light cone. Along the geodesic (2.5.5) becomes $df/du = \alpha(x(u), p(u)) - \beta f$.

The integral is

$$f_o = f_e \exp\left[-\int_{u_e}^{u_0} \beta(u')\,du'\right] + \int_{u_e}^{u_0} du'\,\alpha(u') \exp\left[-\int_{u'}^{u_0} \beta(u'')\,du''\right]. \tag{2.5.7}$$

f_e is the initial value for that geodesic, perhaps due to a discrete source, perhaps due to photons left over from earlier epochs. It is attenuated by absorption with $[\int\beta]^{-1}$ the optical length. The second, emission term is similarly attenuated. Much of astrophysical cosmology consists of staring, in a mildly depressed way, at (2.5.7). The measurable quantity f_o contains a lot of information. But the unknowns, whose values we wish to estimate, are f_e for each direction, the geometric effects that come in *via* (2.5.6), and the values of α and β. A lot of model building is required to work backward from f_o to the desired estimates. For example if f_e is due to a discrete source, one can sometimes approximate it by assuming the source characteristics in the past were similar to those of nearby sources. The optical lengths estimated via (2.5.7) are surprisingly large, probably corresponding to a surprisingly tenuous intergalactic medium (Peebles, 1971; Weinberg, 1972; Longair, 1971).

For $\alpha=0=\beta$ (2.5.7) becomes the surface-brightness theorem (Ellis, 1971); we rewrite it using (2.5.3) and the standard definition $1+z=\nu_e/\nu_o$ of the red-shift z

$$(i_\nu)_o = (i_\nu)_e(1+z)^{-3}. \tag{2.5.8}$$

(2.5.8) indicates the three main effects which are present in evolutionary models: there is a strong cutoff due to the red-shift factor $(1+z)^{-3}$ in any expanding model; the source radiation brightness per unit frequency interval $(i_\nu)_e$ must be taken at the earlier time t so that source evolution has to be estimated; and $(i_\nu)_e$ must be known at the parallel displaced momentum $(p_\mu)_e$ corresponding to a different frequency, so that comparison with nearby sources is difficult. Often these naive effects simply swamp specific geometric effects such as the area effect now to be discussed.

2.6. Lightlike Geodesics

Theoretical expressions for the apparent angle $d\Omega_P$ subtended by a small distant source are of interest for two reasons: first, direct measurements of the angles yield geometric information about spacetime (compare Peebles, 1971 and Longair, 1971 for discussions of recent measurements by Baum); second, for a source which can't be or isn't resolved one needs to compare with observations *via* (2.5.4). In models of high symmetry $d\Omega_P$ can often be obtained by symmetry arguments, as in Section 3. General expressions for $d\Omega_P$ can be obtained by analyzing lightlike curves in a way similar to that used for timelike ones in 2.4. Since the general analysis has various other applications (Hartle, 1973; Hawking and Ellis, 1973; Penrose, 1967) we start with a few comments on it.

Consider a geodesic, lightlike, future-pointing vector field $p^i(x^j)$; thus

$$p^i p_i = 0 = p^i{}_{;j} p^j. \tag{2.6.1}$$

As in 2.4, let $x^i(u)$ be an integral curve and $n^i(u)$ be a connecting vector; thus

$$Dn^i/du = p^i{}_{;j} n^j. \tag{2.6.2}$$

To interpret we shall again split up $p_{i;j}$.

Since lightlike geodesics are rather tricky we need some preliminary results for the tangent space to a point, say the point $x^i(0)$. For vividness we shall regard the tangent space as Minkowski space. To relevant order the integral curve can for the moment be considered as a lightlike geodesic L in Minkowski space with equation

$$\mathbf{x}(u) = \mathbf{p}u, \quad \mathbf{p}\cdot\mathbf{p} = 0. \tag{2.6.3}$$

Neighboring integral curves can for the moment be regarded as parallel lightlike geodesics in Minkowski space each with equation

$$\mathbf{x}(\bar{u}) = \mathbf{p}\bar{u} + \mathbf{x}_o. \tag{2.6.4}$$

One such curve $\bar{L}$ lies in the same lightlike hyperplane as L, namely $(\mathbf{p})^{\perp}$, iff

$$\mathbf{x}_o \cdot \mathbf{p} = 0\,. \tag{2.6.5}$$

One then also says L and $\bar{L}$ meet at lightlike infinity (Penrose, 1967). Suppose L and $\bar{L}$ meet at lightlike infinity in Minkowski space and let $\mathbf{n}$ be a vector from any point on L to any point on $\bar{L}$. Then from (2.6.3)–(2.6.5) there is some τ with

$$\mathbf{n} = \tau\mathbf{p} + \mathbf{x}_o \rightarrow \mathbf{n}\cdot\mathbf{p} = 0, \quad \mathbf{n}\cdot\mathbf{n} = \mathbf{x}_o\cdot\mathbf{x}_o \equiv |\mathbf{x}_o|^2\,. \tag{2.6.6}$$

Thus L and $\bar{L}$ have the peculiar property that any connecting vector is orthogonal to both and has the given length $|\mathbf{x}_o|$ (Figure 2).

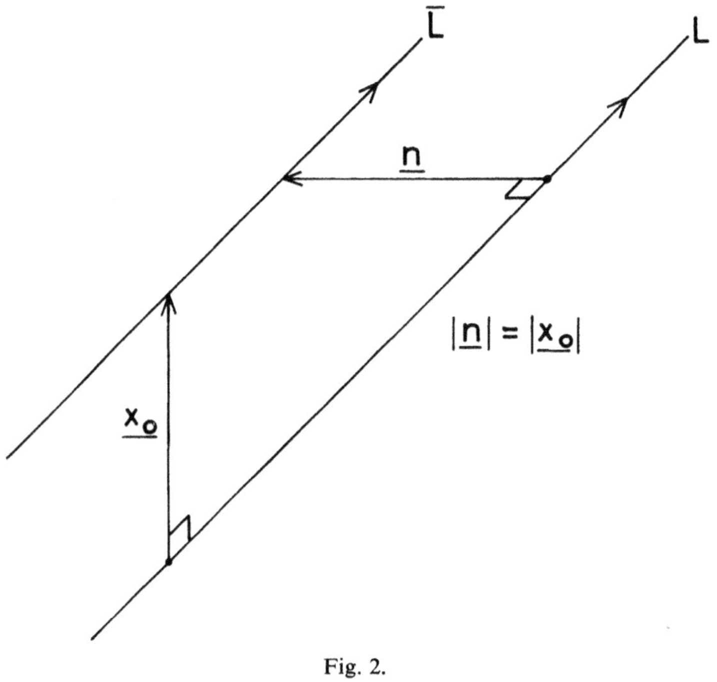

Fig. 2.

$|\mathbf{x}_o|$ is the distance that any observer would measure between the two photons on a photographic plate held normally. For consider any observer with 4-velocity $\mathbf{f}$ at the origin of Minkowski space. Projecting into his rest space gives the spatial direction of L as $\mathbf{p}' = \mathbf{p} + \mathbf{f}(\mathbf{f}\cdot\mathbf{p})$. If he measures, he chooses a point on $\bar{L}$ simultaneous in his frame with the origin, so he gets that unique connection vector $\mathbf{n}$ in (2.6.6) which obeys $\mathbf{n}\cdot\mathbf{f} = 0$, namely

$$\mathbf{n} = -\left(\mathbf{f}\cdot\mathbf{x}_o/\mathbf{f}\cdot\mathbf{p}\right)\mathbf{p} + \mathbf{x}_o\,. \tag{2.6.7}$$

Thus $\mathbf{n}\cdot\mathbf{p}' = \left[-\left(\mathbf{f}\cdot\mathbf{x}_o\right) + \mathbf{f}\cdot\mathbf{x}_o\right]\mathbf{f}\cdot\mathbf{p} = 0$, so $\mathbf{n}$ lies on a photographic plate orthogonal to $\mathbf{p}'$. We may summarize by saying that each of the following conditions on two parallel lightlike geodesics in Minkowski space implies the others: they lie in the same lightlike hyperplane; they meet at infinity; one vector connecting them is orthogonal to both; every vector connecting them is orthogonal to both; every vector connecting them has the same length $|\mathbf{n}|$; and every observer who measures them simultaneously gets a vector in his own rest space which is orthogonal to their common spatial direction and has the observer-independent length $|\mathbf{n}|$.

Now consider a bundle of parallel lightlike geodesics in Minkowski space, all lying in the same lightlike hyperplane. By an easy extension of the above argument one finds the bundle has an observer-independent 2-dimensional area and shape, namely that area and shape which any observer would obtain by letting the photons hit his photographic plate simultaneously (and thus orthogonally in his rest space).

Once these curious algebraic properties are realized the general relativistic analysis is rather straightforward. We first restrict attention to orthogonal connecting vectors in (2.6.2):

$$n_i(u)\, p^i = 0. \tag{2.6.8}$$

By (2.4.3) this is consistent. Now we introduce auxiliary observers* along the integral curve $x^i(u)$; which observers we choose doesn't matter, by the above argument, so take their world velocities $f^i(u)$ to be parallel displaced for simplicity. To replace (2.2.3) introduce the projection operator which projects orthogonal to both p^i and f^i, namely

$$t^{ij} = g^{ij} - f^i p^j/(f_k p^k) - f^j p^i/(f_k p^k) - p^i p^j/(f_k p^k)^2. \tag{2.6.9}$$

Then the Equation (2.6.2) for Lie transport implies

$$D(t^i_j n^j) = (t^i_k t^j_m p^k{}_{;j})\,(t^m_q n^q) \equiv A^i_m\,(t^m_k n^k), \tag{2.6.10}$$

where everything is appropriately projected. The appropriate analogue of (2.4.6) is

$$t^k_i t^m_j n_{k;m} \equiv A_{ij} = \tfrac{1}{2}\,\Theta t_{ij} + \sigma_{ij} + \omega_{ij};\quad \sigma_{ij} = \sigma_{ji};\quad \sigma^i_i = 0;\quad \omega_{ij} = \omega_{ji} \tag{2.6.11}$$

(2.6.11) defines the expansion Θ, shear σ_{ij}, and twist ω_{ij} of the lightlike geodesic vector field. By algebra (2.6.11) is equivalent to

$$\Theta = p^i{}_{;i};\quad \sigma_{ij} = \tfrac{1}{2}\,(A_{ij} + A_{ji}) - \tfrac{1}{2}\,\Theta t_{ij};\quad \omega_{ij} = \tfrac{1}{2}\,(A_{ij} - A_{ji}). \tag{2.6.12}$$

Θ is explicitly independent of the auxiliary observers; if A is the intrinsic 2-dimensional cross sectional area of a small bundle of integral curves centered on $x^i(u)$ then, by a simple local argument

$$(\mathrm{d}A/\mathrm{d}u)/A = \Theta \tag{2.6.13}$$

similar to the result in 2.4. Although σ_{ij} and ω_{ij} do depend on the auxiliary observers,

* One can avoid the auxiliary observers by working with spinors and/or 'flags' (Penrose, 1967); this is much neater but somewhat less intuitive.

the dependence is trivial. Specifically, algebra shows that

$$\sigma_{ij}\sigma^{ij} = \tfrac{1}{2}\,(p^{i;j} + p^{j;i})\,p_{i;j} - \tfrac{1}{2}\,\Theta^2; \quad \omega_{ij}\omega^{ij} = \tfrac{1}{2}\,(p^{i;j} - p^{j;i})\,p_{i;j} \tag{2.6.14}$$

independent of the observers. σ_{ij} is the shearing rate for the two dimensional shape, ω_{ij} its 'twisting' relative to directions parallel propagated along $x^i(u)$.

We can again compute $\mathrm{d}\Theta/\mathrm{d}u$, $D\sigma_{ij}/\mathrm{d}u$, $D\omega_{ij}/\mathrm{d}u$; (2.4.8) applies with f^i replaced by p^i. Algebra gives

$$D\omega_{ij}/\mathrm{d}u = -\,\omega_{ik}\sigma_j^k + \omega_{jk}\sigma_i^k - \Theta\omega_{ij}. \tag{2.6.15}$$

For $\omega_{ij}=0$ more algebra gives

$$D\sigma_{ij}/\mathrm{d}u = -\,\sigma_{ik}\sigma_j^k + \tfrac{1}{2}\,t_{ij}\sigma_{km}\sigma^{km} - \Theta\sigma_{ij} - C_{imjk}p^m p^k\,; \tag{2.6.16}$$

$$\mathrm{d}\Theta/\mathrm{d}u = -\,\sigma_{ij}\sigma^{ij} - \tfrac{1}{2}\,\Theta^2 - R_{ij}p^i p^j\,. \tag{2.6.17}$$

The interpretations are quite similar to those of Subsection 2.4. In particular, to calculate the apparent angle subtended by a small distant source we can proceed as follows (Figure 3).

Consider a lightlike geodesic $x^i(u)$ which comes from somewhere near the center of source to our location x_o^i. Consider all those lightlike geodesics which lie on our past light cone from x_o^i and intercept the source. This is a small bundle of lightlike geodesics. By local arguments near x_o^i one finds that any connecting vector n^i near x_o^i obeys $n^i p_i=0$. Thus our formalism applies. Similarly, $\omega_{ij}=0$ near x_o^i and, by

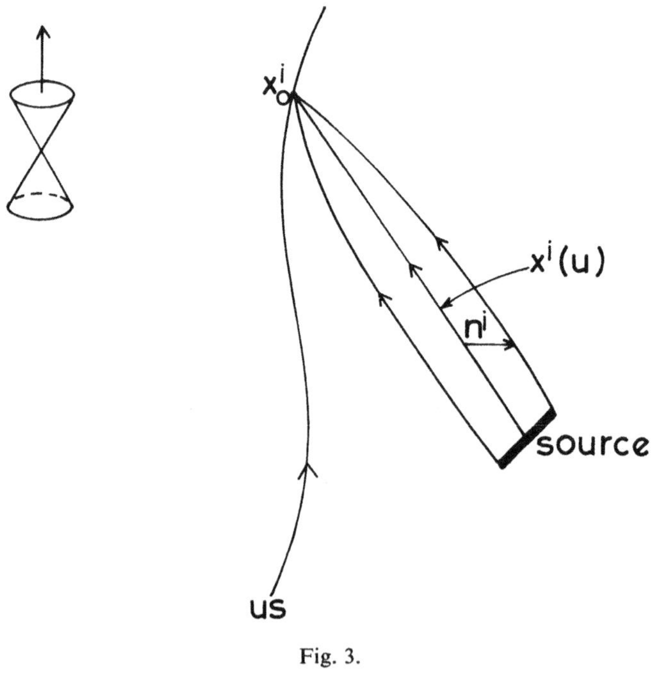

Fig. 3.

(2.6.15) $\omega_{ij}=0$ for all u along $x^i(u)$. The 2-dimensional area of the bundle at the source is the intrinsic cross sectional area of the source. The 2-dimensional area of the bundle near x_o determines the apparent angle subtended, as can be seen special relativistically in our local Lorentz frame. (2.6.13) shows that the way the area changes is governed by Θ. (2.6.17) exhibits the focussing effects of shear and of matter (in the terms $\sigma_{ij}\sigma^{ij}$ and $R_{ij}p^ip^j$). (2.6.16) shows that the shear is driven by the conformal tensor, the anisotropic part of the external gravitational field. In particular, these equations can be used to estimate the following selection effect: if we actually see a distant source there is normally no intervening galaxy, so $R_{ij}p^ip^j$ is abnormally low (Dyer and Roeder, 1973).

3. The Geometry of the Hot Einstein De Sitter Model

In this section we consider the most popular current cosmological model from a geometric viewpoint. 3.1 discussed how the spacetime fits into the classification 2.4. Subsection 3.2 gives a more explicit coordinate version. The remaining subsections are primarily concerned with the behaviour of geodesics. The model is so simple that one can do most calculations explicitly, but sufficiently representative that knowing how to work with it suffices for understanding most of the estimates needed for the applications.

3.1. Robertson-Walker Spacetimes

We shall make the isotropy assumptions of 2.2 on matter. If in addition the first derivatives $f_{i;j}$ of the preferred reference frame do not define a special direction in any local rest frame then $f_{i;j}f^j=0=\omega_{ij}=\sigma_{ij}$. The spacetimes with these properties are called Robertson-Walker spacetimes. (2.4.5) now becomes the equation for a uniform stretching

$$Dn^i/\mathrm{d}s = \tfrac{1}{3}\,\Theta n^i \tag{3.1.1}$$

with s being t of 2.4 evaluated along one observer world line. (3.1.1) identifies $\frac{1}{3}\Theta_o$ as the Hubble constant H. To get the equations in familiar form, we can define $R(s)$ along any observer world line, up to an arbitrary positive multiplicative constant, by

$$\dot{R}/R = \tfrac{1}{3}\,\Theta, \quad R > 0 \tag{3.1.2}$$

here '$\cdot$' means $\mathrm{d}/\mathrm{d}s$.

Thus 2.4.11 becomes the deceleration equation

$$3\ddot{R}/R = -\tfrac{1}{2}(\varrho + 3p). \tag{3.1.3}$$

In the sense of this deceleration equation the active mass of the sources per unit volume is $\varrho+3p$. Moreover we now have by algebra

$$T^{ij}{}_{;j} = 0 \Leftrightarrow \begin{cases} f_iT^{ij}{}_{;j} = 0 \Rightarrow \dot{\varrho} + 3(\varrho + p)\,\dot{R}/R = 0. \\ p_{,i}h^i_j = 0 \end{cases} \tag{3.1.4}$$

Note for later reference the following relations for ϱ and for a conserved quantity $b(s)f^i$

$$p = 0 \Rightarrow \varrho = \varrho_o (R_o/R)^3 \tag{3.1.5}$$

$$p = (\tfrac{1}{3})\, \varrho \Rightarrow \varrho = \varrho_o (R_o/R)^4 \tag{3.1.6}$$

$$(bf^i)_{;i} = 0 \Leftrightarrow \dot{b} + b\Theta = 0 \Leftrightarrow b = b_o (R_o/R)^3 . \tag{3.1.7}$$

(3.1.3) and (3.1.4) together imply the Friedmann equation

$$\tfrac{1}{2}\dot{R}^2 - \tfrac{1}{6}\varrho R^2 = -\tfrac{1}{2}k , \tag{3.1.8}$$

where k is a constant of integration corresponding in Newtonian terms to $-2\times$ the total energy per unit mass of a comoving particle on the surface of a sphere of radius R (Peebles, 1971). One often chooses the multiplicative factor in R such that $k=1$, 0, or -1; however this convention is a little misleading and won't be used here. The sign of k corresponds to the sign of the spatial curvature in the Robertson-Walker models (Ellis, 1971). Since from (3.1.8) $3H^2 - \varrho_o = -3k/R^2$ the sign can in principle be determined directly from observations, with $k<0$ iff $\varrho_o < \varrho_c$. However current observations are not accurate enough to determine this sign. In the remaining subsections we shall take $\varrho_o = \varrho_c$, $k=0$. This gives a representative case consistent with the apparent order of magnitude agreement between ϱ_o and ϱ_c. To see how models with $\varrho_o \neq \varrho_c$ differ from the case we shall consider it is useful to get a qualitative feel for the main effects. For example, (3.1.3) indicates that (for H fixed) a high density model $\varrho_o > \varrho_c$ 'lives faster' owing to the greater deceleration. Moreover, note that in 2.6 the focussing of light due to $R_{ij}\, p^i p^j$ is larger in a high density model. Now imagine an observed, distant, unresolved, discrete source at a given redshift. What difference do we heuristically expect between the predictions of the $\varrho_o = \varrho_c$ model and a $\varrho_o > \varrho_c$ model (with H fixed)? First, there should be more focussing and thus a larger predicted energy flux of photons in the $\varrho_o > \varrho_c$ model. Second, the elapsed cosmic time in a $\varrho_o > \varrho_c$ model should be less, owing to the speedup mentioned, so that source evolution should be somewhat less important. Detailed calculations (Peebles, 1971) bear out both guesses. On the whole, such heuristic arguments are as useful as the standard detailed equations (Weinberg, 1972) for the various Robertson-Walker models.

3.2. Einstein de Sitter Space

Consider a spacetime which obeys the algebraic assumptions of 3.1, namely $G_{ij} = \varrho f_i f_j + p h_{ij}$ and $f_{i;j} = \Theta h_{ij}$. Suppose one also has $\varrho_o = \varrho_c = 3H^2 = \tfrac{1}{3}\Theta_o^2$. Then (Ellis, 1971) the spacetime is (locally) isometric to Einstein de Sitter space, as defined below.

A single coordinate system, with $-\infty < x^\mu < \infty$ and $0 < t < \infty$ covers all spacetime. The metric tensor is determined by

$$g_{ij}\, \mathrm{d}x^i\, \mathrm{d}x^j = -\, \mathrm{d}t^2 + R^2(t)\, \mathrm{d}x^\mu\, \mathrm{d}x^\mu \tag{3.2.1}$$

where $R(t) > 0$ and $R(t) \to 0$ for $t \to 0$. An orientation is assigned by choosing $\eta = \sqrt{-g}\, \mathrm{d}^4x$, rather than $-\eta$, as volume element. From (3.2.1) the Einstein tensor has the required form

$$G_{ij} = \varrho f_i f_j + p h_{ij} \tag{3.2.2}$$

with $f^i = \delta^i_4$, h_{ij} as in Section 2, $\varrho = 3(\dot{R}/R)^2$ and $p = -2(\ddot{R}/R) - (\dot{R}/R)^2$. Thus we have a perfect fluid stress energy tensor. f^i, chosen future-pointing to time-orient the spacetime, is the preferred reference frame (2.2.1). (3.2.1) gives

$$f_{i;j} = \tfrac{1}{3}\,\Theta h_{ij}, \quad \Theta = 3\dot{R}/R \tag{3.2.3}$$

as required.

In particular $\omega_{ij} = 0$, and t is the cosmic time; the additive constant in t is fixed by $\varrho \to \infty$ for $t \to 0$, so t is uniquely defined.

To complete the definition we need the function $R(t)$, obtained by integrating $G_{ij} = T_{ij}$ for suitable matter variables. The complete scheme will be analyzed in Section 4. In this section we need only the following comments. First, R is determined only up to an arbitrary multiplicative constant, since $R \to KR$ with K constant in (3.2.1) gives an isometric spacetime. Second, integrating the inequalities $0 \leqslant p = -2 \times \times (\ddot{R}/R) - (\dot{R}/R)^2 \leqslant \frac{1}{3}\varrho = (\dot{R}/R)^2$ gives for $t \leqslant t_o$

$$(2Ht)^{1/2} \geqslant R(t)/R_o \geqslant (\tfrac{3}{2}Ht)^{2/3}. \tag{3.2.4}$$

In particular

$$\tfrac{2}{3}H^{-1} \geqslant t_o \geqslant \tfrac{1}{2}H^{-1}. \tag{3.2.5}$$

Thus this model (like most models) is roughly consistent with the observed order of magnitude agreement between H^{-1} and the ages of stars and rocks. Next, for the limiting values we have

$$p = 0 \Rightarrow R = R_o(t/t_o)^{2/3} \tag{3.2.6}$$

$$p = \tfrac{1}{3}\varrho \Rightarrow R = R_o(t/t_o)^{1/2}. \tag{3.2.7}$$

Finally, we remark that our assumption in the introduction that ϱ_o is determined by the baryon rest-mass contribution means that in the present epoch and for most of the past the $p=0$ behavior (3.2.6) holds. This assumption involves neglecting the galactic peculiar velocities, which are of order 1000 km s^{-1} at most (thus much less than c), and their corresponding pressure. It also involves assuming negligible neutrino and graviton contribution to ϱ_o. One has no direct observational handle on the neutrinos and gravitons except the very weak limit that if their present energy density is larger than ϱ_c optical effects like those discussed in 3.1 should show up. However, considerations of the early universe (Section 4) suggest that in fact their contribution is now less than the photon contribution (1.1.3). By extrapolating the microwave photons and the baryons backward in time, assuming $T^{ij}{}_{;j} = 0$ for both components, we can estimate when the $p=0$ assumption becomes unreasonable. Using (3.1.6), (3.1.7), and (1.1.3) for the estimate we find that ϱ_γ equals the rest mass energy contribution of baryons when

$$R/R_o \cong 5 \times 10^{-5} \Rightarrow t/t_o \cong (5 \times 10^{-5})^{3/2} \cong 4 \times 10^{-7}. \tag{3.2.8}$$

For t smaller than $10^{-7} t_o$ (3.2.7) is a reasonable estimate except perhaps at very early times (Section 4).

3.3. Symmetries

By inspection, Einstein de Sitter spacetime has the symmetries of Euclidean 3-space (rotations, translations and reflections). We now show that the obvious symmetries are the only symmetries (unlike the case of the sphere with metric determined by $ds^2 = d\theta^2 + \sin^2\theta \, d\phi^2$).

Let D be an isometry (Schmidt, 1973) of Einstein de Sitter spacetime onto itself. Since f^i and t are intrinsically defined D carries each integral curve of f^i onto an integral curve of f^i and carries the hypersurface $t=1$ onto itself. The former statement means that when we know how D acts on the hypersurface $t=1$ we know how D acts even off that hypersurface. The latter means that on $t=1$, D is an isometry of $ds^2 = R^2(1)\, dx^\mu\, dx^\mu$; it thus coincides with one of the obvious symmetries on the hypersurface $t=1$ and, by our above argument coincides with one of the obvious symmetries everywhere. A converse holds (Weinberg, 1972). Locally a spacetime is isometric to that of 3.2 iff it has exactly the same symmetries as Euclidean 3-space and obeys the Einstein field equations for appropriate sources.

3.4. Freely falling particles

'Understanding' a spacetime primarily means knowing a lot about its geodesics. The symmetries of Einstein de Sitter space make the computation of geodesics comparatively simple. If $x^i(u)$, $p^i(u)$ are the world line and 4-momentum of a freely falling particle (2.5.6) in Einstein de Sitter space then $E = -p^i f_i$ is the energy measured by that comoving observer the particle happens to be passing and $P(u) = h_{ij} p^i p^j = R^{-1}(t(u)) \sqrt{p_\mu p_\mu}$ is the measured magnitude for the 3-momentum. We have

$$p^i p_i = -m^2 \Leftrightarrow E^2 = P^2 + m^2. \tag{3.4.1}$$

Since the Lagrangian $(\frac{1}{2})\, g_{ij} (dx^i/du)(dx^j/du)$ for particle motion is independent of x^μ each conjugate momentum p_μ is a constant of the motion*. Specifically,

$$p^i p_i = -m^2 \text{ and } Dp^i/du = 0 \Leftrightarrow p^i p_i = -m^2 \text{ and } dp_\mu/du = 0. \tag{3.4.2}$$

We shall write $q = \sqrt{p_\mu p_\mu}$ for the corresponding 'magnitude' which is also a constant of the motion.

The measured momentum $P = R^{-1} q$ thus decreases as R^{-1}. The standard phrase is that "as the universe expands the particle wavelength λ also expands proportionally so $P \propto R^{-1}$". A more useful picture is phase mixing (Peebles, 1971). Imagine the

* Alternately, we can argue that $\partial/\partial x^1$, say, is a Killing vector because each element of its flow is an isometry. Thus $g(p, \partial/\partial x^1)$ is constant along $x^i(u)$; this gives the same result.

particle shot out from some comoving observer. It overhauls other comoving observers. The other comoving observers are themselves moving outward so the measured 3-momentum decreases. It is basically this effect which is responsible for the standard cooling laws, discussed in Section 4, of a cosmological gas. A special case of $P=R^{-1}q$, that of a photon, gives the standard redshift relation. For $m=0$ we have $1+z=\omega_e/\omega_o=E_e/E_o=P_e/P_o$. Thus

$$1 + z = R_o/R_e. \tag{3.4.3}$$

The four constants of the motion m^2 and p_μ, together with the inequalities (3.2.4) suffice to give a qualitatively complete picture of the geodesics. One integrates $\mathrm{d}x^\mu/\mathrm{d}u=R^{-2}(t(u))\,p_\mu$. The results are presented in Table I.

The following are freely falling particles.

TABLE I

Type	Energy E	Worldline: $x^2=x^3=0$ and $x^1=$	Example $R=t^{2/3}\Rightarrow p=0$	Example $R=t^{1/2}\Rightarrow p=\frac{1}{3}\varrho$
lightlike	$R^{-1}(t(u))$	$\int_0^t \mathrm{d}t'/R(t')$	$x^1=3t^{1/3}$	$x^1=2t^{1/2}$
$m=1,\ q\neq 0$ timelike	$(1+R^2)^{-1/2}$	$q\int^t \mathrm{d}t'(R^4+q^2R^2)^{-1/2}$	$R\gg q\Rightarrow x^1\cong -\frac{1}{3}qt^{-1/3}$ —	$R\ll q\Rightarrow x^1\cong 2t^{1/2} - -(\frac{1}{3})t^{3/2}/q^2$
$m=1,\ q=0$ timelike	1	0	–	–

Moreover, every freely falling particle can be obtained from one of those in the table by a symmetry operation and/or the following rescaling with a constant scale factor $\mathrm{d}>0$: $u\to \mathrm{d}^{-1}u$, $m\to \mathrm{d}m$, $E\to \mathrm{d}E$, $P\to \mathrm{d}P$, $p_\mu\to \mathrm{d}p_\mu$, world line unchanged.

The geodesics can best be visualized on the following spacetime diagram (Figure 4).

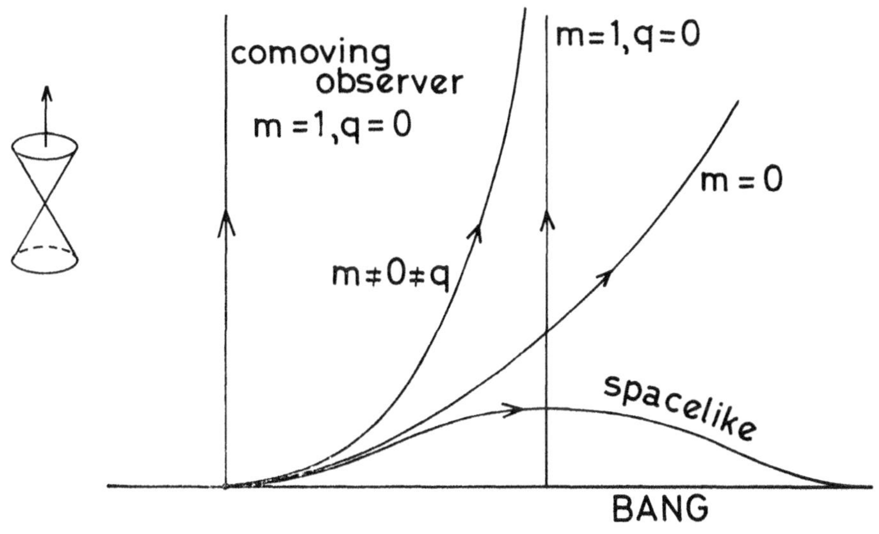

Fig. 4.

The inequalities (3.2.4) imply that every geodesic approaches a definite value of x^μ for $t \to 0$. Moreover the $m \neq 0 \neq q$ geodesics approach a definite value of x^μ for $t \to \infty$.

3.5. Photons

As discussed in Section 2, solving the Boltzmann equation for photons and computing apparent angular diameters are the essential steps in analyzing observations. We now indicate how the use of a particular spacetime, here Einstein de Sitter spacetime, makes explicit computations possible.

The apparent angular diameter of a small distant sphere with intrinsic diameter a can be obtained from a symmetry argument (Figure 5).

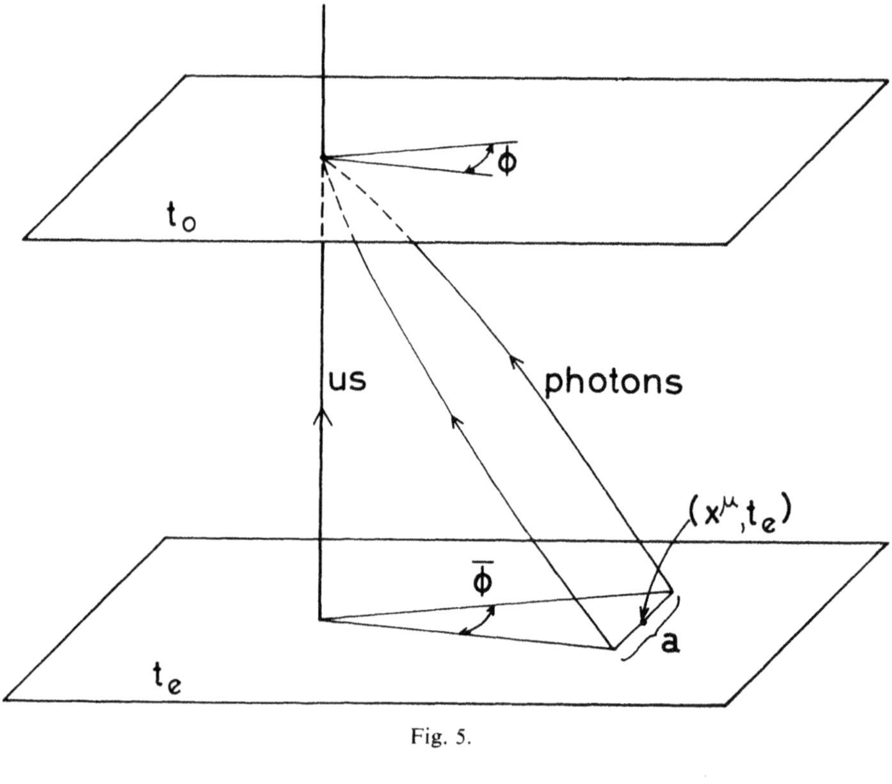

Fig. 5.

The spacetime diagram shows our world line, two photons coming from the edges of a sphere at (x^μ, t_e) and the projections of the photon momenta into our rest space at $(0, t_o)$, which determine the measured angle ϕ. By symmetry $\phi = \bar{\phi}$. By Euclidean geometry within the hypersurface $t = t_e$, $\bar{\phi} = a/R_e\sqrt{(x^\mu x^\mu)}$. In the case $R = t^{2/3}$, corresponding to $p = 0$, we have $H = \frac{2}{3} R_o^{-3/2}$ and, from the equations for lightlike geodesics in 3.4, $\sqrt{(x^\mu x^\mu)} = 3R_o^{1/2} - 3R_e^{1/2}$. Thus $\phi = \bar{\phi} = Haf(z)$, with

$$f(z) = \tfrac{1}{2}(1 + z)\left[1 - (1 + z)^{-1/2}\right]^{-1}. \qquad (3.5.1)$$

The graph of $f(z)$ shows a typical minimum, here at $z=\frac{5}{4}$, 'due to gravitational focussing'. (3.5.1) and (2.5.7) or (2.5.8) lead to the standard relations (Weinberg, 1972) for the standard tests on discrete sources: red shift *vs.* angular diameter, red shift *vs.* luminosity, red shift *vs.* number of sources and flux density *vs.* number of sources (Longair, 1971). Since these have been discussed so often we only summarize the main conclusions (Longair, 1971). The observed results are consistent with the Einstein de Sitter model and with many similar models. There seems to be some evidence from the radio data that there were more and/or brighter sources in the past, at a time corresponding to the range $z=2-3$. There is no convincing evidence for absorption by an intergalactic medium.

To analyze backgrounds one can simplify the Boltzmann equation by assuming that the photon distribution function and the distribution function for any species with which photons interact have the same symmetries as the spacetime. Using the criterion discussed in 2.1, one finds that a distribution function f has the appropriate symmetry iff it can be written as a function of t and q alone. In the rest of this subsection we assume

$$f = f(t, q). \tag{3.5.2}$$

The collision terms automatically have the same form, $c=c(q, t)$. Since q is a constant of the motion (2.1.8) simplifies to

$$p^4(\partial f/\partial t) = R^{-1}q(\partial f/\partial t) = c(q, t). \tag{3.5.3}$$

If in addition we assume $(Rc/q)=\bar{\alpha}(q, t)-\bar{\beta}(q, t)\, c$ corresponding to (2.5.5) we get the integral

$$f(t_o, q) = f_e(q)\exp\left[-\int_{t_e}^{t_o}\bar{\beta}(t')\,dt'\right] + \int_{t_e}^{t_o}\bar{\alpha}(t')\,dt'\exp\left[-\int_{t'}^{t_o}\bar{\beta}(t'')\,dt''\right]. \tag{3.5.4}$$

Using this integral and $E=\omega=R^{-1}$ the absorption and volume emission terms can again be estimated and comparison with observations again suggest that there is little intergalactic gas (Longair, 1971). Moreover, at meter wavelengths attempts to account for the isotropic background by emission over long times from discrete sources, modelled with an $\bar{\alpha}$, again suggest that the sources have evolved.

If $\bar{\alpha}=0=\bar{\beta}$ (3.5.4) becomes

$$f_o = f_e \tag{3.5.5}$$

As shown in 4.2 this gives the standard laws for the energy density ϱ_γ, entropy density s_γ, number density n_γ and spectrum of an undisturbed photon gas: $\varrho_\gamma \propto R^{-4}$, $s_\gamma \propto R^{-3}$, $n_\gamma \propto R^{-3}$ and a Planck spectrum remains Planck with $T \propto R^{-1}$.

Equation (3.5.3) has been applied to estimate the results of scattering by high energy photons or protons off the microwave photons (Longair, 1971; Sciama, 1971). In particular there should be a very strong attenuation in the cosmic ray proton flux above 10^{20} eV due to such scattering if the microwave photons are in fact cosmological; in fact no protons at energies 10^{21} eV or more have yet been observed, though in the absence of satisfactory models for the creation of such high energy protons it is not clear what to make of this result.

3.6. The g-structure of the big bang; horizons

In almost all physically interesting spacetimes information can leak into or out of the spacetime at singularities (Hawking and Ellis, 1973), lightlike infinity (Penrose, 1967), and timelike infinity (Eardley and Sachs, 1973). I don't really see how one can do physics unless one can somehow control this information. In cosmology one must apparently analyze and in some natural way restrict the structure at or near the big bang. Perhaps the best way to try to make sense of the leaky regions is to try to close them up using the g-boundary suggested by Geroch, 1967.

Applied to the big bang of Einstein de Sitters pacetime, Geroch's construction amounts to the following. The geodesics are divided into equivalence classes. The equivalence relation is 'coming arbitrarily close near the big bang'. By considering what happens when a geodesic is varied slightly this equivalence relation can be made precise without reference to any coordinate system (Geroch, 1967). Each equivalence class is regarded as one point on the big bang. As might have been expected from Figure 4, two geodesics are equivalent iff in the coordinates used they approach the same x^{μ} for $t \to 0$. Thus there is one point on the big bang for each comoving observer. Following up Geroch's construction gives the following results. One gets a spacetime-with-boundary, corresponding to the coordinate range $0 \leqslant t < \infty$, $-\infty < x^{\mu} < \infty$. The hypersurface $t=0$ is the big bang, a 3-dimensional spacelike hypersurface attached to the spacetime as its boundary. The coordinates here used are smooth on the whole spacetime-with-boundary.

Given a point p in the Einstein de Sitter spacetime, its past lightcone intersects the big bang in a 2-surface. If p has coordinates $(0, 0, 0, t)$, 3.4 gives for the surface: $t=0$ and $\sqrt{x^{\mu}x^{\mu}} = \eta(t)$ where

$$\eta = \int_0^t \mathrm{d}t'/R(t'). \tag{3.6.1}$$

For example, if $R = t^{2/3}$, $\eta = 3\,t^{1/3}$. The comoving observers who hit the big bang on or within this surface will here be said to be within the horizon for p. Thus a comoving observer is within p's horizon iff he has had a chance to signal p by a signal no faster than light.

In discussing lump formation (Rees, 1971) the interesting questions are how much matter is within a horizon and how big an observed angle a horizon sphere for some earlier time has. For $t > 10^{-7} t_0$ we can estimate these as follows. Assume $R = \mathrm{t}^{2/3}$, $p=0$ and assume that matter is baryons with density $b = b_0\, R_0^3/R^3$ as in (3.1.7). At time t, corresponding to redshift $z = (R_0/R) - 1$, consider the matter within the sphere $\sqrt{x^{\mu}x^{\mu}} \leqslant \eta(t)$. The total number of baryons in the sphere is $B(t) = \int B^i \,\mathrm{d}\sigma_i = (4\pi/3)$

$(bt)\,[R(t)\,\eta(t)]^3$. In particular, the corresponding rest mass now is $(4\pi/3)\,\varrho_o[R_o\eta_o]^3 = 32\pi H^{-1} \cong 4\cdot 10^{23}\,M_\odot$. Moreover $B(t)/B(t_o) = [\eta(t)/\eta(t_o)]^3 = (1+z)^{-3/2}$. Thus the mass within a horizon is

$$M(t) \cong 4\cdot 10^{23} M_\odot (1+z)^{-3/2}. \tag{3.6.2}$$

To get the apparent angle, note that the intrinsic length at t corresponding to $\eta(t)$ is $R(t)\,\eta(t) = 3t = 3(1+z)^{-3/2}\,(2/3H)$. Thus for $z \gg 1$ (3.5.1) gives

$$\phi \cong z^{-1/2}. \tag{3.6.3}$$

For example, at $z = 1000$ (which plays an important role in some arguments since the universe is believed opaque for larger z by the arguments of 4.8) the mass within a horizon is about 10^{19} solar masses and the horizon for that time subtends about 1° of arc.

4. Evolution of the Hot Einstein De Sitter Model

We now consider in more detail the evolution of geometry and the thermal history of matter. The discussion involves extrapolating backward to very early times; of course the earlier the time involved the more speculative the analysis. However, the rather remarkable, if as yet tentative, success of the helium formation theory suggests that perhaps even extrapolating backward 20 orders of magnitude or more in the time may yield qualitatively correct results.

Throughout this section we shall assume a Robertson-Walker model. For discussing early stages the integration constant k in (3.1.8) is not important; as $R \to 0$ both the kinetic energy term $\frac{1}{2}\dot{R}^2$ and the potential energy term $(\frac{1}{6})\,\varrho R^2$ become large and k becomes negligible compared to them. Thus we shall in fact continue to work with the hot Einstein de Sitter model for explicitness. We shall also assume that every distribution function f_A has the symmetries of the spacetime; thus $f_A = f_A(q, t)$ with $q = \sqrt{p_\mu p_\mu}$ throughout this section. The analysis follows that given by Peebles (1971) and Weinberg (1972).

Subsection 4.1 indicates how the symmetry assumption on f_A can be built into the analysis. Subsections 4.2 and 4.3 are comments on the two extreme cases of complete thermal equilibrium and of collision-free propagation. Subsection 3.4 gives some references to models for very early times, $t < 10^{-2}$ s. Subsection 3.5 outlines the consequences of assuming complete thermal equilibrium at $t = 10^{-2}$ s. Subsection 3.6 analyzes the subsequent time development of $R(t)$. The remaining subsections discuss briefly what happens to the various species of particles in this model.

4.1. Evolution Equations

The assumption $f_A = f_A(q, t)$ simplifies the form of the stress-energy tensor. Define

$$V_A(R, t) = 4\pi \int_0^\infty f_A(q, t)\,(R^{-2} q^2 + m_A^2)^{1/2}\, q^2\, dq\,. \tag{4.1.1}$$

Then (2.1.3), (2.2.3) give

$$\varrho_A = R^{-3}(t)\, V_A(R(t), t), \quad p_A = -\left(\tfrac{1}{3}\right)(\partial V_A/\partial R)_{R=R(t)}. \tag{4.1.2}$$

Similarly $N_A = n_i^A f^i$ with $f^i = \delta_4^i$ in the coordinates of Section 3 and with

$$n_A = 4\pi R^{-3} \int_0^\infty f_A q^2 \, dq. \tag{4.1.3}$$

The Boltzmann equation simplifies as in (3.5.3) to

$$\partial f_A/\partial t = (R^{-2}q^2 + m_A^2)^{-1/2}\, c_A(q, t). \tag{4.1.4}$$

Suppose for some initial time t_i we are given each $f_A(q, t_i)$ and choose $R(t_i)$ arbitrarily. Then (4.1.4) and

$$(G^{ij} - T^{ij})\, f_i f_j = 0 \Leftrightarrow 3(\dot{R}/R)^2 - R^{-3} \sum_A V_A = 0 \tag{4.1.5}$$

uniquely determine $f_A(q, t)$ and $R(t)$. Conversely, if (4.1.4) and (4.1.5) hold then the full Einstein field equations also hold, as can be seen by the following standard argument. Equation (1.1.7), the Boltzmann Equation (4.1.4), and the energy momentum conservation built into the collision integrals $c_A(q, t)$ imply $T^{ij}{}_{;j} = 0$ (Ehlers, 1973). Therefore $f_i(T^{ij} - G^{ij})_{;j} = 0 = [f_i(T^{ij} - G^{ij})]_{;j} - (T^{ij} - G^{ij}) f_{i;j}$. Algebra using (2.2.3), (3.2.2) and (3.2.3) now gives $T^{ij} = G^{ij}$. Thus (4.1.4) and (4.1.5) are the only equations needed. The game now is to approximate c_A appropriately and integrate. To find suitable approximations for c_A involves a lot of physics (Peebles, 1971; Weinberg, 1972). The following comments may clarify roughly what is involved.

Consider a particular reaction, for example photon scattering

$$\gamma + e \rightarrow \gamma + e. \tag{4.1.6}$$

With such a reaction one can associate one or more collision times. In the example suppose henceforth that in the preferred reference frame the electrons are non-relativistic and that the photons have average energy $\langle E \rangle$ much less than the electron rest energy m_e. Let f be the photon distribution function. Then in (4.1.4) there will be a term W due to the reaction (4.1.6):

$$\partial f/\partial t = W + \cdots. \tag{4.1.7}$$

Now $\tau = \int f\pi / \int W\pi$ has the dimensions of a time and measures, roughly speaking, how long it takes a photon to collide. An order of magnitude estimate for τ is $c\tau = (n\sigma)^{-1}$, where n is the electron density and σ the Thomson scattering cross section. $\tau' = (\langle E \rangle / m_e)\, \tau$ is another time. The average energy transfer in one non-relativistic scattering (4.1.6) is of order $\langle E \rangle^2 / m_e$, so τ' measures roughly how long an excess energy photon takes to lose a significant amount of its energy by scattering.

By comparing every relevant collision time with the characteristic expansion time $R/\dot{R}$ one gets estimates for which reactions are important at any particular cosmological epoch. If $\tau \ll R/\dot{R}$ for sufficiently many collision times one can use the complete thermal equilibrium approximation discussed in subsection 3.2. If $\tau \gg R/\dot{R}$ for all collision times involving a particular species one can take that species as collision free. Many intermediate situations are conceivable. Subsections 2.5, 3.5, and 6.3 indicate a few of the techniques used when c_A is neither negligible nor dominant. More systematic analyses are given by Peebles (1971) and Weinberg (1972).

4.2. Complete equilibrium

As long as complete local equilibrium (2.3) holds the only unknowns are $R(t)$, $T(t)$ and the basic chemical potentials α_B etc. The time development of the chemical potentials is determined from (3.1.7), for example, we can regard the law $b(t)=b(t_e)(R_e/R)^3$, for baryon number density as determining the baryon chemical potential α_B. $R(t)$ is determined by (3.2.2). The equation for $T(t)$ can be taken as $f_i T^{ij}{}_{;j}=0$ which is equivalent to

$$\dot{\varrho} + 3(\varrho + p)\,\dot{R}/R \equiv 0\,. \tag{4.2.1}$$

Conversely, by the argument of Subsection 4.1, the above relations imply $G^{ij}=T^{ij}$ so they can be taken as the complete set of basic equations. Sometimes one applies these equilibrium relations to a subset of all the species. Note that by (2.3.3) entropy density is conserved:

$$s = s(t_e)\,[R_e/R]^3\,. \tag{4.2.2}$$

4.3. Collision-free behaviour

If $c_A=0$ the solution to (4.1.4) is just $f_A=f_A(q)$. Then $n_A \propto R^{-3}$ by (4.1.3); ϱ_A and p_A obey (4.2.1); s_A is conserved as in (4.2.2).

Suppose that at one time f_A has the equilibrium form (2.3.1). Substituting $E^2=\sqrt{R^{-2}q^2+m_A^2}$ into $f_A=f_A(q)$ shows that in general the collision free time development does not preserve this equilibrium form. However, for m_A negligible $E=R^{-1}q$ shows that (2.3.1) is preserved with

$$T \propto R^{-1}\,. \tag{4.3.1}$$

At the other extreme $E=m_A+\frac{1}{2}m_A R^{-2}q^2$ we again have the form (2.3.1) preserved with

$$T \propto R^{-2}\,. \tag{4.3.2}$$

4.4. Early epochs

We now discuss the evolution of the model. Our basic 'initial' value assumptions will presently be formulated for a time $t_i \cong 10^{-2}$ s. At still earlier times, between the big bang and 10^{-2} s, even speculative results are as yet rather scarce. It is conjectured that for $t<10^{-23}$ s quantum effects may be dominant. Some attempts to deal with these very early epochs are in Parker (1972), Omnes (1971), Ryan (1972) and the references given there.

It has been suggested that for t greater than 10^{-23} s but less than, say, 10^{-4} s the main uncertainty in the model may be due to our lack of sufficient knowledge of hadron physics. One might assume a finite number of elementary particle species – say gravitons, photons, $e^{\pm}$, the various neutrinos, muons and quarks (compare Weinberg, 1972) – treated by kinetic theory. Then the mixture presumably behaves as a $p = \frac{1}{3}\varrho$ gas in thermal equilibrium 4.3 throughout this epoch, with temperatures of order 10^{12} K at the end and much higher temperatures earlier.

Alternately, one can assume a Hagedorn spectrum of particle species. For a Hagedorn spectrum, the number $V(m)$ of species in rest mass interval dm at rest mass m is taken to be

$$V(m) \propto m^{-B} \exp(m/kT_H), \tag{4.4.1}$$

where B is about 3 and the Hagedorn temperature T_H is somewhat less than $2 \cdot 10^{12}$ K. Our previous formalism is assumed applicable with $\sum_A \rightarrow \int V \, dm$. Assuming (4.1.1) and approximate thermal equilibrium one finds that even at very early stages the temperature never exceeds T_H. The dominant contribution to the energy density comes from extremely massive particles; $p/\varrho \rightarrow 0$ as we go backward in time toward $T \rightarrow T_H$. Recent discussions of the Hagedorn model are given by Weinberg (1972) and by Frautschi *et al.* (1972).

4.5. Initial Conditions

It is possible, though by no means certain, that at some time t_i the universe arrives at a state well approximated by assuming complete thermal equilibrium at an 'initial' temperature $T_i = 10^{11}$ K. The relevant constituents are then photons, $e^{\pm}$, ν_e, $\bar{\nu}_e$, ν_μ, $\bar{\nu}_\mu$, protons and neutrons. We shall assume, as an initial condition for the further time development, that there is such a time. The assumption is self consistent: at 10^{11} K, kT is less than the muon rest mass energy so muons or unstable strongly interacting particles need not be present; moreover, if one estimates collision times as in 4.1 one finds that small deviations from thermal equilibrium would quickly be wiped out by collisions. On the other hand, one can envisage many situations where deviations from equilibrium are so large that the collision time estimates (which require approximate knowledge of the densities and energies) are inapplicable.

On our assumption of thermal equilibrium at $T_i = 10^{11}$ K we must merely fix the basic chemical potentials to define the state completely. The simplest assumption is to take the charge density, electron-lepton number density, and muon-lepton number density as zero and choose that value for the baryon number density which leads to (1.1.2) at the end of the integration process described in 4.1. We henceforth specialize to this simplest case. Other cases are considered by Weinberg (1972) and Peebles (1971).

Then our initial state can be qualitatively described as follows. The energy density in photons is aT^4 (Subsection 2.3); the total energy density in e and μ neutrinos and anti-neutrinos is $\frac{7}{4} aT^4$, with each species present in equal numbers to sufficient approximation. The rest mass of electrons and positrons is negligible to good approximation. There is a slight excess of electrons over positrons but to good approximation we can neglect the electron chemical potential. With these two approximations the contribution of $e^{\pm}$ to ϱ and p is

$$\varrho = \tfrac{7}{4}\, aT_i^4, \quad p = \tfrac{1}{3}\,\varrho\,. \tag{4.5.1}$$

Neutrons and protons are non-relativistic but the neutron-proton mass difference is (barely) negligible. We thus have approximately equal numbers of neutrons and protons. The contribution they make to the energy density is approximately the rest-mass contribution

$$\varrho = m_P b(t_i) \lesssim 10^{-6} aT_i^4, \quad b(t_i) \sim 10^{27}\ \text{cm}^{-3} \tag{4.5.2}$$

and is negligible at this initial time. All other species are negligible (we here ignore a possible contribution from gravitons). From these estimates we can assign an initial time t_i as follows: the mixture is a $p=\frac{1}{3}\varrho$ gas to good approximation and for a $p=\frac{1}{3}\varrho$ gas $R\propto t^{1/2}$ (Subsection 3.2). Extrapolating backward to $t=0$ we get $\varrho_i = 3(\dot{R}_i/R_i)^2 = 3/4t_i^2$. This gives

$$t_i = (4\varrho_i/3)^{-1/2} = (6aT_i^4)^{-1/2} \cong 10^{-2}\,. \tag{4.5.3}$$

Whether this t_i is actually in some sense time since the big bang depends on what happened earlier, but for temperature somewhat lower than 10^{11} K the corresponding approximation

$$t \cong (6aT^4)^{-1/2} \cong (T/10^{10}\,\text{K})^{-2} \tag{4.5.4}$$

is useful.

4.6. Evolution of $R(t)$

If we want to compute $R(t)$ for $t>t_i$ we can use simpler approximations than if we try to keep track of the matter in detail. In this subsection we sketch the results. For present purposes we may divide the species into 3 groups: neutrinos; neutrons and protons; and a mixture of photons, electrons, and positrons. Estimates of neutrino cross sections (Weinberg, 1972) indicate that for $t>t_i$ we can approximate the neutrinos as collisionless for present purposes. Thus by 4.3 the neutrino contribution to T^{ij} is determined by $T\propto R^{-1}$ or more explicitly

$$\varrho_\nu = \tfrac{7}{4}\, a\,(T_i R_i/R)^4, \quad p_\nu = \tfrac{1}{3}\,\varrho_\nu \tag{4.6.1}$$

at all later times.

The protons and neutrons, perhaps assembled into nuclei (or galaxies for that matter), contribute an amount given by (1.1.2) and baryon number conservation:

$$\varrho \cong b(t_o)\, m_P (R_o/R)^3, \quad p = 0\,. \tag{4.6.2}$$

They exert no significant influence on the geometry until $R \gtrsim 10^5\, R_i \sim 10^{-5}\, R_o$ and thereafter become dominant.

Finally, the chemical potentials for electrons and positrons can be taken zero for present purposes. Collision time estimates indicate that for present purposes the electron, positron, photon mixture remains in thermal equilibrium at a common temperature $T(t)$ until after the electrons and positrons no longer exert significant influence on the geometry. The methods of (4.2), applied to just these three species, remain applicable throughout. At temperatures of about 5×10^9 K, $(kT \approx m_e c^2)$ the electrons and positrons annihilate, feeding their energy into the photons. Near the start of the process we thus have $p = \frac{1}{3}\varrho \Rightarrow \varrho \propto R^{-4}$ for the mixture and near the end we have the same behaviour for the photons alone. Numerical integrations are required for the transition region. The ultimate photon temperature can be obtained from entropy conservation. 2.2 and 4.2 give

$$T = T_o (R_o/R) \cong 1.4 T_i (R_i/R). \tag{4.6.3}$$

The above equations give a complete scheme for finding $R(t)$ by integrating $\varrho = 3(\dot{R}/R)^2$. The qualitative behaviour is that outlined in 3.2. Peebles (1971), Weinberg (1972), and Rees (1971) give graphs.

Note that with these approximations the entropy per baryon remains strictly constant during the entire time development from 10^{11} K to the present.

4.7. Helium Production

The model predicts primordial helium. Essentially what happens is the following. As the mixture at 10^{11} K cools, neutrons and protons remain in thermal equilibrium with the γ, $e^{\pm}$ components until the electron positron pairs annihilate at about 5×10^9 K. Thereafter the neutron proton ratio remains nearly fixed until $T \sim 10^9$ K since processes like $p + e \Leftrightarrow n + \nu_e$ are no longer effective and neutron decay is fairly slow compared to the time scales (4.5.4). At about 10^9 K deuterium formation, previously negligible because of photodissociation and other high energy collisions, becomes possible. Almost all the remaining neutrons go into deuterium and then almost all the deuterium quickly becomes He^4. Detailed calculations (Peebles, 1971) give a mass fraction of about 25–30% helium. It is the apparent agreement of this fraction with observed values that has encouraged the extrapolations to early times.

4.8. Later Thermal Evolution

On the present assumptions – strict homogeneity, strict isotropy, nearly thermal equilibrium, etc. – calculating the behaviour of the matter for temperatures less than 10^9 K is comparatively very straightforward. At $T < 10^9$ K one has neutrinos completely decoupled from all other matter, a mixture of nuclei (almost entirely hydrogen and helium), enough residual electrons to give charge neutrality, and photons. Apart from the fact that nuclei are not present in the thermal equilibrium abundance (which can of course be taken into account by taking the fraction of each stable nuclear species fixed at the same value as it was at 10^9 K) all species except neutrinos remain in thermal equilibrium until photon temperatures of about 5×10^3 K. Thereafter atomic hydrogen and helium are formed, with a very small residual ionization and a small distortion in the photon spectrum due to recombination radiation.

Thermal equilibrium is broken at temperatures of about 4000 K, corresponding by (4.3.1) to redshifts of order 10^3, when scattering becomes ineffective due to the disappearance of free electrons. Thereafter the photons and the other matter follow the collision-free behaviour 4.3.

This simple predicted behaviour is considerably modified when one tries to put in enough lumps to account for the present lumps in the universe. We now turn to a brief discussion of some of the methods used in trying to analyze lumps.

5. Inhomogeneous Models

In order to get models which show anything like the very rich lump structure we observe, one must consider inhomogeneous spacetimes. The known exact inhomogeneous cosmological spacetimes all have high symmetry; most attempts use perturbation theory. Sections 5.1–5.2 summarize a few features of linearized perturbation spacetimes. Section 5.3 mentions some of the known exact inhomogeneous models. Inhomogeneous models have perhaps not received as much attention from geometrically minded theoreticians as they deserve. Though less elegant than the homogeneous anisotropic spacetimes they are probably more important.

5.1. Perturbation methods

The key advantage of linearized perturbation theory is that one can consider arbitrary deviations from homogeneity and isotropy, rather than having to impose artificial symmetry assumptions to get manageable equations. The drawbacks are that perturbation theory has not yet been formulated in a really geometric way and that one doesn't respect the full non-linearities of the basic equations. The pioneering work is that of Lifshitz (1946). A systematic review of the literature and results is given by Weinberg (1972).

In linearized perturbation theory one considers an unperturbed spacetime M endowed with an unperturbed metric tensor g and an 'infinitesimally neighbouring'* perturbed spacetime M' with metric tensor g'; M' is take to have the same topology and differential structure as M. It is assumed that for at least one diffeomorphism Y of M onto M' (a 1–1 identification smooth both ways; see Schmidt (1973)) we have $Y^*g' = g + k$, where all terms quadratic or higher in k and its derivatives can be neglected to adequate approximation in all calculations. For example, if in Section 3.2 we set $p=0$ we get a unique spacetime (M, g) and if we set $p=10^{-6}\,p$ we get another unique spacetime (M', g'). One identification Y would be defined by identifying those points which have the same coordinates in the coordinate system(s) used. The approximation of neglecting terms quadratic in k then in fact gives sensible results except near $t=0$.

The diffeomorphism Y mentioned is not in any natural sense unique. In our

* The concept of infinitesimally neighboring can be formalized by considering a smooth curve in the infinite dimensional manifold Lorentz (M) of C^∞ Lorentzian structures on a given C^∞ manifold M. We omit the construction here.

example one might as well identify points with the same value of $\dot{R}/R$ as identify points with the same value of t, etc. This arbitrariness in Y corresponds to gauge transformations. If $k = Y^* g' - g$ is one perturbation, a tensor field on the unperturbed manifold, and ξ is any smooth vector field on the unperturbed manifold then $\bar{k} = k + L_\xi g$ is a perturbation with exactly the same physical significance, where L_ξ denotes the Lie derivative. In fact, there will then be a sensible $\bar{Y}$ such that $\bar{k} = \bar{Y}^* g' - g$ to relevant order. Moreover, if we consider any geometric structure on (M', g') such as the Ricci tensor R' the effect of a gauge transformation is to replace $\delta R = Y^* R' - R$ by $\delta\bar{R} = \bar{Y}^* R' - R = \delta R + L_\xi R$, where the last equality holds to relevant order.

In coordinates the perturbation rules amount to the following. We consider various unperturbed tensor fields g_{ij}, ϱ, R_{ijkm}, etc. on spacetime. The actual spacetime of interest has metric $g'_{ij} = g_{ij} + k_{ij}$, density $\varrho' = \varrho + \delta\varrho$, curvature $R_{ijkm} + \delta R_{ijkm}$, etc., where k_{ij}, $\delta\varrho$, δR_{ijkm} are tensor fields on spacetime. All terms quadratic in the deviations k_{ij}, $\delta\varrho$, etc. and their derivatives are neglected. If ξ^i is any smooth vector field on spacetime then

$$\begin{aligned} \bar{k}_{ij} &= k_{ij} + \xi_{i;j} + \xi_{j;i}\,; \\ \overline{\delta\varrho} &= \delta\varrho + \varrho_{,i}\xi^i\,; \\ \overline{\delta R}_{ijkm} &= \delta R_{ijkm} + L_\xi R_{ijkm}\,; \end{aligned} \tag{5.1.1}$$

etc., describe the same perturbed situation as the set $(k_{ij}, \delta\varrho, \delta R_{ijkm}, \text{etc.})$.

5.2. Simple Perturbation Results

The simplest example which shows some of the methods and results of linearized perturbation theory is that of $p = 0$ perturbations on a $p = 0$ Einstein de Sitter spacetime (Sachs and Wolfe, 1967).

By Section 3 the metric is

$$ds^2 = -\,dt^2 + t^{4/3}\,dx^\mu\,dx^\mu + k_{ij}\,dx^i\,dx^j\,. \tag{5.2.1}$$

The restriction on k_{ij} is that the Einstein tensor calculated from (5.2.1) by linearizing have the algebraic form

$$G^{ij} = \varrho f^i f^j \tag{5.2.2}$$

with $f^i f_i = -1$. By using comoving coordinates for the perturbed spacetime one can insist that

$$f^i = \delta^i_4 \Rightarrow k_{44} = 0\,. \tag{5.2.3}$$

Then the remaining allowed gauge transformations are restricted by the condition $L_\xi f^i = 0 \Leftrightarrow \partial\xi^i/\partial t = 0$. Moreover,

$$G^{44} = \varrho \tag{5.2.4}$$

$$G^{\mu\nu} = 0 = G^{\mu 4} \tag{5.2.5}$$

(5.2.4) simply determines ϱ. Because of the form of the unperturbed metric and the linearization, (5.2.5) are linear partial differential equations in k_{ij} with coefficients that depend only on t, not on x^μ. They can be solved explicitly to get the following results.

Up to a gauge transformation every solution k_{ij} is a linear combination of gravitational radiation modes, vorticity modes, and density perturbation modes. To characterize the latter we introduce the density contrast

$$\varepsilon = \delta\varrho/\varrho = [\varrho - (\tfrac{4}{3}t^2)]/(\tfrac{4}{3}t^2). \tag{5.2.6}$$

We shall also use the horizon parameter $\eta(t) = 3t^{1/3}$ introduced in Subsection 3, and write $\nabla^2 = \partial^2/\partial x^\mu \partial x^\mu$.

The gravitational wave modes are 'generated' by an arbitrary solution $D_{\mu\nu}(\eta, x^\nu)$ of

$$(\partial^2/\partial\eta^2 - \nabla^2)\, D_{\mu\nu} = 0, \quad D_{\mu\nu,\,\nu} = 0 = D_{\mu\mu}, \quad D_{\mu\nu} = D_{\nu\mu}. \tag{5.2.7}$$

Specifically, given any such $D_{\mu\nu}$ setting

$$k_{44} = 0 = k_{4\mu}, \quad k_{\mu\nu} = \eta^3\,(\partial/\partial\eta)\,(\eta^{-1} D_{\mu\nu}) \tag{5.2.8}$$

gives a solution of (5.2.2) with $\varepsilon = 0$. The corresponding conformal tensor is gauge invariant by the transformation laws mentioned in 5.1 and the conformal flatness of the background metric. By comparing (5.2.8) with solutions of Maxwell's equations on the unperturbed spacetime one finds that the linearized gravitational waves have properties wholly analogous to electromagnetic waves. In particular, gravitational waves with wavelengths small compared to the horizon parameter show the standard red shift.

The vorticity modes are generated by an arbitrary transverse time independent 'vector' field $C_\mu(x^\nu)$, i.e. any solution of

$$C_{\mu,\,\mu} = 0 = \partial C_\mu/\partial t. \tag{5.2.9}$$

Given such a C_μ, setting

$$k_{44} = 0, \quad k_{\mu 4} = \nabla^2 C_\mu, \quad k_{\mu\nu} = (\tfrac{1}{9})\,(8\eta^{-3} - \eta^{-1}\nabla^2)\,(C_{\mu,\,\nu} + C_{\nu,\,\mu}) \tag{5.2.10}$$

gives a linearized solution of (5.2.2.) with $\varepsilon = 0$. The corresponding gauge invariant quantity is the vorticity $\tfrac{1}{2}\omega_{ij}\omega^{ij} \equiv \omega^2$. (5.2.10) shows that along the world line of any comoving observer

$$\omega^2 \propto R^{-4} = t^{-8/3} \tag{5.2.11}$$

in agreement with Newtonian estimates based on angular momentum conservation (Peebles, 1971) and with the linearized version of (2.4.9).

Finally, the density perturbations are determined by two 'potentials' $A(x^\mu)$, $B(x^\mu)$ which are time independent but otherwise arbitrary. Setting

$$k_{44} = 0 = k_{4\mu}, \quad k_{\mu\nu} = \eta A_{,\,\mu\nu} + \eta^4\,[5\delta_{\mu\nu} - (\tfrac{1}{2})\,\eta^2\,(\partial^2/\partial x^\mu \partial x^\nu)]\,B \tag{5.2.12}$$

gives a linearized solution of (5.2.2) with

$$\varepsilon = \text{constant}\left[2\eta^{-3}\nabla^2 A - \eta^2\nabla^2 B\right]. \tag{5.2.13}$$

Perturbation calculations, such as the above and its generalizations, have been applied in two main ways, summarized by Rees (1971) and by Sunyaev and Zeldovich (1972). First, attempts have been made to explain the present lumps by following the growth of perturbations. For example, since the era of decoupling at $z \sim 1000$ (Subsection 4.8) the main matter in the universe has presumably behaved roughly as a $p=0$ gas so the above calculations apply. We can ask what value ε_e one needs at decoupling to account for a present observed lump of, say, $10^{17}\ M_\odot$ with a density contrast ε_o of order unity. By (5.2.13) ε grows as $\eta^2 \propto t^{2/3} \propto (1+z)^{-1}$ so one finds that $\varepsilon_e \sim 10^{-3}$ is needed. By 3.6 the scale of the perturbation at t_e would have to be somewhat less than a horizon scale and its observed angle now would be less than 1°. A peculiar feature of such estimates is that they seem always to lead us further back into the past in the search for an explanation. In the example, one should ask why there should be density perturbations of a particular magnitude and length scale at t_e.

A second application has been to try to place upper limits on the magnitude of the perturbations by estimating their observable effects on microwave and other photons. The perturbations in the geometry influence photons *via* their effects on the Liouville operator L. Moreover, the emission, absorption, and scattering properties of a perturbed region must be somewhat different from those of the background model. The main conclusion has been that on the largest scales, those comparable to the present horizon, the perturbation amplitudes must be quite small (say $\varepsilon \ll 1\%$) to get compatibility with the isotropy of the microwave radiation.

5.3. Other Inhomogeneous Models

Only a few exact spatially inhomogeneous cosmological spacetimes are known. For $p \neq 0$, only spherically symmetric 'similarity' solutions (Cahill and Taub (1971) and references there) and plane symmetric $p=\varrho$ solutions (Tabensky and Taub, 1972) seem to have been investigated. For $p=0$ there are the Swiss cheese models (Kantowski (1969) and references there), and some plane and spherically symmetric models.

The most intensively investigated models are the $p=0$ spherically symmetric ones (Tolman, 1934; Bondi, 1947). Like the perturbation models these have the disturbing feature that by adjusting the big bang one can get almost any behavior one likes. For example, a subclass of the spherically symmetric models has the property that the big bang is spacelike (in Geroch's, 1967 sense) but one can have a delayed little bang (Ne'eman and Tauber, 1967) of arbitrary mass and arbitrary delay time.* In comoving coordinates the metric is

$$\mathrm{d}s^2 = -\,\mathrm{d}t^2 + X^2\,\mathrm{d}M^2 + Y^2(\mathrm{d}\theta^2 + \sin^2\theta\,\mathrm{d}\phi^2). \tag{5.3.1}$$

Here M is essentially the mass within a given sphere. One can choose any bang time

* I am indebted to D. Eardley for discussions of these models.

$\tau(M)$ subject to

$$\mathrm{d}\tau/\mathrm{d}M \leqslant 0, \quad (\mathrm{d}\tau/\mathrm{d}M)_o \geqslant -\tfrac{4}{3}, \quad \tau \underset{M\to\infty}{\to} \text{constant}. \tag{5.3.2}$$

Then setting

$$Y = (9M/2)^{1/3}\,[t-\tau]^{2/3},\ X = Y\left[\frac{1}{3M} - \frac{2}{3}\frac{\mathrm{d}\tau}{\mathrm{d}M}(t-\tau)^{-1}\right] \tag{5.3.3}$$

gives a solution of $G^{ij} = \varrho f^i f^j$, with $f^i = \delta^i_4$. The singularity is three dimensional and spacelike; for large M (corresponding roughly to large radius) the solution becomes an Einstein de Sitter spacetime. But near the center $M=0$ the bang can go off as late as desired.

Finally, we mention that Lifshitz, Khalatnikov and Belinskii have introduced approximations specifically designed to be applicable near the big bang for an inhomogeneous model (Belinskii and Khalatnikov, 1970).

6. Homogeneous Anisotropic Spacetimes

Anisotropic, (spatially) homogeneous spacetimes have been analyzed very intensively during the last five years. Interest has centered on various issues raised by Misner. His 'chaotic cosmology' suggestions were the following. Perhaps 'most' spacetimes highly chaotic near the big bang, in particular highly anisotropic, can acquire an overall isotropy consistent with observation by damping processes. Perhaps 'most' cosmological spacetimes effectively have no horizons, so that inhomogeneities can at least communicate at early times and perhaps settle down to the large scale homogeneity which we appear to observe. Insofar as one can judge from the homogeneous anisotropic models these suggestions now seem to be overoptimistic. But their analysis has clarified a key point: an explanation for the large scale homogeneity and isotropy is both required and difficult. In addition, Misner pointed out that the homogeneous anisotropic spacetimes furnish an excellent way to test ideas about superspace and quantization. Because the mathematics is relatively manageable, one can hope to relate the ideas to actual observations or at least well defined conceptual experiments. Theoreticians have also been fascinated by having available a rich set of examples for analyzing non-linearities in the Einstein field equations, analyzing vorticity, and analyzing the structure of the big bang. Some very pretty formal techniques have been developed for handling the spacetimes. Summaries of the work on such spacetimes have been given by Ryan (1972) and by MacCallum (1973). References to recent work are given by Hawking and Collins (1973). In 6.1–6.3 a brief outline of some, though by no means all, of the main ideas is given.

It should perhaps be emphasized that homogeneous anisotropic spacetimes have severe limitations as models. We seem to observe isotropy and homogeneity in the large, and certainly observe inhomogeneities on smaller scales. Thus to model the universe assuming large-scale anisotropy and assuming homogeneity on all scales is rather artificial. From the point of view of perturbation theory, assuming homogeneity and allowing anisotropy amounts to restricting attention to the long wavelength limit. There is no convincing reason for assuming the long wavelength limit is the one of main interest. Moreover, by assuming homogeneity one automatically loses the chance to analyze causality, to see how dynamical influences spread out. Finally, detailed investigations have not yet uncovered results which are common to all homogeneous anisotropic spacetimes. For example, in many cases assuming a perfect fluid stress-energy tensor results in models which eventually approach a Robertson-Walker model. But this is by no means true in all cases. Since the homogeneous anisotropic spacetimes are at best approximations, the lack of general results is rather disconcerting if we want to decide what conclusions to draw from the spacetimes for observable quantities.

6.1. Group Theoretical Techniques

An 'orbit' $\mathcal{O}$ of an isometry (symmetry) group G is the set of all points reachable from a given point by the operations of G; thus for any point p in spacetime the orbit $\mathcal{O}$ through p is $\mathcal{O}=\{q \mid \exists g\in G \text{ such that } gp=q\}$. A (spatially) homogeneous spacetime is one whose isometry group has 3-dimensional spacelike orbits. Schmidt (1973) and MacCallum (1973) give the definitions and results needed to exploit this definition. In the present subsection let us assume for brevity that G 'acts freely' (is 'simply transitive'); this assumption is equivalent to demanding that for any two points p and q in spacetime there is at most one $g\in G$ such that $gp=q$. We sketch how computations involving quantities like the curvature tensor can be reduced to ordinary differential equations without explicitly using a full coordinate system.

Let $\mathcal{O}$ be any orbit. Label $\mathcal{O}$ as the hypersurface $t=1$. Consider the unique geodesics perpendicular to $\mathcal{O}$. (Locally) one such geodesic goes through each spacetime point. For any (sufficiently small) real t define a transformation f_t of spacetime onto itself by the rule "push each point a distance t toward the future along the geodesic through the point". We shall first prove the following properties: (6.1.1), $f_t\mathcal{O}=\mathcal{O}'$ is an orbit; (6.1.2), $f_tg=gf_t$ for all g in G; (6.1.3), the set $H=\{f_tg \mid g \text{ in } G,\ t \text{ real}\}$ is a group (though not a group of isometries) which acts freely and (locally) has spacetime as its orbit; (6.1.4), the geodesics are orthogonal to $\mathcal{O}'$.

In fact, if p and q are points in $\mathcal{O}$ there is a g in G for which $gp=q$, since $\mathcal{O}$ is an orbit. Let L be the geodesic through p. Then gL is a geodesic since g is an isometry and gL goes through q. Indeed, gL is the orthogonal geodesic through q since g preserves orthogonality and the image of g_* contains the tangent space to $\mathcal{O}$ at q. Moreover, as an isometry, g carries a distance t along L to a distance t along gL. Thus if $p'=f_tp$ and $q'=f_tq$ then $gp'=q'$. This implies $\mathcal{O}'$ is an orbit. Moreover, we have $gf_tp=q'=f_tgp$ for any p, any q and any t. Thus (6.1.2) also holds. (6.1.3) follows by closure: $f_tgf_{-\bar{t}}\bar{g}^{-1}=f_tf_{-\bar{t}}g\bar{g}^{-1}=f_{t-\bar{t}}\tilde{g}$ with $\tilde{g}$ in G for any t, $\bar{t}$, g, $\bar{g}$.

Finally, for any vector n Lie transported along a unit, geodesic vector field f one has $g(n,f)=\text{constant}$ (subsection 1); this implies that the geodesics are orthogonal to $\mathcal{O}'$. Thus the original choice of our orbit makes no essential difference.

Now let p be any point in spacetime and $v\equiv v^i(\partial/\partial x^i)$ be a non-zero vector at p.

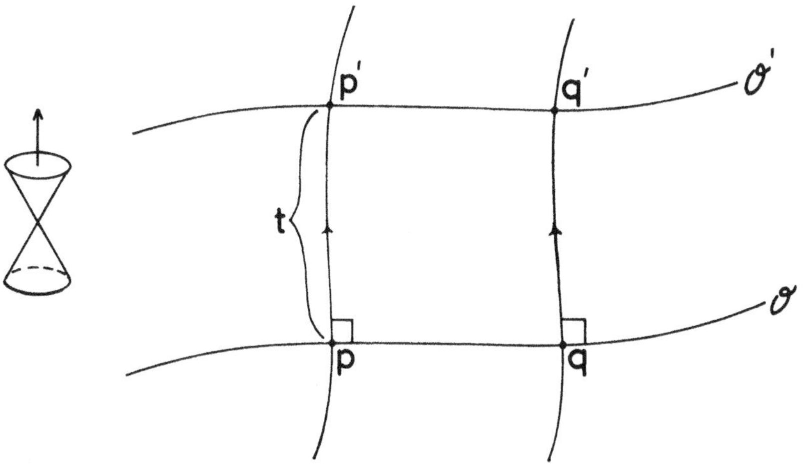

Fig. 6.

Since H acts freely there is a unique 'pushed forward' vector field V on spacetime, defined by $(h_* V)(p) = V(hp)$ for all p and all h in H and by $V(p) = v$ (Schmidt, 1973). Similarly, choosing a basis (v_1, v_2, v_3, v_4) for the tangent space at p gives a unique basis (v_1, v_2, v_3, v_4) for vector fields and a dual basis $(\omega^1, \omega^2, \omega^3, \omega^4)$ for one-forms on spacetime. Define $g_{ij} = g(V_i, V_j)$. Since G is an isometry group the functions g_{ij} are constant on each orbit. Assigning each orbit its appropriate value of t we can write $g_{ij} = g_{ij}(t)$. It is now straightforward to show (MacCallum, 1973) that the connection forms and curvature forms for $(\omega^1, \omega^2, \omega^3, \omega^4)$ depend only on $g_{ij}(t)$, its derivatives, and the structure constants for H. By (6.1.2) the structure constants for H are determined by those for G. All relevant calculations can thus be reduced to working with the structure constants of G and the ten functions $g_{ij}(t)$.

6.2. Bianchi type I spacetimes

The simplest example is obtained by choosing G Abelian. Then the spacetime is called a Bianchi type I spacetime. (Locally) each orbit must be intrinsically flat, since it is a 3-dimensional Riemannian space which is the orbit of an Abelian isometry group. Now choose an orbit and introduce local coordinates by the following requirements: $t = x^4$ is as above; in the orbit chosen x^μ are the ordinary coordinates for flat 3-space; and (x^μ, t) are comoving coordinates. Then the metric takes the form

$$ds^2 = -\,dt^2 + g_{\mu\nu}\,dx^\mu\,dx^\nu, \quad g_{\mu\nu} = g_{\mu\nu}(t). \tag{6.2.1}$$

For example, the absence of cross terms $dt\,dx^\mu$ corresponds to the facts that the geodesics in 6.1 are orthogonal to the orbits and that we can choose and have chosen $g_{\mu\nu} = g(\partial/\partial x^\mu, \partial/\partial x^\nu)$ in the present case.

6.3. Neutrino viscosity

One application of homogeneous anisotropic spacetimes is estimating the effects of neutrino viscosity on anisotropy. We sketch how the field equations and Boltzmann equation are set up in a special case analyzed by Misner (1968) and Matzner and Misner (1972). They obtain controllable approximations for the full range of possibilities, from collision dominated neutrinos to collision free ones.

The spacetime is taken as a Bianchi type I spacetime with additional, reflection symmetries. Suppose that in (6.2.1) $(x^1, x^2, x^3, t) \to (-x^1, x^2, x^3, t)$ and the similar reflections involving x^2 or x^3, are isometries. This occurs iff $g_{\mu\nu}$ is diagonal. It is convenient to rewrite (6.2.1) in the form

$$ds^2 = -\,dt^2 + e^{2\alpha}(e^{2\beta})_{\mu\nu}\,dx^\mu\,dx^\nu\,. \tag{6.3.1}$$

Here $\alpha(t)$ is a function and $\beta(t)$ is a real, diagonal, trace free matrix; $e^{2\beta}$ is the usual matrix exponential. α and β are uniquely determined by $g_{\mu\nu}(t)$. However the latter can be set equal to $\partial_{\mu\nu}$ at any one time by 'stretching' the coordinates x^μ. Thus α and β have no direct significance. Their interpretation is that if $g_{\mu\nu}(t_e)=\partial_{\mu\nu}$ then by (1.4) $2\dot{\alpha}$ is the expansion and $2\dot{\beta}$ is essentially the shear of the reference frame $f^i=\partial^i_4$. It is the non-zero shear which destroys isotropy.

The neutrinos are modelled by a distribution function f which has the same symmetries as the spacetime (Subsection 2.1). The condition that f have the Bianchi type I symmetries is equivalent to the requirement that it depends only on t and p_μ, not on x^μ; thus $f=f(t, p_\mu)$. f has the reflection symmetries mentioned above iff it is an even function of each p_μ. From the reflection symmetry we get directly that f^i is the preferred reference frame (1.4) and that

$$T_{\mu 4} = 0 = G_{\mu 4}; \quad T^{ij}{}_{;j} = 0 \quad \text{iff} \quad f_i T^{ij}{}_{;j} = 0\,. \tag{6.3.2}$$

In the case we shall consider here, Matzner and Misner (1972) specialize further by assuming a particular form for f and replacing the Boltzmann equation by a collision-time approximation ansatz. We state the formal assumptions and then some motivations. It is assumed that for all t, f has the form

$$f = f(t, [e^{-2\psi}]_{\mu\nu}\,p_\mu p_\nu)\,, \tag{6.3.3}$$

where $\psi=\beta$ at some initial time t_e. The initial value assumption is equivalent to assuming isotropy of f at t_e, since $e^{-2\alpha}(e^{-2\beta})_{\mu\nu}p_\mu p_\nu = E^2$ with $E=-f_i p^i$ the energy measured by a comoving observer. The time development equations for f and $g_{\mu\nu}$ are taken to be

$$\dot{\psi} = -\,(1/\tau)\,(\psi - \beta)\,; \tag{6.3.4}$$

$$f_i T^{ij}{}_{;j} = 0\,; \tag{6.3.5}$$

$$G^{\mu\nu} = T^{\mu\nu}\,. \tag{6.3.6}$$

Here the collision time $\tau(t)$ is to be estimated from microphysics. (6.3.4)–(6.3.6) actually give a unique time development consistent with energy-momentum conservation from appropriate initial data. This can be seen, roughly, by imagining $g_{\mu\nu}$ given for the moment. Then (6.3.4)–(6.3.5) determine f from f_e and $T^{ij}{}_{;j}=0$ holds by (6.3.2). On the other hand, if T^{ij} is given (6.3.6) determines $g_{\mu\nu}$ from the initial values $\dot{\alpha}(t_e)$, $\dot{\beta}(t_e)$ and the (irrelevant) initial values of α and β. $G^{ij}=T^{ij}$ as in (4.1).

The particular feature of (6.3.4) is that it extrapolates smoothly between a collision free situation $\tau = \infty$ and a collision dominated situation $\tau \to 0$. For $\tau = \infty$ (6.3.4) and (6.3.5) are equivalent to the collision free Boltzmann equation. For the general solution of the collision free Boltzmann equation is $f = f(p_\mu)$, by the same argument as in Subsection 3.5 and $f = f(p_\mu) \Rightarrow T^{ij}{}_{;j} = 0$. A solution of (6.3.3) and (6.3.4) with $\tau = \infty$ is $f = f[(e^{-2\beta(t_e)})_{\mu\nu} p_\mu p_\nu] \Rightarrow Lf = 0 \Rightarrow T^{ij}{}_{;j} = 0$. By uniqueness the solution of $Lf = 0$ with the indicated initial values and the solution of (6.3.3)–(6.3.5) with $\tau = \infty$ coincide.

In the other limit $\tau \to 0$ we have in zeroth order that $\psi = \beta$ for all times. This gives an isotropic f for all times, by the argument used above at t_e, and thus duly corresponds to the collision dominated perfect fluid case. In first order, for τ small, we have

$$\psi = \beta - \tau\dot{\psi} \cong \beta - \tau\dot{\beta}. \tag{6.3.7}$$

Matzner and Misner (1972) show that (6.3.7) yields the usual viscosity approximation (Weinberg, 1972) for imperfect fluids, with the coefficient of viscosity determined in terms of the collision time τ.

Matzner and Misner integrate the ordinary integro-differential equations (6.3.4)–(6.3.6) by an ingenious combination of Lagrangian and numerical techniques. Two main conclusions are drawn. First, under appropriate circumstances an initial shear can be damped by many orders of magnitude by the neutrino collisions. Second, in agreement with the comments by Stewart (1969) and others, the damping is not so effective that arbitrarily large amounts of shear near the big bang can be dissipated.

7. Conclusion

Four comments might be added. Perhaps the most striking feature of current theoretical cosmology is that the most naive self consistent model we can think of, the hot Einstein de Sitter model, is as good as any other. A discouraging feature is that none of the models suggested during the last fifty years, with the honorable exception of the steady state model, have been sufficiently explicit to give observers reasonable targets to shoot at. The most fascinating puzzles at present seem to be those connected with the initial conditions. The most important immediate task for theoreticians may be that of coming to grips with inhomogeneous models.

Acknowledgements

The author is grateful to the NSF, grant GP-29257, for financial support and to D.A.M.T.P., Cambridge University, for hospitality. Discussions with D. Eardley, G. Ellis, W. Kundt, E. Liang, M. MacCallum, R. Matzner and A. Taub were very useful.

References

Belinskii, V. and Khalatnikov, I.: 1970, *JETP* **30**, 1174.
Bishop, R. and Goldberg, S.: 1968, *Tensor Analysis on Manifolds*, MacMillan.
Bondi, H.: 1947, *Monthly Notices Roy. Astron. Soc.* **107**, 410.

Cahill, M. and Taub, A.: 1971, *Comm. Math. Phys.* **21**, 1.
Collins, B. and Hawking, S.: 1973, *Astrophys. J.* **180**, 317.
Dyer, C. and Roeder, R.: 1973, *Astrophys. J.* **180**, L31.
Eardley, D. and Sachs, R.: 1973, *J. Math. Phys.* **14**, 209.
Ehlers, J.: 1971, in *General Relativity and Cosmology*, Academic Press; Varenna Course XLVII.
Ehlers, J.: 1973, this volume, p. 1.
Ellis, G.: 1971, in *General Relativity and Cosmology*, Academic Press; Varenna Course XLVII.
Frautschi, S., Steigman, G., and Bahcall, J.: 1972, *Astrophys. J.* **175**, 307.
Geroch, R.: 1967, Thesis, Princeton University.
Harlte, J.: 1973, this volume, p. 153.
Hawking, S. and Ellis, G.: 1973, *The Large Scale Structure of Spacetime*, Cambridge U.P.
Kantowski, R.: 1969, *Astrophys. J.* **155**, 89.
Lifshitz, E.: 1946, *J. Phys. U.S.S.R.* **10**, 116.
Mitton, S. and Reinhardt, M.: 1972, *Astron. Astrophys.* **20**, 337.
MacCallum, M.: 1973, *Cargese Lectures* **6**, Gordon and Breach.
Matzner, R. and Misner, 1972, *Astrophys. J.* **171**, 415.
Misner, C.: 1968, *Astrophys. J.* **158**, 431.
Ne'eman, Y. and Tauber, G.: 1967, *Astrophys. J.* **150**, 766.
Omnes, R.: 1971, *Astron. Astrophys.* **11**, 450.
Novikov, I. D. and Zeldovich, Ya. B.: 1973, *Relativistic Astrophysics* **2**, Chicago Univ. Press (to be published).
Parker, L.: 1972, *Phys Rev. Letters* **28**, 11, 705.
Peebles, J.: 1971, *Physical Cosmology*, Princeton U.P.
Penrose, R.: 1967, in *Battelle Rencontres*, Benjamin.
Rees, M.: 1971, in *General Relativity and Cosmology*, Academic Press, Varenna Course XLVII.
Ryan, M.: 1972, *Hamiltonian Cosmology*, Springer.
Sachs, R. and Wolfe, A.: 1967, *Astrophys. J.* **147**. 73.
Schmidt, B.: 1973, this volume, p. 289.
Sciama, D.: 1971, in *General Relativity and Cosmology*, Academic Press, Varenna Course XLVII.
Stewart, J.: 1969, *Monthly Notices Roy. Astron. Soc.* **145**, 347.
Stewart, J.: 1971, *Non-Equilibrium Relativistic Kinetic Theory*, Springer.
Sunyaev, R. and Zeldovich, Ya. B.: 1972, *Astron. Astrophys.* **20**, 189.
Tabensky, R. and Taub, A.: 1973, *Comm. Math. Phys.* **29**, 61.
Tolman, R.: 1934, *Proc. Nat. Acad. Sci.* **20**, 169.
Weinberg, S.: 1972, *Gravitation and Cosmology*, Wiley and Sons.

CANONICAL QUANTIZATION OF GRAVITY

KAREL KUCHAŘ
Joseph Henry Laboratories, Princeton University, Princeton, N.J. 08540, U.S.A.

imprimatur

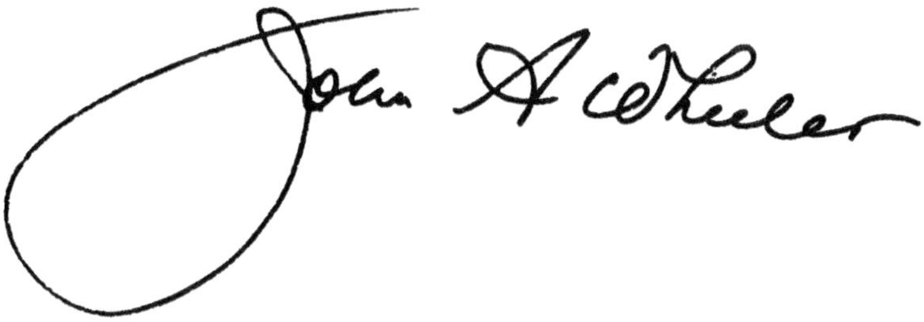

EXPLANATION CONCERNING THE CANON LAW*

From Volume 12, Encyclopaedia Britannica, 1971, p. 10

Imprimatur. In Roman Catholic canon law permission of the bishop is required for the publication of any work on Sacred Scripture, dogmatic, moral or ascetical theology, or, in general, writings containing anything of peculiar significance to religion or morality (can. 1385, sec. 1).

Strictly speaking, the *imprimatur* ('let it be printed') is nothing more than this permission. But because its concession must be preceded by the favourable judgment of a censor (*nihil obstat*, 'nothing hinders [it from being printed]'), the term has come to imply ecclesiastical approbation of the publication itself. This does not mean episcopal endoursement of the content, nor is it a guarantee of doctrinal integrity. It does indicate that ordinarily prudent means have been taken to assure the worthiness of the book, and that nothing offensive to faith or morals has been discovered (cf. can. 1393, sec. 2).

By the Licensing Act of 1662, a 'royal imprimatur' was required for all books printed in or imported into England; the act expired in 1695.

Bibliography. – E. Jombart, 'Censure des livres', *Dictionnaire de droit canonique*, vol. iii, col. 157–169 (1942); D. Wiest, *Precensorship of Books* (1953); T. L. Bouscaren and A. C. Ellis, *Canon Law*, 3rd ed., pp. 750–759 (1957).

* Not to be confused with the canonical formalism which is the subject of these lectures.

W. Israel (ed.), Relativity, Astrophysics and Cosmology, 237–288.

Contents

1. Parametrized Theories

Time present and time past
Are both perhaps present in time future,
And time future contained in time past.
If all time is eternally present
All time is unredeemable.

T. S. Eliot: *Four Quartets*
Burnt Norton, 1–5.

Two questions haunt anybody who sets out to quantize gravity: Why and how. Concerning the first, it is not at all logically necessary that the gravitational field be quantized because other fields which interact with it are. As one alternative, Rosenfeld and Møller proposed that the gravitational field should be treated as a classical field produced by the expectation value of the energy-momentum tensor of all other (quantized) fields according to Einstein's gravitational law. However, it is bad policy to put yourself on the defensive at the very beginning of your lecture by inventing rationalizations why you are doing what you are doing. Let me therefore answer the first question in the words of George Leigh Mallory when he was asked why he wanted to climb Mount Everest: "Because it is there!"

The second question is unfortunately much more specific, and therefore much more difficult to answer. Different investigators have proposed different methods how to quantize the gravitational field, and at present it is unknown whether these methods are equivalent, and unclear which one is the best. To give a few examples, we have a covariant quantization by DeWitt, the constraint quantization by Dirac, the ADM method, Mandelstam's quantization through the gauge-invariant quantities, Schwinger's source-theory formalism, quantization by summing over histories, quantization on light cones, Penrose's twistor formalism, and at least half a dozen other major methods. To decide which method, if any, seems worth pursuing is largely a matter of personal faith. I shall lecture here about the canonical quantization of gravity, though I would describe myself as a quantization polytheist, reluctant to pray my 'credo in unum Deum'. I shall explain my preference for the canonical quantization in the course of these lectures.

In the canonical approach, one tries to cast the action functional of the system into Hamiltonian form and then use the Poisson bracket – commutator relationship to quantize it. This seems straightforward, but it turns out that there are variations on this theme. A year ago, Bryce DeWitt counted seven of them, and I am sure that others have since been composed. Whithin one creed, there are different denominations. I shall mostly follow the Dirac constraint version of the canonical quantization.

Two features of the gravitational field make it special for any method of quantization: its intrinsic non-linearity, and its invariance with respect to an infinitely dimensional group. None of these features is, of course, an exclusive attribute of gravity. Concerning non-linearity, any theory of interacting fields is non-linear. These non-linearities, however, are usually introduced in a highly phenomenological way, adding some coupling terms to the free field Langrangians. On the other hand, the coupling of the gravitational field to itself is something dictated by the geometrical principles of its description, and therefore very fundamental. One finds no coupling constant on the left-hand side of Einstein's gravitational law (though Newton's gravitational constant couples gravity to other fields), and any attempt to divide the pure gravitational action into two terms, one describing the free gravitational field, another the interaction of that field with itself, seems unnatural. I believe that any quantization of the gravitational field should take this message seriously. One should treat the gravitational field as a single entity, and not decompose it into a background which remains classical, and a deviation from that background, which is quantized. Here is one of the reasons why I prefer the canonical quantization to covariant methods. Even if one could prove that the choice of the background does not alter results, one would think that the split into the background and the quantized part of the field is an unecessary procedure.

Connected with this point is a matter of emphasis which questions in quantized gravity does one find more interesting: those concerning scattering experiments in an asymptotically Euclidean spacetime, or those concerning phenomena in strongly nonlinear regions, for example, during the final stage of the gravitational collapse of a star or of a closed universe? The covariant methods are ideal for answering the first type of questions, being geared to an S-matrix approach. For myself, the fascinating question is whether quantization can prevent the singularities predicted by the classical theory. I believe that canonical methods have a good chance to articulate this question, though I would avoid any prediction what the answer may be.

One more remark is appropriate at this place. Classical relativity does not seem to have any special preference for asymptotically Euclidean spacetimes, allowing solutions with different topologies. This suggests that one should prefer those methods of quantization which do not conceptually depend on the asymptotic structure at infinity and which are applicable, with appropriate modifications, to closed as well as to open universes. Canonical quantization is such a method. On the other hand, it is difficult to conceive how to use some other methods of quantization as, for example, the quantization on light cones, in the spaces which are not asymptotically Euclidean.

The second feature of the gravitational field, namely, the existence of an infinitely dimensional invariance group, is also not special to gravity. Electrodynamics, as well as the theory of Yang-Mills fields, is invariant with respect to a gauge group. However, the gauge group of electrodynamics is too simple, being Abelian, and if one is going to climb the Yang-Mills fields, one does not have the certainty which inspired Mallory when he attempted to climb Everest: one does not know if they are there! Moreover, the invariance group of relativity is related to the relativity of space and time description itself, which poses special problems.

An infinitely dimensional gauge group is particularly troublesome to the canonical quantization. It implies that the theory contains more variables than we need to describe the physical state of the system. For example, in electrodynamics we use the components A_ι of the four-potential as our field variables, but we actually need only the equivalence classes $\{\bar{A}_\iota : \bar{A}_\iota = A_\iota + \Lambda_{,\iota}\}$. Similarily, in relativity we use the metric ${}^4g_{\iota\kappa}$ as field variables, but we need only the equivalence classes of metrics which describe the same geometry, $\{{}^4\bar{g}_{\iota\kappa}(\bar{x}^\lambda) : {}^4\bar{g}_{\iota\kappa} = {}^4g_{\alpha\beta}(\partial x^\alpha/\partial\bar{x}^\iota)(\partial x^\beta/\partial\bar{x}^\kappa)\}$. It would be ideal to use as field variables only gauge-invariant quantities, i.e., such function(al)s of A's or g's which are not changed by transformations induced by the invariance group. In free field electrodynamics, such quantities are known: use the transversal part $A^T{}_\iota$ of the four-potential as field variables. In gravitation, we do not know how to separate the gauge-invariant variables from the superfluous variables in a simple way, and this is one of our major troubles. We are forced to use redundant variables as canonical coordinates, and if we use them, there are constraints imposed on canonical momenta. My main concern in these lectures is how to quantize a theory expressed in terms of redundant variables. General relativity is a very complicated example of such a theory, and I shall use a series of simple models to illustrate its basic dynamical structure. I hope that when one understands these models, one has a better feeling for what is going on when quantizing gravity. In fact, an alternative and much more fitting title for my lectures would be:

a concise summation and encompassment of the true theory

of the

quantization of the gravitational field

in which the right proper

canonical quantization

is examin'd, explain'd, expostulat'd, exhibit'd, elucidat'd, execut'd, and verified

in four discourses

and further, in which ye shall find the **notable problems** therein described, illustrated, and made clear by a compacted

series of simple models:

in strict conformity to all known references to the same in the teachings

of the

hebraic fathers, apostles, and all divines of true religion

The problem of redundant variables takes on a very interesting aspect in the general theory of relativity. In picking up redundant variables, the theory lost Old Father Time! To see how this happened, think of a good method of losing something. A possibility is: mix it with something else! A good way to lose a needle is to mix it with a stock of hay. A good way to make it difficult for Cinderella to collect poppy is to mix it with ashes. However, the method is not fool-proof, and Cinderella was not a fool. There were doves that helped her to distinguish poppy from ashes, and there are magnets that help us to distinguish needles from hay. So what is a sure method to lose a needle or a poppy seed? The answer is, I believe, to mix it with something very much the same, the needle with needles and the poppy seed with poppy seed. There is a lovely story by G. K. Chesterton about a would-be murderer who is confronted with the problem which frequently confronts murderers, at least in detective stories: what to do with the body? Fortunately, he has a convenient profession: he is a general. So he starts a battle around his single body, the only battle he ever lost, because he is a good general, and the battle serves his purpose: it creates many more bodies around the one he would like to lose. How simple.

Taking this lesson, the special theory of relativity looked for something very similar to Time when trying to lose Time, and having decided that the nearest thing to Time is Space, she mixed them together. As everybody knows, she made it quite difficult for us to separate Time from Space unambiguously, each observer trying his best, but their answers differing. However, she still kept Space and Time pretty distinct from the rest of physics. In particular, she kept them distinct from the dynamical variables; a field $\phi(T, X)$ evolves in Time and Space, but it is not mixed with Time and Space.

This oversight was promptly corrected by the general theory of relativity. In special relativity, the Minkowskian coordinates T and X have a double function; they serve as labels, but at the same time they inform us about the time durations and space distances. In the general theory of relativity, we must strictly distinguish coordinate labels from the proper time intervals. We work at two totally different levels: that of the differential topology, and that of the metric geometry. To predict what a physical clock will show, we must know the metric, not only the labels distinguishing one event from another. But the metric changing along a worldline describes at the same time a dynamical quantity – the gravitational field. So Time is mixed up with dynamics. You may say: it does not mean it is lost – it only means that the same quantity which describes Time also describes Gravitation. Time has a double face, as the Time-god, Janus, has in the ancient representations. One quantity has two interpretations – chronometrical, and dynamical.

Unfortunately, I have used the term 'dynamical variable' rather loosely. I noticed that ${}^4g_{IK}$ describes gravitation and that it changes along a worldline – so I called it a dynamical variable. However, there is a much more specific use of this term, and it is this use which is important for quantum mechanics. This use is related to spacelike hypersurfaces rather than to timelike lines. 'Dynamical quantity' in this sense is something which can be freely specified on a given spacelike hypersurface ('at a given moment') together with its rate of change, and which is afterwards uniquely determined by the equations of motion. Or, in the canonical language, dynamical variables are the canonical coordinates and momenta which can be initially freely specified, and which are afterwards uniquely determined by the Hamilton equations.

In this sense, the metric tensor ${}^4g_{IK}$ is *not* a dynamical variable. One knows from the study of the gravitational Cauchy problem that neither can ${}^4g_{IK}$ and ${}^4\dot{g}_{IK}$ be initially freely specified, nor do the equations of motion afterwards determine these quantities uniquely. One would somehow like to recover the old comforts of a Hamiltonian-like scheme: a system of hypersurfaces stacked in a well defined way in spacetime, with a system of dynamical variables distributed over these hypersurfaces and developing uniquely from one hypersurface to another. This is not at all easy in the general theory of relativity, because the dynamical variables and the 'kinematical variables' (space and time) are too intimately interconnected.

If it is difficult to treat the gravitational field according to the canonical pattern of ordinary field theories, is it perhaps not easier to treat the ordinary field theories according to the pattern of the gravitational field? If the mountain won't come to Mohammed, Mohammed must go to the mountain. This is the route I shall take. Now, time can be lost essentially in the same way, though far less successfully, by theories other than the general theory of relativity. These (simpler) theories will serve as models indicating how to interpret the hidden dynamics of the gravitational field.

Start with the action functional

$$S = \int \mathrm{d}T \mathring{L}\left(T, X^i, \frac{\mathrm{d}X^i}{\mathrm{d}T}\right) \tag{1}$$

of an elementary dynamical system with a finite number of degrees of freedom X^i. The time T may be a Newtonian time in classical physics, e.g., when we investigate the motion of a non-relativistic particle in a potential field,

$$\mathring{L} = \frac{1}{2} m\delta_{ik} \frac{\mathrm{d}X^i}{\mathrm{d}T} \frac{\mathrm{d}X^k}{\mathrm{d}T} - V(T, X^i), \tag{2}$$

or a Minkowskian time in relativistic physics, e.g., when we investigate the motion of a single free relativistic particle,

$$\mathring{L} = -m\sqrt{1 - \delta_{ik}\frac{dX^i}{dT}\frac{dX^k}{dT}}. \tag{3}$$

Introduce an arbitrary label time t,

$$T = T(t), \quad t = t(T) \tag{4}$$

and denote the derivatives with respect to this label time by dot. Change the time variable in the action functional (1) and find in this way the new Lagrangian $\mathscr{L}$,

$$S = \int dt \, \underbrace{\dot{T}\mathring{L}(T, X^i, \dot{X}^i\dot{T}^{-1})}_{\equiv L(T, \dot{T}; X^i, \dot{X}^i)}. \tag{5}$$

The expressions $T(t)$ and $\dot{T}(t)$ are treated as prescribed functions of t, not as dynamical variables. Only the physical coordinates X^i are to be varied. Define the momentum π_i relative to the label velocity $\dot{X}^i$,

$$\pi_i \equiv \frac{\partial L}{\partial \dot{X}^i} = \dot{T}\frac{\partial \mathring{L}}{\partial \dot{X}^i} = \dot{T}\frac{\partial \mathring{L}}{\partial\left(\frac{dX^i}{dT}\right)}\dot{T}^{-1} = \mathring{\pi}_i. \tag{6}$$

One sees that it does not differ from the physical momentum

$$\mathring{\pi}_i \equiv \frac{\mathring{L}}{\partial\left(\frac{dX^i}{dT}\right)}. \tag{7}$$

Similarly, the new Hamiltonian differs from the physical Hamiltonian merely by the factor $\dot{T}$,

$$H = \pi_i\dot{X}^i - \mathring{L} = \mathring{\pi}_i\frac{dX^i}{dT}\dot{T} - L\dot{T} = \mathring{H}\dot{T}. \tag{8}$$

What is most important, $\mathring{H}$ when expressed as a function of new canonical variables π_i and X^i does not depend on $\dot{T}$:

$$\mathring{H} = \mathring{H}(T, X^i, \mathring{\pi}_i) = \mathring{H}(T, X^i, \pi_i). \tag{9}$$

Writing the action functional (1) in the Hamiltonian form,

$$S = \int dt(\pi_i\dot{X}^i - \mathring{H}(T, X^i, \pi_i)\,\dot{T}), \tag{10}$$

we notice that it is linear not only in the 'dynamical velocities' $\dot{X}^i$, but also in the 'time velocity' $\dot{T}$. It is thus tempting to introduce the physical time T as one of the canonical coordinates $X^\prime = (T, X^i)$, and to identify $-\mathring{H}(T, X^i, \pi_i)$ as the conjugate momentum π_T, $\pi_\iota = (-\mathring{H}, X^i)$ bringing the action functional (10) into a homogeneous form

$$S = \int \mathrm{d}t\, \pi_\iota \dot{X}^\iota = \int \mathrm{d}t\, (\pi_T \dot{T} + \pi_i \dot{X}^i)\,. \tag{11}$$

Unfortunately, the canonical variables X^ι, π_ι cannot be varied freely, but only subject to the constraint

$$\mathscr{H} \equiv \pi_T + \mathring{H}(T, X^i, \pi_i) = 0\,. \tag{12}$$

However, we may take the constraint (12), adjoin it by means of a Lagrange multiplier $N(t)$ to the action,

$$S = \int \mathrm{d}t\, (\pi_\iota \dot{X}^\iota - N\mathscr{H})\,, \tag{13}$$

and then vary all the variables X^ι, π_ι, N freely.

The physical meaning of the Lagrange multiplier $N(t)$ may be obtained from the Hamilton equations which follow from the action functional (13). Varying the energy π_T, we get

$$\dot{T} = \frac{\partial N\mathscr{H}}{\partial \pi_T} = N\,. \tag{14}$$

Thus, the Lagrange multiplier N gives the rate of change of the physical time T with respect to the label time t. As it determines the lapse of the physical time T, the Lagrange multiplier N is called the lapse function.

In the action functional (13), the expression $N\mathscr{H}$ plays the role of the Hamiltonian. Varying the action functional (13) with respect to the lapse function N, we get the constraint (12) which tells us that the Hamiltonian $N\mathscr{H}$ vanishes. It is usual to refer to the constraint function $\mathscr{H}$ as to the super-Hamiltonian. I think it is a particularly well chosen term, because it reminds us how easily a superentity is converted into zero. Memento mori! The process of passing from the action functional (1) to the action functional (13) is called parametrization, and the inverse process is called deparametrization. It is straightforward to deparametrize the action functional (13). Solve the constraint (12) with respect to the canonical momentum π_T, substitute it into the action, and then choose T as a privileged time label,

$$T = t\,. \tag{15}$$

This yields the Hamiltonian form of the physical action functional.

In the parametrized form (13) of the action, we see the time T for the first time mixed with the dynamical variables X^i. At any single instant of the label time t, the physical time T and the conjugate energy $-\pi_T$ enter into the super-Hamiltonian: as in Eliot's lines, all time is eternally present there.

Fortunately, it is easy to distinguish the energy $-\pi_T$ from the dynamical momenat π_i, as it enters into the super-Hamiltonian linearly, and then proceed with the deparametrization. Time is mixed with other variables, but not lost among them. Notice that by adjoining the physical time and energy to other canonical variables, we have obtained an action functional which is invariant with respect to the relabeling of the t-time,

$$t \rightarrow \bar{t} = \bar{t}(t). \tag{16}$$

The super-Hamiltonian constraint is thus the price we pay for the invariance with respect to the gauge group (16).

Parametrizing the motion of a non-relativistic particle (2), we get the super-Hamiltonian constraint

$$\mathscr{H} \equiv \pi_T + \frac{1}{2m} \delta^{ik}\pi_i\pi_k + V(T, X^i) = 0. \tag{17}$$

Similarly, parametrizing the motion of a free relativistic particle (3), we get the super-Hamiltonian constraint

$$\tilde{\mathscr{H}} \equiv \pi_T + \sqrt{\delta^{ik}\pi_i\pi_k + m^2} = 0. \tag{18}$$

In this case, it is more advantageous to take as a super-Hamiltonian constraint the squared form of (18),

$$\begin{aligned} \mathscr{H} &= -\pi_T{}^2 + \delta^{ik}\pi_i\pi_k + m^2 = \eta^{\iota\kappa}\pi_\iota\pi_\kappa + m^2 = 0, \\ \eta^{\iota\kappa} &= \eta_{\iota\kappa} = diag(-1, 1, 1, 1), \end{aligned} \tag{19}$$

getting rid of an inconvenient square root, and gaining an explicit Lorentz invariance. This slightly changes the meaning of the lapse function N. Varying the action functional (13) containing the new super-Hamiltonian (19) with respect to π_T, we get

$$N = \frac{1}{2} \frac{\dot{T}}{\sqrt{m^2 + \delta^{ik}\pi_i\pi_k}} = \frac{1}{2m} \dot{\tau}. \tag{20}$$

The Lagrange multiplier N is thus connected with the lapse of the proper time τ rather than with the lapse of the Minkowskian time T.

One may also write the parametrized form of the action functional of a relativistic particle moving in a prescribed Riemannian spacetime, using the super-Hamiltonian

$$\mathscr{H} \equiv {}^4g^{\iota\kappa}(x^\lambda)\,\pi_\iota\pi_\kappa + m^2. \tag{21}$$

Now take the last step and parametrize a field, i.e., a system with an infinite number of degrees of freedom. For definiteness, take a scalar field $\phi(X^\iota)$ propagating in a flat spacetime with Minkowskian coordinates $X^\iota \equiv (T, X^i)$. However, the same procedure may be easily applied to more complicated tensor fields. Introduce the curvilinear coordinates $x^\iota = (t, x^i)$,

$$X^\iota = X^\iota(x^\kappa). \tag{22}$$

Equation (22) describes a one-parameter system of hypersurfaces $t = \text{const.}$, labeled by the value of a time parameter t, with a spatial system of coordinates x^i defined in each of them (Figure 1). Denote the partial derivatives with respect to the labels x^κ by the standard symbol $_{,\kappa}$.

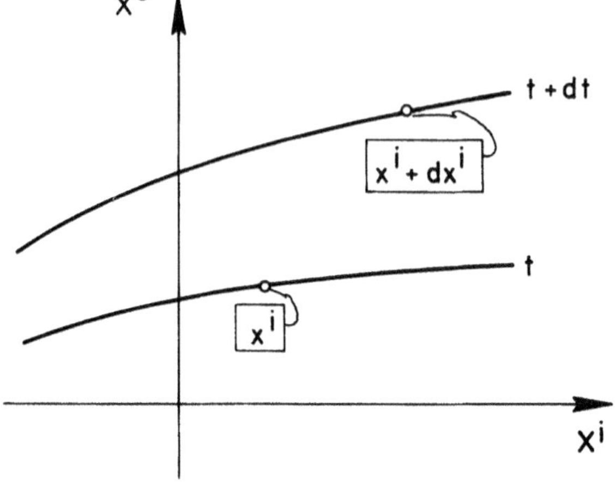

Fig. 1.

Expressing the action functional in curvilinear coordinates, we get

Proper volume element — Minkowskian coordinates — Proper volume element — Jacobian — Curvilinear coordinates

$$S = \int d^4X \mathring{\mathscr{L}}\left(\phi, \frac{\partial \phi}{\partial X^\kappa}\right) = \int d^4x \, |X^\lambda{}_{,\mu}| \, \underbrace{\mathring{\mathscr{L}}\left(\phi, \phi_{,\iota} \frac{\partial x^\iota}{\partial X^\kappa}\right)}_{\mathscr{L}(\phi, \phi_{,i}, \dot{\phi}; X^\iota{}_{,k}, \dot{X}^\iota)} \tag{23}$$

Action — Flat spacetime Lagrangian density — Lagrangian density in curvilinear coordinates

We may define the scalar field momentum π_ϕ relative to the label velocity $\dot{\phi} = \phi_{,0}$ in the standard way

$$\pi_\phi \equiv \frac{\partial \mathscr{L}}{\partial \dot{\phi}}, \tag{24}$$

and pass to the Hamiltonian density

$$\mathfrak{H} = \pi_\phi \dot{\phi} - \mathscr{L} \tag{25}$$

by the steps analogous to those used in Equation (8). The resulting relation is of course more complicated:

$$\mathfrak{H} = \underbrace{\frac{\partial x^{\circ}}{\partial X^{\iota}} |X^{\lambda}{}_{,\mu}|} \cdot \mathring{T}^{\iota}{}_{\kappa} \cdot \dot{X}^{\kappa}. \tag{26}$$

$\mathfrak{H}$	$\frac{\partial x^{\circ}}{\partial X^{\iota}} \lvert X^{\lambda}{}_{,\mu}\rvert$	$\mathring{T}^{\iota}{}_{\kappa}$	$\dot{X}^{\kappa}$
Hamiltonian density in curvilinear coordinates	cofactor $(X^{\iota}{}_{,0})$: does not depend on 'kinematical velocities' $\dot{X}^{\kappa}$	Canonical energy-momentum tensor in Minkowskian coordinates	Kinematical velocities

We have four 'kinematical velocities' $\dot{X}^{\kappa}$ instead of one 'time velocity' $\dot{T}$. We also need the entire energy-momentum tensor $\mathring{T}^{\iota}{}_{\kappa}$ in Minkowskian coordinates to get the single Hamiltonian density $\mathfrak{H}$ in curvilinear coordinates. The first term on the right of Equation (26), cofactor $(X^{\iota}{}_{,0})$, has the geometrical meaning of the surface element of the hypersurface $t = \text{const}$. More important for us at this moment is the purely algebraical fact that it depends only on the spatial derivatives $X^{\iota}{}_{,i}$, not on the kinematical velocities $\dot{X}^{\iota}$. In analogy with Equation (9), though in a slightly more complicated manner, one can prove that the energy momentum tensor $\mathring{T}^{\iota}{}_{\kappa}$, when expressed as a function of canonical variables ϕ and π_{ϕ}, does not depend on the kinematical velocities,

$$\mathring{T}^{\iota}{}_{\kappa} = \mathring{T}^{\iota}{}_{\kappa}(\phi, \phi_{,i}, \pi_{\phi}; X^{\lambda}{}_{,l}). \tag{27}$$

This shows that the Hamiltonian density (26) is a linear function of kinematical velocities. We may thus cast the action functional (23) into the form homogeneous in the dynamical *and* kinematical velocities,

$$S = \int \mathrm{d}^4 x \, (\pi_{\phi} \dot{\phi} + \Pi_{\iota} \dot{X}^{\iota}), \tag{28}$$

but we must vary it under four constraints

$$\mathscr{H}_{\iota} \equiv \Pi_{\iota} + \frac{\partial x^0}{\partial X^{\kappa}} |X^{\lambda}{}_{,\mu}| \, \mathring{T}^{\kappa}{}_{\iota} = 0. \tag{29}$$

Instead of taking the constraints (29) as they are, it is more geometrical to project them into directions perpendicular and parallel to the hypersurfaces $t = \text{const.}$,

$$\mathscr{H} \equiv \mathscr{H}_{\perp} = \mathscr{H}_{\iota} n^{\iota}, \quad \text{and} \quad \mathscr{H}_i \equiv \mathscr{H}_{\iota} X^{\iota}{}_{,i}. \tag{30}$$

The unit normal n^{ι} to these hypersurfaces is defined by the equations

$$\eta_{\iota\kappa} n^{\iota} n^{\kappa} = \varepsilon \begin{cases} -1 & \text{for a timelike } \mathbf{n}\text{, which is our case} \\ +1 & \text{for a spacelike } \mathbf{n}\text{,} \end{cases}$$

$$n_{\iota} X^{\iota}{}_{,i} = 0; \tag{31}$$

it is a function of kinematical variables only and does not depend on kinematical velocities. I retain the indicator ε for further purposes. Each constraint thus has a 'kinematical part',

$$\Pi_\perp \equiv \Pi_\iota n^\iota \quad \text{or} \quad \Pi_i \equiv \Pi_\iota X^\iota{}_{,i},$$

which has the same structure for all fields, and a 'field part', $\mathscr{T}_\perp$ or $\mathscr{T}_i$, which depends on the chosen Lagrangian. The kinematical parts are linear in the 'kinematical momenta'.

In the unparametrized action functional, the Minkowskian coordinates X^ι assumed an Olympian position as arguments of the ordinary field variables ϕ. In the parametrized action functional (28), they stepped down from their heights and mingled with ordinary field variables as Olympians used to mingle with mortal men (or, more often, with mortal women). Even in their most democratic mood, Olympians were still recognizable by their size and beauty; so I at least use the capital symbols to keep track of the supernatural origin of the kinematical coordinates X and their canonically conjugate momenta Π. But even if I did not, one could still recognize these privileged variables from the beautiful linear manner in which they enter the constraint function $\mathscr{H}$. The kinematical variables X^ι, representing Minkowskian time and space coordinates, and their conjugate momenta Π_ι, representing energy and momentum densities, are thus once again (as in particle dynamics) mixed with the dynamical variables, but not lost among them.

We call $\mathscr{H}$ the super-Hamiltonian and $\mathscr{H}_i$ the supermomentum, and adjoin them to the action functional (28) by means of Lagrange multipliers $N(x^\kappa)$ and $N^i(x^\kappa)$, giving

$$S = \int \mathrm{d}^4x \, (\pi_\phi \dot{\phi} + \Pi_\iota \dot{X}^\iota - N\mathscr{H} - N^i \mathscr{H}_i) . \tag{32}$$

We learn the geometrical meaning of these multipliers by varying the action functional (32) with respect to the kinematical momenta Π_ι,

$$\dot{X}^\iota = N n^\iota + N^i X^\iota{}_{,i} . \tag{33}$$

The displacement vector $\delta X^\iota = \dot{X}^\iota \delta t$ connecting two points with the same label x^i on two neighboring hypersurfaces t and $t+\delta t$ is decomposed into the normal and tangential components (Figure 2). The multiplier N of the super-Hamiltonian $\mathscr{H}$ gives the proper distance (in spacetime, the proper time separation) between the hypersurfaces, and is accordingly called the lapse function. The multiplier N^i of the supermomentum $\mathscr{H}_i$ tells us the label $x^i + N^i \delta t$ of the point from which the normal must be erected to pierce the point with the label x^i on the displaced hypersurface $t+\delta t$. It thus characterizes the mutual shift of the spatial systems of coordinates x^i and is called the shift function.

The action functional (28) is invariant with respect to the change

$$x^0 \to \bar{x}^0 = x^0 + F(x^k),$$
$$x^i \to \bar{x}^i = \bar{x}^i(x^k)$$

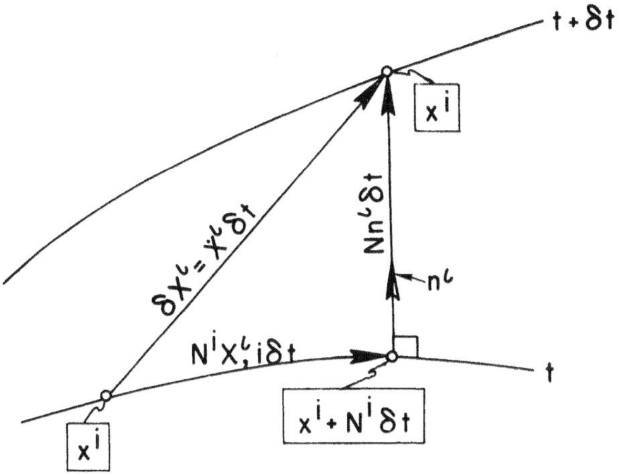

Fig. 2.

of the curvilinear coordinates x^i when we transform the canonical coordinates ϕ, X^ι as scalars and the canonical momenta as scalar densities of weight 1,

$$\bar{\pi}_\phi = \left|\frac{\partial x^\lambda}{\partial \bar{x}^\mu}\right| \pi_\phi, \quad \bar{\Pi}_\iota = \left|\frac{\partial x^\lambda}{\partial \bar{x}^\mu}\right| \Pi_\iota . \tag{34}$$

Similarly as in the parametrized particle dynamics, the super-Hamiltonian and supermomentum constraints are the price we pay for the invariance with respect to this gauge group.

To end with something concrete, let me write down the super-Hamiltonian and supermomentum constraints for a massless scalar field in one spatial dimension. The Minkowskian Lagrangian for such a field is

$$\mathring{\mathscr{L}} = -\frac{1}{2}\eta^{\iota\kappa}\frac{\partial\phi}{\partial X^\iota}\frac{\partial\phi}{\partial X^\kappa}, \quad \iota, \kappa = 0,1 , \tag{35}$$

and the projected constraints have the form

$$\begin{array}{lrl} \mathscr{H} = & \dfrac{1}{\sqrt{X'^2 - T'^2}} (X'\Pi_T + T'\Pi_X & + \tfrac{1}{2}(\pi_\phi{}^2 + \phi'^2)), \\ \mathscr{H}_1 = & T'\Pi_T + X'\Pi_X & + \phi'\pi_\phi \\ & \text{Kinematical parts} & \text{Dynamical parts} \end{array} \tag{36}$$

where prime denotes the derivative with respect to x. The kinematical and dynamical parts are easily recognizable, because the kinematical momenta Π_T and Π_X enter linearly into the super-Hamiltonian $\mathscr{H}$, whereas the dynamical momentum π_ϕ is squared. This simple model will introduce us later on to the quantization of a parametrized field theory.

2. Geometrodynamics

There is only the fight to recover
what has been lost
And found and lost again and again:
T. S. Eliot: *Four Quartets*
East Coker, 186–7.

Last time, I brought the action functional of a field propagating on the Minkowskian background to the form

$$S = \int \mathrm{d}t \int \mathrm{d}^3x \, (\pi_\phi \dot{\phi} + \Pi_l \dot{X}^l - N\mathscr{H} - N^i \mathscr{H}_i), \tag{1}$$

where

$$\begin{aligned} \mathscr{H} &= \Pi_\perp + \mathscr{T}_\perp(\phi, \phi_{,k}, \pi_\phi; X^\kappa{}_{,k}), \\ \mathscr{H}_i &= \Pi_i + \mathscr{T}_i(\phi, \phi_{,k}, \pi_\phi; X^\kappa{}_{,k}). \end{aligned} \tag{2}$$

In the action functional (1), both the field variables ϕ, π_ϕ and the kinematical variables X^l, Π_l are varied independently. The resulting Hamiltonian equations determine the evolution of these canonical variables with the t-time. Also the lapse and shift functions N and N^i are varied independently, yielding the constraints

$$\mathscr{H} = 0, \quad \mathscr{H}_i = 0. \tag{3}$$

To get a consistent theory, the constraints (3) must be preserved in t-time. This means that their Poisson bracket with the Hamiltonian must vanish by virtue of the original constraints themselves. Because the Hamiltonian

$$H = \int \mathrm{d}^3x (N\mathscr{H} + N^i \mathscr{H}_i) \tag{4}$$

is itself a linear combination of the constraints, this implies that the Poisson bracket of any two constraints taken at any two points of the hypersurface must be expressible as a linear combination of the original constraints. To see what this combination is, we need calculate only the Poisson bracket of the kinematical parts of the constraints; the field parts necessarily follow suit. This is due to the fact that the kinematical momenta enter into the kinematical parts $\Pi_\perp$ and Π_i linearly, and do not enter at all into the field parts. Therefore, if any Π arises out of the commutator of two $\mathscr{H}$'s, it must arise out of the commutator of the kinematical parts. Once such a Π arises, we know that a corresponding $\mathscr{T}$ must accompany it, to yield a complete $\mathscr{H}$. This enables us to calculate the Poisson bracket of two $\mathscr{H}$'s, by calculating the Poisson bracket of their kinematical parts. This was done a long time ago by Dirac, with the result ($\varepsilon = -1$)

$$[\mathscr{H}(x), \mathscr{H}(x')] = \mathscr{H}^i(x) \, \delta_{,i}(x, x') + \mathscr{H}^i(x') \, \delta_{,i}(x, x'), \tag{5}$$

$$[\mathscr{H}_i(x), \mathscr{H}(x')] = \mathscr{H}(x) \, \delta_{,i}(x, x'), \tag{6}$$

$$[\mathscr{H}_i(x), \mathscr{H}_k(x')] = \mathscr{H}_k(x) \, \delta_{,i}(x, x') + \mathscr{H}_i(x') \, \delta_{,k}(x, x'). \tag{7}$$

Here, δ is the three-dimensional delta function and the spatial labels x^k are represented by the single letter x.

The supermomenta $\mathscr{H}^i$ on the right hand side have the indices raised by the spatial metric tensor

$$g_{ik} = \eta_{\iota\kappa} X^{\iota}{}_{,i} X^{\kappa}{}_{,k} \tag{8}$$

of the surface $t = \text{const}$. This is the only combination through which the kinematical coordinates ever explicitly enter into the commutation relations. This strongly suggests that the commutation relations (5)–(7) have a deep geometrical meaning which is independent of the original assumption that the dynamics takes place in a flat Minkowskian background. Claudio Teitelboim has realized that and he has shown that Dirac's argument does not in fact depend on the particular structure (2) of the constraints. I shall follow his ideas, casting them into a group theoretical language. I shall thus rederive the relations (5)–(7) in a way revealing their true geometrical meaning and permitting an immediate transition to the general theory of relativity.

Let us start by removing the assumption of a flat Minkowskian background and work instead in an arbitrary Riemannian spacetime. Any dynamics, be it the dynamics of a field propagating in this Riemannian spacetime or be it the dynamics of the gravitational field itself, must tell us how the field develops when we slice the spacetime by a spacelike hypersurface and then deform this hypersurface forward into the future or backward into the past. The set of all spacelike hypersurfaces in a Riemannian space forms itself a sort of a space. The deformations of hypersurfaces into each other form a set of operations acting on this space of hypersurfaces. Any dynamics taking place in a Riemannian spacetime must mirror the structure of this set; on the other hand, the structure of the set is independent of any particular dynamics which we may happen to investigate. Let us therefore first study the set of deformations of hypersurfaces in a Riemannian space for its own sake. We may leave undecided for the moment the signature of the space and whether the hypersurfaces are spacelike or timelike. In fact, one question we would like to answer is precisely how the signature enters into the structure of the commutators of two infinitesimal deformations.

Introduce an arbitrary system of coordinates X^{ι} in the Riemannian space and describe a hypersurface by giving four functions $X^{\iota}(x^i)$ of three coordinates x^i;

$$X^{\iota} = X^{\iota}(x^i). \tag{9}$$

Our hypersurface is thus a labeled hypersurface, i.e., a hypersurface together with a definite intrinsic coordinate system x^i defined in it. Equation (9) tells us that the point of the hypersurface carrying the intrinsic label x^i is located in spacetime at the point carrying the spacetime lable X^{ι} (Figure 3).

We may change a labeled hypersurface either by leaving it fixed in the embedding spacetime but relabeling its points, or (intuitively for the moment!) by deforming it into another hypersurface while leaving its labeling fixed. These two operations generate all others, for one can show that an arbitrary change of a labeled hypersurface may be decomposed into these two changes. How do we know, however, that the

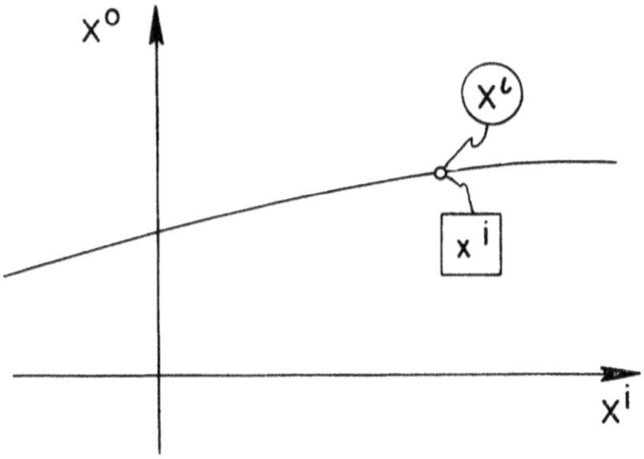

Fig. 3.

intrinsic system of coordinates on a deformed hypersurface is the same as the intrinsic system of coordinates on the old hypersurface? Fortunately, we may use the Riemannian structure of the embedding spacetime to identify two intrinsic systems of coordinates (Figure 4).

Start from the hypersurface $X^\iota(x^i)$. Draw geodesics perpendicular to that hypersurface. Move along the geodesic which starts from the point x^i, eventually meeting a point of the deformed hypersurface $\bar{X}^\iota$. Attach to that point the same label x^i as that of starting point. Describe the displacement of the second hypersurface with respect to the first hypersurface by giving the proper distance $\tau(x^i)$ (for spacelike hypersurfaces in spacetime, the proper time $\tau(x^i)$) measured along the geodesic. Repeat this at each point of the hypersurface $X^\iota(x^i)$. A single function $\tau(x^i)$ unambiguously describes the operation $\gamma[\tau(x^i)]$, which we may call "pure deformation of a hypersurface in a Riemannian spacetime without change of labeling".

It is difficult to write down the action of this operation $\gamma[\tau(x^i)]$ on the hypersurface

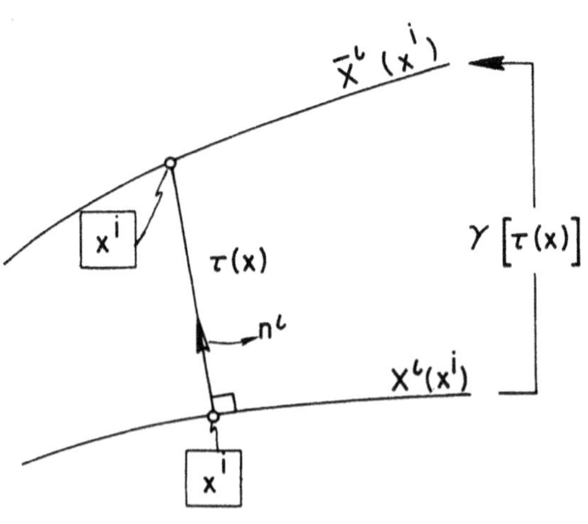

Fig. 4.

$X^{\prime}(x^i)$ explicitly. Luckily, we need to know only the action of an infinitesimal deformation, characterized by $\tau = \delta N(x^i)$. This action is

$$\gamma[\delta N(x^i)]\, X^{\prime}(x^i) = \bar{X}^{\prime}(x^i) = X^{\prime}(x^i) + n^{\prime}(x^i)\, \delta N(x^i), \tag{10}$$

where $n^{\prime}$ is the unit normal to the surface.

Similarly, 'relabelling' $\gamma[\bar{x}^i(x^k)]$ is the operation which takes the label x^k from a fixed spacetime point $X^{\prime}$ and reattaches it to the point $\bar{X}^{\prime}$ which originally had the label $\bar{x}^i(x^k)$ (Figure 5).

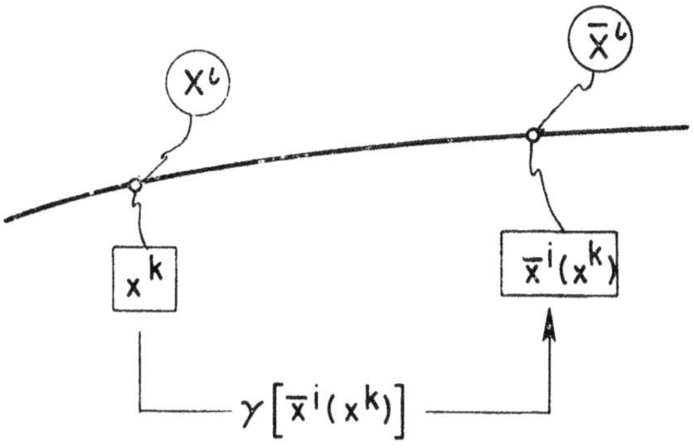

Fig. 5.

Obviously,

$$\gamma[\bar{x}^i(x^k)]\, X^{\prime}(x^k) = \bar{X}^{\prime}(x^k) = X^{\prime}(\bar{x}^i(x^k)).$$

The infinitesimal relabeling is characterized by three functions $\delta N^i(x^k)$,

$$\bar{x}^i = x^i + \delta N^i(x^k),$$

and we have

$$\gamma[\delta N^i(x^k)]\, X^{\prime}(x^k) = \bar{X}^{\prime}(x^k) = X^{\prime}(x^k) + X^{\prime}{}_{,i}(x^k)\, \delta N^i(x^k). \tag{11}$$

Strictly speaking, the set of deformations of hypersurfaces as characterized by four functions $\tau(x^k)$, $\bar{x}^i(x^k)$ of three labels x^k does not form a group, because the composition of two deformations depends also on the position of the hypersurface in the embedding spacetime; one cannot write a composition law involving the functions $\tau(x^k)$, $\bar{x}^i(x^k)$ themselves without reference to the space of hypersurfaces $X^{\prime}(x^i)$ on which the deformations act and to the spacetime metric ${}^4g_{\iota\kappa}(X^{\lambda})$. In spite of this, many notions of group theory, particularly those of the generators of infinitesimal transformations and their commutators, are still useful.

The set of deformations of hypersurfaces is an infinitely dimensional set, because to characterize its elements, we need functions $\tau(x^k)$, $\bar{x}^i(x^k)$ instead of continuous parameters τ, $\bar{x}^i$, as in the theory of Lie groups. An important notion in the theory of Lie groups is that of a generator of an infinitesimal transformation. Consider a space with points $X^{\prime}$ on which a Lie group operates, the elements $\gamma[N^i]$ of the group being labeled by continuous parameters N^i. Let the action of the group on the space be

$$\gamma[N^i]\, X^\iota = \bar{X}^\iota(X^\kappa, N^i).$$

Write down the infinitesimal transformation

$$\gamma[\delta N^i]\, X^\iota = X^\iota + \underbrace{\left.\frac{\partial \bar{X}^\iota(X^\kappa, N^i)}{\partial N^i}\right|_{N^i=0}}_{\xi_i^\iota(X^\kappa)} \delta N^i$$

in the neighborhood of the identity $N^i = 0$. Then, the operators

$$X_i \equiv \xi_i^\iota(X^\kappa)\frac{\partial}{\partial X^\iota} \tag{12}$$

are the generators of the Lie group. In the theory of infinitely dimensional groups, we may define the generators in exactly the same way, except that we need continuous labels x in addition to discrete indices, and we must turn the partial derivatives into variational derivatives, as the generators of an infinitely dimensional group operate on functionals rather than on functions. We shall add these labels among indices and generalize the summation convention so that the integration over a repeated label is implied. To give an example, $\mathscr{H}_{ix}$ is another notation for the function $\mathscr{H}_i(x^k)$, and the Hamiltonian (4) could be written as $N^x \mathscr{H}_x + N^{ix}\mathscr{H}_{ix}$. Thus, if

$$\gamma[N^{ix}]\, X^{\iota x'} = \bar{X}^{\iota x'}[X^{\kappa x''}, N^{ix}] \tag{13}$$

expresses the action of the group on the function space $X^{\iota x'}$, we identify the generators by writing down the infinitesimal transformation

$$\gamma(\delta N^{ix})\, X^{\iota x'} = X^{\iota x'} + \underbrace{\left.\frac{\delta \bar{X}^{\iota x'}[X^{\kappa x''}, N^{ix}]}{\delta N_{ix}}\right|_{N^{ix}=0}}_{\xi_{ix}^{\iota x'}} \cdot \delta N^{ix} \tag{14}$$

and taking

$$X_{ix} = \xi_{ix}^{\iota x'} \frac{\delta}{\delta X^{\iota x'}}. \tag{15}$$

Comparing (14) with (11), we get the 'vectors' $\xi_{ix}^{\iota x'}$,

$$\xi_{ix}^{\iota x'} = X^\iota{}_{,i}(x')\,\delta(x, x'),$$

and the generators of relabelings

$$X_{ix} = \xi_{ix}^{\iota x'}\frac{\delta}{\delta X^{\iota x'}} = \int d^3x' X^\iota{}_{,i}(x')\,\delta(x, x')\frac{\delta}{\delta X^\iota(x')} = X^\iota{}_{,i}(x)\frac{\delta}{\delta X^\iota(x)}. \tag{16}$$

Similarly, the generators of 'pure deformations' are

$$X_x = n^\iota(x)\frac{\delta}{\delta X^\iota(x)}. \tag{17}$$

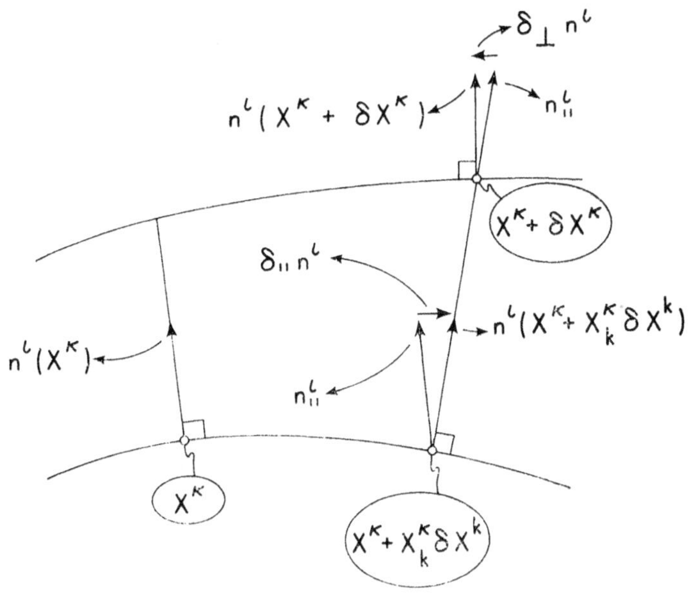

Fig. 6.

In the theory of Lie groups, the commutation relations of generators determine the structure constants of the group, from which the whole group may be reconstructed,

$$[X_i, X_k] = C^l{}_{ik} X_l . \tag{18}$$

Similarly, we may determine the structure constants of an infinitely dimensional group from the commutation relations of its generators. For the set of deformations of hypersurfaces, we want to know the commutation relations between any two generators (16), (17). Let us calculate, as in illustration, the commutator between two generators (17), X_x and $X_{x'}$, which is the most interesting of the whole set. In order to proceed, we need to know how the normal of a hypersurface changes, if we displace the hypersurface by the amount $\delta X^\iota(x)$ (Figure 6). I shall write down the final expression for this change of the normal and discuss its geometrical meaning, rather than trying to derive it:

$$\begin{aligned}\delta n^\iota &\equiv n^\iota(X^\kappa + \delta X^\kappa) - n^\iota(X^\kappa) = \\ &= \underbrace{\underbrace{n^\iota(X^\kappa + X^\kappa_k \,\delta X^k) - n^\iota(X^\kappa)}_{n^\iota{}_{.i}\,\delta X^i}}_{\underbrace{\frac{{}^4\nabla n^\iota}{\partial x^i}\delta X^i}_{\text{Bending term}} - \underbrace{\Gamma^\iota{}_{\kappa\lambda} n^\kappa X^\lambda_i \,\delta X^i}_{\text{transport correction}}} + \underbrace{n^\iota(X^\kappa + \delta X^\kappa) - n^\iota(X^\kappa + X^\kappa_k \,\delta X^k)}_{\underbrace{-\,\varepsilon X^{i\iota}\,\delta X^\perp{}_{,i}}_{\text{Tilting term}} - \underbrace{\Gamma^\iota{}_{\kappa\lambda} n^\kappa n^\lambda \,\delta X^\perp}_{\text{transport correction}}}\end{aligned} \tag{19}$$

Here, $X^\iota_i \equiv X^\iota{}_{,i}$ are the tangent vectors to the hypersurface. The Greek indices are raised and lowered by the spacetime metric tensor, and the Latin indices by the metric tensor

$$g_{ik} = {}^4g_{\iota\kappa} X^\iota_i X^\kappa_k \tag{20}$$

of the hypersurface. Let me explain now the geometrical significance of formula (19). We pass from the point X^ι on the original hypersurface to the point $X^\iota + \delta X^\iota$ on the deformed hypersurface in two steps, decomposing the displacement vector into parts parallel and perpendicular to the hypersurface,

$$\delta X^i \equiv X^i_\iota\, \delta X^\iota, \quad \delta X^\perp \equiv \varepsilon n_\iota\, \delta X^\iota . \tag{21}$$

As we go along the hypersurface, the normal simply changes through its derivative with respect to the internal coordinates: $n^\iota{}_{,i}\, \delta X^i$. This change itself is not a vector, because it is a difference of vectors at two different spacetime points. However, we may compare the vector $n^\iota(X^\kappa + X^\kappa_i\, \delta X^i)$ with the vector $n^\iota(X^k)$ parallel transported to the point $X^\iota + X^\iota_i\, \delta X^i$, decomposing $n^\iota{}_{,i}\, \delta X^i$ into a real vector

$$\delta_\| n^\iota = \frac{{}^4\nabla n^\iota}{\partial x^i}\, \delta X^i$$

and a correction to the parallel transport. The vector $\delta_\| n^\iota$ tells us how the normal changes due to the bending of the original hypersurface. Because the length of a vector is preserved under parallel transport, ${}^4\nabla n^\iota/\partial x^i$ has no component perpendicular to the hypersurface,

$$n_\iota \frac{{}^4\nabla n^\iota}{\partial x^i} = 0 .$$

The tangential components of ${}^4\nabla n^\iota/\partial x^i$ are known as the extrinsic curvature of the hypersurface,

$$K_{ij} \equiv -\, X_{i\iota} \frac{{}^4\nabla n^\iota}{\partial x^j} . \tag{22}$$

Next, we go from the point $X^\iota + X^\iota_i\, \delta X^i$ to the point $X^\iota + \delta X^\iota$ by the displacement $\delta X^\perp n^\iota$ perpendicular to the original hypersurface. We again decompose the change of the normal into a real vector change $\delta_\perp n^\iota$ and the correction to the parallel transport. The real vector change $\delta_\perp n^\iota$ is this time determined by the tilting of the displaced hypersurface with respect to the original hypersurface (Figure 7). This explains why it has only a component tangential to the hypersurface (proportional to $X^{i\iota}$) and no component normal to the hypersurface (proportional to n^ι). The tilting of the displaced hypersurface is determined by the gradient of $\delta X^\perp$, as seen from the picture. If the distance between the two hypersurfaces increases as we go along the coordinate line x^i, the normal is tilted in the direction opposite to the tangent vector of this coordinate line, which explains the minus sign when the interval between the two hypersurfaces is spacelike. We thus see the meaning of all terms in the formula (19).

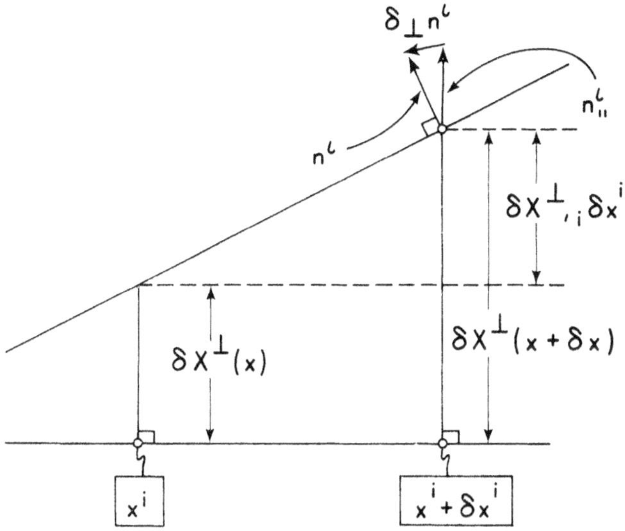

Fig. 7.

I shall now use (19) to calculate the commutator of X_x and $X_{x'}$. We have

$$[X_x, X_{x'}] = - n^\kappa(x') \frac{\delta n^\iota(x)}{\delta X^\kappa(x')} \frac{\delta}{\delta X^\iota(x)}$$

$$+ \text{ the same term with } x \text{ and } x' \text{ interchanged.} \qquad (23)$$

The antisymmetrization in x and x' kills all terms in $\delta n^\kappa(x)/\delta X^\iota(x')$ which are proportional to the delta function $\delta(x, x')$. Therefore, only the tilting term contributes by the amount

$$\frac{\delta_\perp n^\iota(x)}{\delta X^\kappa(x')} = - \varepsilon X^{i\iota}(x)\, \varepsilon n_\lambda(x) \underbrace{\frac{\delta X^\lambda{}_{,i}(x)}{\delta X^\kappa(x')}}_{\delta^\lambda_\kappa \delta_{,i}(x, x')}$$

$$= - X^{i\iota}(x)\, n_\kappa(x)\, \delta_{,i}(x, x').$$

Substituting it into the commutator (23), we get

$$[X_x, X_{x'}] = n^\kappa(x')\, n_\kappa(x)\, X^{i\iota}(x)\, \delta_{,i}(x, x') \frac{\delta}{\delta X^\iota(x)}$$

$$- \text{ the same term with } x \text{ and } x' \text{ interchanged}.$$

We may now use the formula

$$f(x)\, \delta_{,i}(x, x')\, g(x') = f(x')\, \delta_{,i}(x, x')\, g(x') - f_{,i}(x)\, \delta(x, x')\, g(x)$$

to get all functions multiplying the $\delta_{,i}$ function to be evaluated at the same point.

The terms proportional to $\delta(x, x')$, however, again drop out by antisymmetrization. Therefore, because

$$n^{\kappa}(x')\, n_{\kappa}(x') = \varepsilon\,,$$

$$X^{i\iota}(x')\, \frac{\delta}{\delta X^{\iota}(x')} = X^{ix'}\,,$$

we get

$$[X_x, X_{x'}] = \varepsilon\,(X^{ix}\,\delta_{,i}(x, x') + X^{ix'}\,\delta_{,i}(x, x'))\,. \tag{24}$$

Following this pattern, we may calculate the remaining commutators, with the result

$$[X_{ix}, X_{x'}] = -\, X_x\, \delta_{,i}(x, x')\,, \tag{25}$$

$$[X_{ix}, X_{kx'}] = -\, X_{kx}\, \delta_{,i}(x, x') - X_{ix'}\, \delta_{,k}(x, x')\,. \tag{26}$$

Comparing them with Equations (5)–(7), we see that these commutators differ (for $\varepsilon = -1$) only by signs from the corresponding Poisson brackets between the super-Hamiltonian and supermomenta. This means that the super-Hamiltonian and super-momenta of a field theory in parametrized form (1) represent the set of deformations of spacelike hypersurfaces*. We should remember, however, that the indices on the right-hand side of Equation (5) are raised by the spatial metric tensor contravariant to

$$g_{ik} \equiv \eta_{\iota\kappa} X^{\iota}{}_{,i} X^{\kappa}{}_{,k}\,, \tag{27}$$

whereas the indices on the right-hand side of Equation (24) are raised by an arbitrary spatial metric tensor (20). Equation (27) tells us that the hyper-surfaces are embedded in a flat Minkowskian spacetime rather than in an arbitrary Riemannian spacetime, so that Equations (5)–(7) represent only the set of deformations of spacelike hyper-surfaces in the Minkowskian spacetime.

From the point of view of Riemannian geometry, the supplementary condition (8) is unnatural. One would like to know if it is possible to discover a super-Hamiltonian and a supermomentum representing the generators of the set of deformations of hypersurfaces in a Riemannian spacetime without imposing this condition. We have seen that the only quantity which enters into the commutation relations explicitly is the spatial metric tensor g_{ik}. A minimal representation would, therefore, use only this spatial metric tensor g_{ik} and a conjugate momentum π^{ik} as canonical variables. This means that we would like to get the dynamics of pure Riemannian spacetime itself, revealed by the spatial metric on all the different spacelike slices which we can cut across this spacetime. In fact, because the set we are representing contains the group of relabelings as a subset, what matters actually is the three-geometry $\mathscr{G}$, not a particular spatial metric g_{ik} which describes it. This brings us to a standpoint which John Wheeler called geometrodynamics. Geometrodynamics studies how the spatial

* The change in sign is due to the conventions in the definition of the Poisson bracket and the commutator; when the dynamical variable A represents the generator X, and the dynamical variable B represents the generator Y, then the Poisson bracket $[A, B]$ represents the commutator $[Y, X]$, not the commutator $[X, Y]$.

geometry of a slice changes if we push this slice through a spacetime. As we proceed, the three-geometry moves along a path in superspace, which is defined as the set of all possible spatial geometries. To characterize superspace in a slightly more formal way, take at first the set of all spatial Riemannian metrics defined over a (three-dimensional) manifold M and call it Riem (M). The equivalence class of all metrics in Riem (M) which differ only by a three-dimensional diffeomorphism forms a three-geometry $\mathscr{G}$, and the set of all these three-geometries is superspace,

$$\mathscr{S}(M) = \text{Riem}\,(M)/\text{Diff}\,(M)\,.$$

The above definitions provide several useful concepts and images which help us to grasp the intuitive meaning of the formalism.

The variables g_{ik} and π^{ik} which we shall now use are still redundant, being subject to the constraints $\mathscr{H} = \mathscr{H}_i = 0$. This is because they represent not only the truly dynamical degrees of freedom, but also space and time together with momentum and energy, treated as canonical variables. At this moment, we do not try to separate dynamical and kinematical variables, though it is an extremely interesting (and extremely difficult) problem in itself. In fact, it is so difficult that one is almost tempted to give up and say that what God has joined together, let no man tear asunder!

Our task now is to find the super-Hamiltonian $\mathscr{H}$ and supermomentum $\mathscr{H}_i$ constructed entirely out of the spatial metric g_{ik} and its conjugate momentum π^{ik} and satisfying the Poisson brackets relations (5)–(7). The Hamiltonian (4) corresponding to the geometrodynamical action functional

$$S = \int \mathrm{d}^4x\,(\pi^{ik}\dot{g}_{ik} - N\mathscr{H} - N^i\mathscr{H}_i) \tag{28}$$

then determines the change δF of an arbitrary functional F of the geometrodynamical variables g_{ik}, π^{ik} induced by the deformation $\delta N = N\delta t$, $\delta N^i = N^i\delta t$ of the hypersurface. Under such a deformation,

$$\delta F = [F, \mathscr{H}_{x'}\,\delta N^{x'} + \mathscr{H}_{lx'}\,\delta N^{lx'}]\,. \tag{29}$$

This is what we mean by saying that geometrodynamics represents the set of deformations of hypersurfaces.

The supermomentum $\mathscr{H}_i$ may be determined directly from Equation (29) specialized to pure relabelings ($\delta N = 0$),

$$\delta F = [F, \mathscr{H}_{lx'}\,\delta N^{lx'}]\,. \tag{30}$$

The trick is that we know beforehand how the canonical variables g_{ik} and π^{ik} behave under the relabelings: g_{ik} transforms as a tensor, and π^{ik} as a tensor density of weight 1 (one sees it from the fact that the integrand $\pi^{ik}\dot{g}_{ik}$ of the action functional (28) must be a scalar density of weight 1). The change in g_{ik} and π^{ik} are thus given by the Lie derivatives of a tensor and a tensor density, respectively. On the other hand, the same change is determined by Equation (30) applied to the canonical variables g_{ikx}, π^{ikx}

in place of F. The comparison gives us two sets of equations for $\mathscr{H}_{lx'}$,

$$\underbrace{[g_{ikx}, \mathscr{H}_{lx'}\,\delta N^{lx'}]}_{\dfrac{\delta \mathscr{H}_{lx'}}{\delta \pi^{ikx}}\,\delta N^{lx'}} = g_{ik,l}\,\delta N^l + g_{il}\,\delta N^l{}_{,k} + g_{lk}\,\delta N^l{}_{,i}, \tag{31}$$

$$\underbrace{[\pi^{ikx}, \mathscr{H}_{lx'}\,\delta N^{lx'}]}_{-\dfrac{\delta \mathscr{H}_{lx'}}{\delta g_{ikx}}\,\delta N^{lx'}} = (\pi^{ik}\,\delta N^l)_{,l} - \pi^{il}\,\delta N^k{}_{,l} - \pi^{lk}\,\delta N^i{}_{,l}. \tag{32}$$

It is straightforward to prove that they have a unique solution for $\mathscr{H}_{ix}$, namely,

$$\mathscr{H}_{ix} = -\,2g_{ik}\pi^{kl}{}_{|l}. \tag{33}$$

The supermomenta $\mathscr{H}_{ix}$ in Equation (33) are easily shown to satisfy the commutation relation (7). Going up the ladder, we pass to the commutation relation (6). Similarly as Equation (32) tells us that the geometrodynamical momentum π^{ik} transforms as a tensor density of weight 1 under relabelings, Equation (6) tells us nothing more and nothing less than that the super-Hamiltonian $\mathscr{H}_x$ is a scalar density of weight 1,

$$\delta\mathscr{H}_x = [\mathscr{H}_x, \mathscr{H}_{ix'}\,\delta N^{ix'}] = (\mathscr{H}_x\,\delta N^i)_{,i}.$$

It is thus easy to satisfy the commutation relations (7) and (6). The really difficult one is the commutation relation (5) which implies that the theory can be interpreted as a slicing of a Riemannian spacetime, and, through the sign of its right-hand side coming from the indicator ε in Equation (24), that this spacetime has the signature $(-, +, +, +)$. It is hard to find out the general solution of Equation (5) for the super-Hamiltonian, but we can at least find a general solution of a certain form. Let us be guided by the super-Hamiltonian of a typical field theory in flat spacetime. Such a super-Hamiltonian contains terms linear in the kinematical momenta, terms quadratic in the dynamical momenta, and a potential term coupling the dynamical degrees of freedom at neighbouring points, the potential containing the spatial derivatives of dynamical coordinates. In geometrodynamics, we do not know how to separate the kinematical and dynamical variables. However, make the *ansatz* that the geometrodynamical super-Hamiltonian has the form

$$\mathscr{H}_x = G_{x\,iklm}\pi^{ik}\pi^{lm} + H_{x\,ik}\pi^{ik} + V_x. \tag{34}$$

$$G_{x\,iklm} = G_{x\,ikml} = G_{x\,kilm} = G_{x\,lmik}, \; H_{x\,ik} = H_{x\,ki},$$

with the position-dependent coefficients G_x and H_x of the momenta depending locally on the spatial metric, but not its derivatives, and the potential term V_x depending locally on the metric and its (spatial) derivatives up to the second order. We already know that V_x is a scalar density of weight 1, which further restricts the possible form of these terms. In fact, G_x is then determined up to two arbitrary constants, α and β,

$$G_{x\,iklm} = g^{-\frac{1}{2}}\left[\alpha(g_{il}g_{km} + g_{im}g_{kl}) + \beta g_{ik}g_{lm}\right], \tag{35}$$

$H_{x\,ik}$ up to one arbitrary constant,

$$H_{x\,ik} = \gamma g_{ik}\,, \tag{36}$$

and V_x, by a well-known theorem of Riemannian geometry, is restricted to have the form

$$V_x = g^{\frac{1}{2}}\, f_x(g_{ik}, R_{ik})\,, \tag{37}$$

where f_x is an arbitrary scalar constructed locally solely from the three-dimensional metric tensor and the Ricci tensor. We impose only one additional condition on the constants α and β; the 'supermetric' G_{iklm} should not kill any degree of freedom carried by π^{ik}. This excludes the singular cases $\alpha=0$ ($G_{iklm}\pi^{ik}\pi^{lm}$ depends then only on the trace $\pi \equiv g_{ik}\pi^{ik}$ of π^{ik}, and not on its traceless part $\pi^{ik}-\frac{1}{3}\pi g^{ik}$) and $2\alpha+3\beta=0$ ($G_{iklm}\pi^{ik}\pi^{lm}$ depends then only on the traceless part of π^{ik}). Under these conditions, Sergio Hojman and I were able to show that the commutation relation determines the constants α, β, γ uniquely (up to the choice of units and gauging of momentum π^{ik}),

$$\alpha = \tfrac{1}{2},\ \beta = -\tfrac{1}{2},\ \gamma = 0\,, \tag{38}$$

and forces $f(g_{ik}, R_{ik})$ into the form

$$f = -R + 2\lambda, \quad \lambda = \text{const}\,. \tag{39}$$

This yields the super-Hamiltonian

$$\mathscr{H} = \tfrac{1}{2} g^{-\frac{1}{2}} \left(g_{il}g_{km} + g_{im}g_{kl} - g_{ik}g_{lm}\right) \pi^{ik}\pi^{lm} - g^{\frac{1}{2}}R + 2\lambda g^{\frac{1}{2}}\,. \tag{40}$$

Because the coefficients α and β have opposite signs, the 'supermetric' (35) depending at a given point x^i on six components of the metric tensor g_{ik} as canonical coordinates is indefinite, with the signature having one minus sign and five pluses. People often conjecture that the supermetric signature reflects in the geometrodynamical language the indefinite signature of the spacetime metric. It is rather surprising to learn that it does not. If one takes the 'spacetime' with signature $(+, +, +, +)$, the only change in the commutation relations (5)–(7) is the opposite sign on the right-hand side of the Poisson bracket between two super-Hamiltonians. If we again make the *ansatz* (34) and go through all the steps, we get exactly the same constants α and β, but a different sign of f in Equation (39). The signature of spacetime is thus reflected in the relative sign of the 'kinetic' and 'potential' terms in the super-Hamiltonian. Unfortunately, I have no time to discuss the significance of this result.

Now, one cannot get something really new in the general theory of relativity. All roads lead to Rome! The natural geometrical requirements are so strong that one always finally ends with Einstein's law of gravitation. There is only the fight to recover what has been lost and found and lost again and again. That the action functional (28) with the super-Hamiltonian (40) and the supermomentum (33) leads to Einstein's law of gravitation was shown by Arnowitt, Deser, and Misner, who started from the Hilbert action functional (I retain the cosmological term $-2\lambda\sqrt{-{}^4g}$)

$$S = \int \mathrm{d}^4x \sqrt{-{}^4g}\,({}^4R - 2\lambda)$$

and, by a series of rearrangements, throwing away some boundary terms and introducing new variables, cast it exactly into the form (28). The method I have sketched here thus amounts to the rederivation of Einstein's law from purely geometrodynamical principles.

To return to where relativity starts one goes in the direction opposite to Arnowitt, Deser, and Misner. The first thing one would like to know, of course, is the geometrical meaning of the geometrodynamical momentum π^{ik}. This is obtained by varying the geometrodynamical action functional with respect to the metric, which gives

$$\pi_{ij} = g^{\frac{1}{2}}(Kg_{ij} - K_{ij}), \tag{41}$$

with K_{ij} introduced as an abbreviation for

$$K_{ij} = \frac{1}{2}\left(-\frac{\delta g_{ij}}{\delta N} + \frac{\delta N_{(i|j)}}{\delta N}\right), \tag{42}$$

$$\delta N = N\,\delta t,\ \delta N_i = N_i\,\delta t\,.$$

It can be checked that K_{ij} is identical with the extrinsic curvature (22). One shows first that the lapse and shift functions N and N_i in a Riemannian spacetime are connected to the components of the spacetime metric tensor ${}^4g_{\iota\kappa}$ by the formulas

$${}^4g_{oi} = N_i,\ {}^4g^{oo} = -N^{-2} \tag{43}$$

and then uses the geometrical definition of the extrinsic curvature.

The geometrodynamical momentum therefore characterizes the bending of the slice carrying the intrinsic geometry g_{ij} in the embedding spacetime. Algebraically, Equations (41), (42) can be viewed as expressing the momentum π_{ik} in terms of the metric g_{ik}, its rate of change with respect to a label time t, and the lapse and shift functions. Using the other set of Hamilton's equations

$$\dot{\pi}^{ik} = \frac{\delta}{\delta g_{ik}}\,(N^{x'}\mathscr{H}_{x'} + N^{lx'}\mathscr{H}_{lx'})$$

and substituting into them the above expressions for momentum, we obtain the Einstein's equations $G_{ik}=0$. The super-Hamiltonian constraint $\mathscr{H}_x=0$ gives the equation $G^o_o=0$, and the supermomentum constraints $\mathscr{H}_{ix}=0$ the equations $G^o_i=0$. Welcome back to Einstein's Rome, the capital of relativistic Christendom!

3. Quantization

These are only hints and guesses,
Hints followed by guesses; and the rest
Is prayer, observance, discipline, thought and action.

T. S. Eliot: *Four Quartets*
The Dry Salvages, 212–214.

In trying to understand the dynamical structure of Einstein's gravitational law, we collected a number of models with redundant degrees of freedom. Now, on the one hand, we know how to quantize the motion of non-relativistic particles in a potential field, or of free relativistic particles; we also know how to quantize a free scalar field. On the other hand, we have cast these theories into a parametrized form in which they take on the garments of geometrodynamics, acquiring its bad habit of mixing time, energy, position and momentum with the true dynamical variables. Clearly, then, these models can give us hints how to quantize geometrodynamics. It is in this way that I shall proceed: by 'hints and guesses, hints followed by guesses'. I leave to you the 'prayer, observance, discipline, thought and action' needed for a more formal development of quantum geometrodynamics.

I won't get too far, even by hints and guesses. In this lecture I shall write down the evolution equation for the geometrodynamical state, analogous to the Schrödinger equation of particle dynamics. Parametrized theories will tell me how to write it. (Sometimes, if we are not cautious, we may be misled by their hints. We'll see this at the end of the lecture, when I will briefly touch on the factor ordering problem). Unfortunately, most of the formal problems of quantum geometrodynamics are too difficult to be adequately discussed in two lectures. Instead of trying to attack them in their full generality, I will show in my last lecture how to construct simple geometrodynamical models. I'll use one of them to illustrate such general questions as the definition of a conserved inner product and the probabilistic interpretation of quantum geometrodynamics.

I start with the parametrized theories of my first lecture. First, there was the parametrized particle dynamics, non-relativistic and relativistic. The parametrized action functional had the form

$$S = \int dt \, (\pi_\iota \dot{X}^\iota - N\mathscr{H}), \qquad (1)$$
$$X^\iota = (T, X^i), \qquad \pi_\iota = (\pi_T, \pi_i),$$

with the super-Hamiltonian (I.17)

$$\mathscr{H} \equiv \pi_T + \frac{1}{2m} \delta^{ik} \pi_i \pi_k + V(X^\iota) \qquad (2)$$

for a non-relativistic particle under the influence of a potential V, and the super-Hamiltonian (I.19).

$$\mathscr{H} \equiv \eta^{\iota\kappa} \pi_\iota \pi_\kappa + m^2 \qquad (3)$$

for a free relativistic particle. More complicated was the parametrized field theory in Minkowskian spacetime, where both time and space were introduced as canonical coordinates. For a massless scalar field ϕ in one spatial dimension described by the Lorentz-invariant Lagrangian (I.35)

$$\mathring{\mathscr{L}} = -\frac{1}{2} \eta^{\iota\kappa} \frac{\partial \phi}{\delta X^\iota} \frac{\partial \phi}{\delta X^\kappa}, \qquad \eta^{\iota\kappa} = (-1, 1),$$

I obtained the parametrized action

$$S = \int \mathrm{d}t \int \mathrm{d}x \left(\pi_\phi \dot{\phi} + \Pi_T \dot{T} + \Pi_X \dot{X} - N\mathscr{H} - N^1 \mathscr{H}_1\right), \tag{4}$$

with the super-Hamiltonian

$$\mathscr{H} = (X'^2 - T'^2)^{-\frac{1}{2}} [X'\Pi_T + T'\Pi_X + \tfrac{1}{2}(\pi_\phi{}^2 + \phi'^2)] \tag{5}$$

and the supermomentum (I.36)

$$\mathscr{H}_1 = T'\Pi_T + X'\Pi_X + \phi'\pi_\phi . \tag{6}$$

Compare the parametrized action (4) with the deparametrized action

$$S = \int \mathrm{d}T \int \mathrm{d}X \left(\pi_\phi \frac{\partial \phi}{\partial T} - \mathfrak{H}\right) \tag{7}$$

in which the true non-vanishing Hamiltonian reappears,

$$H = \int \mathrm{d}X\, \mathfrak{H} = \int \mathrm{d}X \frac{1}{2}\left(\pi_\phi{}^2(X) + \left(\frac{\partial \phi(X)}{\partial X}\right)^2\right). \tag{8}$$

An easy way to recover the standard Schrödinger equation from the parametrized action functional for a single non-relativistic particle is as follows: Take all the canonical coordinates X^ι and all the canonical momenta π_ι and regard them as operators satisfying the commutation relations

$$[X^\iota, X^\kappa] = 0, [\pi_\iota, \pi_\kappa] = 0, [X^\iota, \pi_\kappa] = i\delta^\iota_\kappa . \tag{9}$$

Choose the X^ι-representation and replace the momenta by differentiation operators,

$$\pi_\iota = - i \frac{\partial}{\partial X^\iota} . \tag{10}$$

The state in the X^ι-representation is described by a state function $\Psi(X^\iota)$. Substitute the momentum operators into super-Hamiltonian (2) and impose the super-Hamiltonian constraint $\mathscr{H} = 0$ as a restriction on the state function $\Psi(X^\iota)$:

$$\mathscr{H}\Psi = - i \frac{\partial \Psi}{\partial T} - \frac{1}{2m} \Delta\Psi + V(X^\iota)\,\Psi = 0 . \tag{11}$$

This is just the Schrödinger equation.

The same prescription also works for the parametrized action functional (1) of a relativistic particle (3), yielding the Klein-Gordon equation for the state function $\Psi(X^\iota)$. There is no need to introduce a state functional $\Psi(X^\iota, t)$ depending on the label time t and ask how it evolves in t. In fact, trying that one finds out that it does not evolve at all. One obtains what some people call the 'frozen formalism'. There is no need, however, to proceed that way; all the dynamics in a parametrized theory is hidden in the constraints and the constraints themselves lead to the correct Schrödinger or Klein-Gordon equations.

Encouraged by these examples we apply the same method of quantization to the scalar field theory described by the action functional (4) with the super-Hamiltanian (5) and the supermomentum (6). Take all the canonical coordinates $\phi^{\iota} \equiv (T, X, \phi)$, kinematical as well as dynamical, and all the conjugate momenta $\pi_{\iota} \equiv (\Pi_T, \Pi_X, \pi_\phi)$, and turn them into operators satisfying the commutation relations

$$[\phi^{\iota}(x), \phi^{\kappa}(x')] = 0, [\pi_{\iota}(x), \pi_{\kappa}(x')] = 0, \quad [\phi^{\iota}(x), \pi_{\kappa}(x')] = i\delta^{\iota}_{\kappa}\,\delta(x, x'). \tag{12}$$

Choose the $\phi^{\iota}(x)$-representation in which the state of the scalar field is described by a state functional $\Psi[\phi^{\iota}(x)]$ and replace the canonical momenta $\pi_{\iota}(x)$ by variational derivatives,

$$\pi_{\iota}(x) = -i\frac{\delta}{\delta\phi^{\iota}(x)}. \tag{13}$$

Finally, substitute these operators into the super-Hamiltonian (5) and the supermomentum (6) and impose the constraints as restriction on the state functional. You get the functional differential equations

$$\mathscr{H}\Psi = (X'^2 - T'^2)^{-\frac{1}{2}}\left(-iX'(x)\frac{\delta\Psi}{\delta T(x)} - iT'(x)\frac{\delta\Psi}{\delta T(x)} + \right. \\ \left. + \frac{1}{2}\left(-\frac{\delta^2\Psi}{\delta\phi(x)\,\delta\phi(x)} + \phi'^2(x)\,\Psi\right)\right) = 0, \tag{14}$$

$$\mathscr{H}_1\Psi = -iT'(x)\frac{\delta\Psi}{\delta T(x)} - iX'(x)\frac{\delta\Psi}{\delta X(x)} - i\phi'(x)\frac{\delta\Psi}{\delta\phi(x)} = 0. \tag{15}$$

Questions of factor ordering arise when we write the constraints (14) and (15) as I did, with momenta to the right and coordinates to the left. They are not serious in this case because all canonical coordinates are differentiated with respect to the labels x. The terms with the momenta on the right therefore differ from the terms with the momenta on the left only by a differentiated δ-function with equal label arguments, e.g.,

$$T'(x)\frac{\delta\Psi}{\delta T(x')} = \frac{\delta}{\delta T(x')}(T'(x)\,\Psi) - \Psi\delta'(x, x').$$

The difference $\Psi\delta'(x, x)$ is to be put equal to zero, due to the antisymmetry of $\delta'(x, x')$.

Looking at the constraints (14), (15), we may hesitate about the correctness of the whole procedure. In particle dynamics, we had no trouble in identifying the super-Hamiltonian constraint with the Schrödinger equation. On the other hand, the constraints (14), (15) look totally different from the Schrödinger equation corresponding to the deparametrized form of the action functional (7), (8). The deparametrized state functional $\Psi(T)[\phi(X)]$ is a functional of the scalar field $\phi(X)$ (which is itself a function of the Minkowskian coordinate X), depending on the Minkowskian time T as a parameter. This functional satisfies the single Schrödinger equation

$$i\frac{\partial\Psi}{\partial T}=\frac{1}{2}\int \mathrm{d}X\left(-\frac{\delta^2\Psi}{\delta\phi(X)\,\delta\phi(X)}+\left(\frac{\partial\phi}{\partial X}\right)^2\Psi\right). \tag{16}$$

In comparison, the state functional of the parametrized field depends on three function variables $T(x)$, $X(x)$, $\phi(x)$, and satisfies a twofold infinity of Equations (14), (15) – two equations per each label x.

Is an infinity of equations too much for a single poor state functional? To see that they are *not* too much, in fact that they are just enough, we recall the meaning of the state functional $\Psi(T)[\phi(X)]$ of the deparametrized theory and of the functional $\Psi[T(x), X(x), \phi(x)]$ of the parametrized theory. In the deparametrized theory we are interested in the state of the scalar field on flat hypersurfaces $T=\text{const}$. The state functional $\Psi(T)[\phi(X)]$ is interpreted as the probability amplitude that the scalar field ϕ has the definite distribution $\phi(X)$ at the moment T. The parametrized theory broadens our interests; we want to know the state of the scalar field on an *arbitrary* curved spacelike hypersurface. Two functions

$$T=T(x),\ X=X(x) \tag{17}$$

describe how this hypersurface is drawn in Minkowskian spacetime. The third function, $\phi=\phi(x)$, gives the distribution of the scalar field on the hypersurface (17). The state functional $\Psi[T(x), X(x), \phi(x)]$ is interpreted as the probability amplitude that the scalar field ϕ has the definite distribution $\phi(x)$ on the curved hypersurface defined by the functions (17). We see why one needs a single Schrödinger equation in the deparametrized field theory, but an infinity of equations in the parametrized field theory. In the deparametrized theory, we want to know how the state changes when we pass from one flat hypersurface $T=T_0$ to another flat hypersurface $T=T_0+\delta T_0$. Such hypersurfaces form a one-parameter family and a single partial differential equation with respect to the T_0 labelling is sufficient to describe the changes of state. In the parametrized theory, we want to know the state on any spacelike hypersurface. These hypersurfaces are too numerous to fit into a single one-parameter family; they differ from each other by an arbitrary deformation. To describe how the state changes, we must know how it changes when we deform the original hypersurface by a bump at one place and by another bump at another place, i.e., we must know how it changes under an infinite number of elementary bumps. An infinite number of equations is just appropriate for this purpose. To compare the parametrized formalism with the deparametrized formalism, we must restrict the former to flat hypersurfaces. No such need arose in the particle theory because, no matter which way one parametrizes the time line, there is no difference between 'flat points' and 'curved points' on the line.

We may now be ready to admit that an infinity of Schrödinger equations is needed in parametrized field theory, but why a twofold infinity? A twofold infinity, because the supermomentum constraint (15) carries the information that the state functional Ψ does not depend on the labelling of the hypersurface on which the state is defined. Under an infinitesimal relabelling $x\to\bar{x}=x+\delta N^1(x)$ of the hypersurfaces, all canonical coordinates $T(x)$, $X(x)$, and $\phi(x)$ behave as scalars,

$$T(x) \to T(x - \delta N^1(x)) = T(x) - T'(x)\,\delta N^1(x), \text{ etc}.$$

The condition that the state functional is unchanged by the relabeling is

$$\Psi[T - T'(x)\,\delta N^1(x), \text{etc.}] = \Psi[T, \text{etc.}].$$

To the first order in $\delta N^1(x)$, we get

$$\int \mathrm{d}x \left[T'(x) \frac{\delta \Psi}{\delta T(x)} + \text{etc.} \right] \delta N^1(x) = 0. \tag{18}$$

Due to the arbitrariness of $\delta N^1(x)$, Equation (18) leads directly to the supermomentum constraint (15).

The entire *dynamical* content of the theory is thus carried in the super-Hamiltonian constraint (14) which governs the evolution of the state functional Ψ from one labelled hypersurface to another. We can eliminate the variational derivative $\delta\Psi/\delta X(x)$ from the super-Hamiltonian constraint through the supermomentum constraint. Moreover, because we already know that the state functional does not depend on the labeling of the hypersurface, we are entitled to choose whatever labelling we wish. Choosing the Minkowskian coordinate itself as a label,

$$X = x, \tag{19}$$

the super-Hamiltonian constraint becomes

$$i \frac{\delta \Psi}{\delta T(X)} = \frac{1}{2}\left(1 - \left(\frac{\partial T}{\partial X}\right)^2\right)^{-1} \left(-i \frac{\delta}{\delta \phi(X)} - \frac{\partial T}{\partial X}\frac{\partial \phi}{\partial X}\right)^2 \Psi + \frac{1}{2}\left(\frac{\partial \phi}{\partial X}\right)^2 \Psi. \tag{20}$$

In this equation any reference to an arbitrary labelling has disappeared and the hypersurface is given in the deparametrized form $T = T(X)$. If the state functional $\Psi[T(X), \phi(X)]$ is prescribed on a hypersurface $T = T(X)$, Equation (20) determines it on a slightly displaced hypersurface $T = T(X) + \delta T(X)$. Moreover, we can displace the hypersurface further and further, until Equation (20) determines the state functional on any hypersurface we wish.

To pass finally to the deparametrized formalism, select from all possible hypersurfaces a one-parameter family of flat hypersurfaces

$$T(X) = T_0 \in (-\infty, \infty), \qquad \frac{\partial T}{\partial X} = 0. \tag{21}$$

The state functional $\Psi[T(X), \phi(X)]$ on this family becomes a functional of $\phi(X)$ depending on T_0 as a parameter. Asking how it changes from one flat hypersurface to another, we realize that

$$\frac{\partial \Psi}{\partial T_0} = \int \mathrm{d}X \frac{\delta \Psi}{\delta T(X)} \underbrace{\frac{\partial T(X)}{\partial T_0}}_{=1},$$

and use Equation (20). Because $\partial T/\partial X=0$, the resulting equation reduces to the single Schrödinger equation (16) of the deparametrized theory.

This shows how the quantization by constraints works in a parametrized field theory. Going further, perhaps a little hesitantly, let us apply the same method to geometrodynamics. The geometrodynamical action had the same general form as the action of parametrized field theories,

$$S=\int \mathrm{d}t \int \mathrm{d}^3x\,(\pi^{ik}\dot{g}_{ik}-N\mathscr{H}-N^i\mathscr{H}_i), \tag{22}$$

differing from them only by the specific structure of the super-Hamiltonian

$$\begin{aligned}\mathscr{H} &= G_{iklm}\pi^{ik}\pi^{lm}-g^{\frac{1}{2}}R\,,\\ G_{iklm} &\equiv \tfrac{1}{2}g^{-\frac{1}{2}}(g_{il}g_{km}+g_{im}g_{kl}-g_{ik}g_{lm})\,.\end{aligned} \tag{23}$$

(I have omitted the cosmological term for simplicity), and the supermomentum

$$\mathscr{H}_i=-\,2\pi^k_{i|k}\,. \tag{24}$$

One is therefore tempted to turn the metric g_{ik} and the momentum π^{ik} into operators satisfying the commutation relations

$$\begin{aligned}&[g_{ik}(x),g_{lm}(x')]=0,\ [\pi^{ik}(x),\pi^{lm}(x')]=0\,,\\ &[g_{ik}(x),\pi^{lm}(x')]=\tfrac{1}{2}i\,(\delta^l_i\delta^m_k+\delta^m_i\delta^l_k)\,\delta(x,x')\,.\end{aligned} \tag{25}$$

In the metric representation, the state functional becomes a functional of the metric, $\Psi=\Psi[g_{ik}]$, and the momenta are replaced by the variational derivatives with respect to the metric,

$$\pi^{ik}(x)=-\,i\,\frac{\delta}{\delta g_{ik}(x)}\,. \tag{26}$$

Following the previous pattern, we substitute these operators into the super-Hamiltonian and the supermomentum and impose the constraints as restrictions on the state functional,

$$-\,G_{iklm}(x)\,\frac{\delta^2\Psi}{\delta g_{ik}(x)\,\delta g_{lm}(x)}-g^{\frac{1}{2}}(x)\,R(x)\,\Psi=0\,, \tag{27}$$

$$\left(\frac{\delta\Psi}{\delta g_{ik}(x)}\right)_{|k}=0\,. \tag{28}$$

As in the parametrized field theories, the $3\infty^3$ Equations (28) imply that the state functional $\Psi[g_{ik}]$ is unchanged by a relabeling of the hypersurface carrying the metric. Indeed, under $x^i\to\bar{x}^i=x^i+\delta N^i(x^k)$ the metric changes into

$$\bar{g}_{ik}(x)=g_{ik}(x)-\delta N_{i|k}(x)-\delta N_{k|i}(x)\,.$$

For the functional to remain the same,

$$\Psi[g_{ik}-\delta N_{i|k}-\delta N_{k|i}]=\Psi[g_{ik}]\,,$$

we must have

$$\int \mathrm{d}^3x \frac{\delta \Psi}{\delta g_{ik}(x)} \delta N_{i|k}(x) = 0 .$$

After an integration by parts, the arbitrariness of $\delta N_i(x)$ leads to the supermomentum constraint (28).

The state functional thus depends only on the spatial geometry and not on the particular metric g_{ik} chosen to represent it. In Wheeler's notation, $\Psi = \Psi[\mathscr{G}]$. The proper domain of the state functional is thus superspace, not the space of Riemannian three-metrics. While six functions $g_{ik}(x)$ are necessary to specify a metric, only three functions are necessary to specify a spatial geometry (though a choice of such functions may be cumbersome). Moreover, our study of the parametrized scalar field indicates that one of these functions should represent an intrinsic time, two others remaining to describe true dynamical degrees of freedom of the gravitational field. Unfortunately, we do not know how to separate the dynamical variables from the time. We thus know that the super-Hamiltonian constraint governs the time evolution of the dynamical state without knowing what is time evolution and what is dynamical state!

The quantum version (27) of the super-Hamiltonian constraint was called the Einstein-Schrödinger equation by John Wheeler, and the Wheeler-DeWitt equation by others. Because I cannot recall any two other physicists who would more strenuously object to the idea of quantizing gravity than Einstein and Schrödinger, and because I am not John Wheeler, I shall use the second name.

Everything seemed so straightforward when writing down the quantum constraints (27), (28) that I am really sorry to remind you that everything is not so straightforward. We have the ordering problems. If we write the supermomentum $\mathscr{H}_i$ with the momentum operators π^{ik} on the right, the labeling independence of the state functional Ψ immediately follows, so we may be inclined to use this ordering rather than others. We may show, however, that the interchange of momenta π^{ik} and coordinates g_{ik} in the covariant form $\mathscr{H}_i$ of the supermomentum leads only to the appearance of terms with $\delta_{,i}(x, x)$ which can be put equal to zero, as in the parametrized scalar field theory. The order of factors in the supermomentum $\mathscr{H}_i$ thus really does not matter. It is the factor ordering of the super-Hamiltonian which is tricky. The following argument is evoked to determine the factor ordering of a parametrized theory like geometrodynamics: the quantum constraints must be closed, i.e., their commutators must not lead to new constraints. In classical theory, the closing relations are just Equations (II.5)–(II.7). In order that the corresponding commutator relations be closed in the quantum theory, we must get on the right-hand side of Equation (II.5) the metric tensor g^{ik} to the left, and the covariant form of the supermomentum to the right, i.e.,

$$\begin{aligned}[\mathscr{H}(x), \mathscr{H}(x')]_{\text{commutator}} = i\,(&g^{ik}(x)\,\mathscr{H}_k(x)\,\delta_{,i}(x, x') \\ &+ g^{ik}(x')\,\mathscr{H}_k(x')\,\delta_{,i}(x, x')) . \end{aligned} \tag{29}$$

This results in general only from a definite ordering of the super-Hamiltonian.

Schwinger offered a solution of the factor ordering problem along these lines and DeWitt proposed another solution. In my opinion, neither of them is geometrical enough to provide the final answer to the problem. To explain what I mean, let me raise another objection against the Wheeler-DeWitt equation written in the form (27): this equation is not invariant with respect to the change of coordinates in Riem (M). This means that if we decide to use some six functionals of the metric

$$\bar{g}_{ik}(x) = \bar{g}_{ik}(x)\,[g_{ab}]$$

(e.g., the eigenvalues and the eigenvectors of the spatial Ricci tensor R_{ik}) as new canonical coordinates, find the canonical momenta $\bar{\pi}^{ik}(x)$ conjugate to these new coordinates, and base the quantization on the commutation relations (25) written down for these new canonical variables, we come out with a different answer for Ψ. We would therefore like to see an invariant equation replacing the original Wheeler-DeWitt equation.

To connect this problem with that of the factor ordering, we recall the ordinary Klein-Gordon equation for a relativistic particle moving in a curved background,

$$({}^4g^{\iota\kappa}(x^\lambda)\,\pi_\iota\pi_\kappa + m^2)\,\Psi = 0\,. \tag{30}$$

To make this equation invariant with respect to changes of the canonical coordinates we replace the momenta π_ι by covariant derivatives with respect to the metric ${}^4g^{\iota\kappa}$,

$$\pi_\iota = -\,i\,{}^4\nabla_\iota\,, \tag{31}$$

and not by the partial derivatives as in Equation (10). The whole difference, however, boils down to the factor ordering, because we may write Equation (30) in the form

$$\frac{1}{\sqrt{-\,{}^4g}}\,\pi_\iota(\sqrt{-\,{}^4g}\;{}^4g^{\iota\kappa}\pi_\kappa)\,\Psi + m^2\Psi = 0\,, \tag{32}$$

and still replace π_ι by partial derivatives.

The super-Hamiltonian constraint (23) is analogous to Equation (30) because the coefficients G_{iklm} of the quadratic form in the momenta π^{ik} depend on the canonical coordinates g_{ik}, just as the coefficients ${}^4g^{\iota\kappa}$ of the quadratic form of the momenta π_ι in Equation (30) depend on the canonical coordinates x^ι. In fact, DeWitt interpreted the coefficients G_{iklm} as a metric in Riem (M). Because the variational derivatives $\delta/\delta g_{ik}$ are analogous to the partial derivatives $\partial/\partial x^\iota$, the Wheeler-DeWitt equation looks like the Klein-Gordon equation (30) in a curved space written with partial derivatives in place of covariant derivatives. One may now think that the universal remedy needed to solve the factor ordering problem and the invariance problem at the same time is to replace the variational derivatives in the Wheeler-DeWitt equation by appropriate covariant variational derivatives with respect to the metric in Riem (M). This is easy to do. We pick up some $\delta(0)$'s by the way, but we can persuade ourselves (e.g., by working out a similar problem in parametrized scalar field theory treated in 'curvilinear coordinates in ϕ-space') that such terms should arise to compensate singularities arising from the succeeding variations of the state functional Ψ at the same space point x. Unfortunately, this simple solution of our problems is still (using Einstein's saying) not the true Jacob. The difficulty is that the particle moving in the curved background is described by physical coordinates x^{ι}, but the geometry moving in a curved superspace is described by redundant coordinates g_{ik} as a metric moving on a curved Riem (M). This is reflected by the fact that no linear constraint restricts the motion of the particle, but the linear supermomentum constraint (24) restricts the motion of the metric. If I had time, I could present a simple example from particle dynamics, where a single free particle moves in a flat spacetime but is described by redundant coordinates as moving in a higher dimensional curved spacetime, restricted in its motion by a linear constraint. The proper way to quantize the motion of this particle in the curved spacetime is to write down the Klein-Gordon equation (30) with the covariant derivatives in place of the redundant momenta, but with the projection operator $\tilde{g}^{\iota\kappa}$ into the physical flat spacetime replacing the curved metric $g^{\iota\kappa}$. An infinite dimensional analogue of such a procedure is apparently needed in geometrodynamics, and I hope that I am close to the solution of the problem along these lines.

4. Midi-superspace

> What we call the beginning is often the end
> And to make an end is to make a beginning.
> The end is where we start from.
>
> T. S. Eliot: *Four Quartets*
> *Little Gidding*, 214–216.

Last time, I wrote down the Wheeler-DeWitt equation governing the evolution of the geometrodynamical state and briefly discussed the difficulties connected with the factor ordering problem. Writing down an analogue of the Schrödinger equation does not in itself complete the quantization of a system. Many more questions remain to be asked and answered. First of all, one would like to have an inner product defined in the set of state functionals which would be conserved during the dynamical evolution of the system and which would impose on the set of state functionals the structure of a Hilbert space. One would like to know what operators constructed out of the metric g_{ik} and the geometrodynamical momentum π^{ik} represent significant dynamical quantities and how to calculate their expectation values. Finally, one would like to have some intuitive understanding and physical interpretation of the complex formalism of quantum geometrodynamics.

Our progress is hindered all along by the fact that we do not know what the geometrodynamical time is. How could one define, for example, the inner product which is conserved in time, if we do not know the time to conserve the inner product in? Nevertheless, one can do something. Observing that the metric G_{iklm} in Riem (M) has an indefinite signature, with one minus sign and five pluses, we can specify a superspacelike super-hypersurface (too many super's and hyper's, I fear!), and define an infinitely dimensional geometrodynamical analogue of the Klein-Gordon product, as DeWitt did. We have some problems, because we normalize over redundant coordinates, counting one geometry infinitely many times; but there are techniques for overcoming these difficulties. More significantly, we unleash the negative probabilities and begin to wonder if there are such things as geometrodynamical and antigeometrodynamical states. Misner, quantizing simple cosmological models, concluded that there are not, the positive and negative frequency solutions corresponding merely to the same universe in the expanding or recontracting stages. I am not sure, however, if this is the final word on the problem.

For the lack of time and the lack of any definite results, I am not going to answer the questions which I have raised in their full generality. Instead, I shall follow my program of elucidating difficult questions by means of simple models. One of the pleasing features of the canonical approach is that it is easily converted (some people would say, perverted) into such models. The general technique is to restrict oneself to a mini-superspace.

What is mini-superspace? Superspace is the full configuration space of geometrodynamics; it is the set of all possible three-dimensional geometries. In the canonical formalism, we work in the corresponding geometrodynamical phase space. A point of the geometrodynamical phase space is the class of couples g_{ik}, π^{ik}; two couples belong to the same class if they can be transformed into each other by a spatial transformation of coordinates. In practice, we describe points in superspace by redundant coordinates, giving a metric in Riem (M). Similarly, we describe points in the geometrodynamical phase space by redundant coordinates, giving a metric g_{ik} and a momentum π^{ik}. The dynamical evolution of a metric can start from an arbitrary point g_{ik} in Riem (M). On the other hand, the metric plus the momentum are restricted by the constraints $\mathscr{H}=\mathscr{H}_i=0$. The dynamical trajectory in phase space is thus confined to the constraint 'surface' $\mathscr{H}=\mathscr{H}_i=0$.

In spite of the famous bet between John Wheeler and Rainer Sachs nine years ago, nobody knows how to find the general solution of Einstein's equations. All that we have today are various particular solutions, or classes of such solutions, characterized either by their symmetries, or by other geometrical properties. Superspace is a bewilderingly large dynamical arena for such solutions. Like animals adapted to their environment, they remain only in certain limited regions of superspace which are favorable to their geometrical properties. The same tendency can also be observed in phase space; the symmetric solutions are to be found only in certain limited regions of the constraint surface. It seems a real waste of energy to apply the formidable canonical formalism of full geometrodynamics to such classes of solutions; one is tempted to think that a limited dynamical arena and the canonical formalism restricted to such a limited arena should suffice.

To construct such an arena, one starts with a spacetime which admits an intransitive group of motions generated by spacelike Killing vectors ${}^4\xi^{\iota}_A$. The minimal invariant varieties of such a group are spacelike surfaces. For example, if our group is that of spatial rotations around a center, the minimal invariant varieties are the two-dimensional spheres drawn in four-dimensional spacetime (the spheres $r=r_0=$const., $t=t_0=$const., θ and φ arbitrary, in the curvature coordinates form of the Schwarzschild geometry). Now allow only such spatial slices which contain the minimal invariant varieties and are themselves invariant varieties (not necessarily minimal). In Schwarzschild spacetime, the hypersurfaces $t=t_0=\text{const.}$ are such slices, but not the only ones; any slice $t=t(r)$ in the curvature coordinates is also an invariant variety. The Killing vectors of the group are tangential to such slices and may therefore be represented as three-vectors in the intrinsic coordinates x^i labeling the points of a slice. Now, on the permissible slices allow only such metrics g_{ik} and extrinsic curvatures π^{ik} which are symmetric with respect to the group of motions generated by these intrinsic Killing vectors,

$$\pounds_{\xi_A{}^i}\, g_{mn} = 0\,, \tag{1}$$

$$\pounds_{\xi_A{}^i}\, \pi^{mn} = 0\,. \tag{2}$$

We say that the geometries represented by the symmetric metrics (1) form a mini-superspace and the couples (g_{mn}, π^{mn}) restricted by both conditions (1) and (2) form a mini-phase-space. If we want to impose additional reflection symmetries on our geometries, supplementary conditions are needed, but let me not write them down explicitly.

The momentum π^{ik} must have the symmetry (2) to preserve the symmetry (1) of the geometry. The symmetry of the momentum alone, however, is insufficient to do the job. The spatial symmetries could not be maintained if we decided to cut the next slice across the symmetrical spacetime in an arbitrary way, disregarding the spacetime symmetry. How do we know, for example, that the next slice is a good one if the spacetime through which this slice is to be cut is as yet not constructed, but is only to be built up step by step in the process of integrating the Hamilton equations? The answer is that we must proceed from one slice to the next in such a way that the proper time between the two slices measured in the direction normal to the first slice has the same value at all points of the slice which are equivalent under our group of motions. The proper time between the two slices is proportional to the lapse function N. We are thus led to the condition

$$\pounds_{\xi_A{}^i}\, N = 0 \tag{3}$$

which ensures that the symmetry will not be broken because of a bad choice of slicing. While the symmetry (3) of the lapse function is necessary to keep the dynamical trajectory within our mini-superspace, no symmetry requirements are imposed on the shift functions N^i. This is intuitively clear because the shift functions fix only the spatial system of coordinates on a new hypersurface and an intrinsic symmetry remains an intrinsic symmetry in all coordinates. Of course, it is advantageous to use a spatial system of coordinates geared to the intrinsic symmetry and then preserve it on the next slice by restricting the shift functions through the symmetry requirement

$$\pounds_{\xi_A{}^i}\, N^k = 0\,. \tag{4}$$

This is what we shall do. The Killing equations (1) reduce the number of independent canonical coordinates. The best way to get rid of the surplus variables is to choose a standard system of spatial coordinates x^l reflecting the symmetries. In it, the canonical coordinates $g_{ik}(x^l)$ are expressed as functions of some independent quantities g^A,

$$g_{ik} = g_{ik}(g^A). \tag{5}$$

In general, the g^A are independent of some of the coordinates x^l, and their number is less than the number of algebraically independent components of the metric tensor g_{ik}, i.e., less than six. Equations (2) reduce the number of independent canonical momenta π^{ik} in the same way as Equations (1) reduce the number of independent canonical coordinates. If we write them down in the standard system of coordinates, the canonical momenta are expressible as functions of some independent quantities π_B,

$$\pi^{ik} = \pi^{ik}(g^A, \pi_B),$$

where, in contrast to (5), π^{ik} (in general) depends on both the g^A and the π_B. Our main task is to choose the new quantities π_A in such a way that the variables g^A and π_A are canonically conjugate. This means that the geometrodynamical action functional, when expressed by means of the variables g^A and π_A, assumes canonical form. In the standard system of coordinates, the super-Hamiltonian $\mathscr{H}$ and the supermomentum $\mathscr{H}_i$, as well as the lapse function N and the shift functions N^i do not depend on certain of the coordinates x^l by virtue of the symmetry conditions. The action may then easily be integrated over these coordinates and the canonical form of the action becomes

$$S = \int \mathrm{d}t \int \mathrm{d}x \left[\pi_A \dot{g}^A - N\mathscr{H}(g^A, \pi_B) - N^i \mathscr{H}_i(g^A, \pi_B)\right]. \tag{6}$$

The integration $\int \mathrm{d}x$ runs over the remaining coordinates and the summation $N^i\mathscr{H}_i$ over the components of the supermomentum which are not killed by the reflection symmetries.

The above reduction of the canonical formalism to a mini-phase-space was used for the first time by DeWitt and Misner to investigate homogeneous cosmological models. The term mini-superspace itself was coined by Misner. Homogeneous cosmologies are very special models, because the minimal invariant varieties of their respective groups of motions are themselves three-dimensional hypersurfaces fitted into a one-parameter family. The homogeneous universe looks homogeneous only when one takes its snapshots at a constant cosmological time. The one-parameter family of hypersurfaces provides a privileged slicing of spacetime which is used in the reduced formalism. This simplifies the formalism enormously. But this very same simplification makes it impossible to illustrate by means of these models one of the most prominent features of the full geometrodynamical formalism: the existence of a many-fingered time, allowing the deformation of a hypersurface by one amount at one point and by another amount at another point. A comprehensive account of the work done on homogenous cosmological models by canonical methods, mostly by

Misner and his students, was given by Ryan. Because I want to exhibit the many-fingered time formalism of the full geometrodynamics, I shall proceed directly to a more complicated model.

Homogeneous cosmologies have another disadvantage in that they drastically reduce the number of the gravitational degrees of freedom; from infinitely many degrees of freedom, only a finite number is left. The reason is that the homogeneity ties the corresponding degrees of freedom at different points rigidly together, forcing a typical degree of freedom at one point to imitate the behavior of the corresponding degree of freedom at any other point, so that they finally dance together like well-disciplined Rockettes. The field aspect of gravity thus completely disappears from these models. To study the propagation of the gravitational field, one must study a wave, not a homogeneous universe. After all, it was not a universe which was first quantized by the quantum field theory, but the electromagnetic wave! There are exact wave solutions known in Einstein's theory, and I shall apply the minisuperspace methods to one of them, namely, to the Einstein-Rosen cylindrical wave. This wave is symmetric with respect to translations along an axis of symmetry and rotations around it, and also with respect to reflections in the planes containing the axis of symmetry and perpendicular to it. It has only one polarization, the other one being killed by the reflection symmetry. However, there are ∞^1 degrees of freedom contained in this polarization, one degree of freedom for each cylindrical surface drawn around the axis of symmetry. The degrees of freedom on one cylindrical surface still dance in unison, but those on another surface may lag behind them, giving rise to the radial propagation of the wave. Furthermore, the slicing of spacetime is not completely fixed by the symmetry. The minimal invariant varieties of the group of cylindrical waves are two-dimensional cylindrical surfaces drawn around the axis of symmetry, just as the minimal invariant varieties of the rotation group were the two-dimensional spherical surfaces. All slices containing these two-dimensional cylindrical surfaces are allowed, and they are too numerous to be fitted into a one-parameter family. An arbitrary cylindrically symmetrical deformation of an allowed slice leads to another slice, the deformation being an ∞^1 fingered time. We have cut $2\infty^2$ fingers from our time, but it still has an infinite number of fingers left! Our mini-superspace is thus infinitely richer than the mini-superspaces of DeWitt and Misner. Let me borrow once more a term from the world of fashion and call it 'midi-superspace'. It is still not the full-length imperial robe that general geometrodynamics wears, but it certainly uses more fabric than the younger models. Curiously enough, truth is better revealed dressed than unclothed in geometrodynamics.

Reducing the canonical formalism for the Einstein-Rosen cylindrical waves, I use as standard coordinates the 'Killing coordinates', φ and and z, defined naturally by the rotational and translational Killing vectors through the requirement that the Killing vectors have the components

$$\xi^i_{(\varphi)}=(0, 1, 0),\ \xi^i_{(z)}=(0, 0, 1) \tag{7}$$

in the coordinates $x^i=(r, \varphi, z)$. The radial coordinate r is not fixed by the symmetry, and we may change it arbitrarily,

$$r \to \bar{r} = f^{-1}(r), \tag{8}$$

subject only to certain boundary restrictions at the axis of symmetry ($r=0$) and at spatial infinity ($r\to\infty$). It is well-known that the symmetry (1) of the metric with respect to the rotations and translations implies that the metric g_{ik} depends only on r, not on φ or z. Moreover, the requirement of symmetry with respect to the reflections

$$\varphi \to -\varphi,\ z \to -z \tag{9}$$

kills the non-diagonal components $i\neq k$ of the metric. The metric is thus reduced to three functions of one coordinate r, which may be (and traditionally are) chosen as $\gamma(r)$, $\phi(r)$, $R(r)$:

$$g_{11} = e^{\gamma-\phi},\ g_{22} = R^2 e^{-\phi},\ g_{33} = e^{\phi}.$$

Similarly, the symmetries (3), (4) of the lapse and shift functions restrict them to be functions of r only, and the reflection symmetry (9) kills the non-radial components of the shift function,

$$N = N(r);\ N^1 = N^1(r),\ N^2 = N^3 = 0. \tag{10}$$

The symmetry (2) of the momentum implies that $\pi^{ik}=\pi^{ik}(r)$, and the reflection symmetry kills its non-diagonal components. The momentum is thus expressible by three functions of one coordinate r, and if we choose these independent functions as π_γ, π_R, π_ϕ

$$\pi^{11} = \pi_\gamma e^{\phi-\gamma},\ \pi^{22} = \tfrac{1}{2}R\pi_R e^{\phi},\ \pi^{33} = (\pi_\gamma + \tfrac{1}{2}R\pi_R + \pi_\phi)\, e^{-\phi}, \tag{11}$$

the action functional (III.22) assumes after the reduction the canonical form

$$S = 2\pi \int_{-\infty}^{\infty} \mathrm{d}t \int_0^{\infty} \mathrm{d}r(\pi_\gamma\dot{\gamma} + \pi_R\dot{R} + \pi_\phi\dot{\phi} - N\mathscr{H} - N^1\mathscr{H}_1). \tag{12}$$

I have limited the integration over z by two 'planes', $z=z_0$, and $z=z_0+1$, to avoid an infinite action. The constant 2π, arising out of the integration over φ, could have been absorbed in the momenta π_γ, π_R, π_ϕ, but would then have messed up our coefficients; hence 2π in front of the integral! The super-Hamiltonian and supermomentum can be calculated directly from their definitions (III.23), (III.24), giving

$$\begin{aligned} \mathscr{H} &= e^{\frac{1}{2}(\phi-\gamma)}(-\pi_\gamma\pi_R + \tfrac{1}{2}R^{-1}\pi_\phi{}^2 + 2R'' - \gamma' R' + \tfrac{1}{2}R\phi'^2),\\ \mathscr{H}_1 &= -2\pi_\gamma{}' + \gamma'\pi_\gamma + R'\pi_R + \phi'\pi_\phi. \end{aligned} \tag{13}$$

The best way to describe the structure of this super-Hamiltonian and supermomentum is to say that it is messy. I would discourage anybody from imposing the constraints (13) on the state functional, because one does not know how to solve the resulting functional differential equations. The best way to unmess a mathematical mess is to use a messy transformation! Because I want to stay within the canonical formalism, it should be a messy canonical transformation. One can arrive at it in a natural way, by using what is known about the cylindrical waves described in a non-canonical language. Because I do not have enough time to proceed in such an inductive way, I shall simply write down the final formulas:

$$\begin{aligned} T(r) &= T(\infty) + \int_{\infty}^{r} - \pi_\gamma(r)\,\mathrm{d}r, \\ \Pi_T(r) &= -\gamma'(r) + \{\ln[R'^2(r) - \pi_\gamma^{\,2}(r)]\}', \\ R(r) &= R(r), \\ \Pi_R(r) &= \pi_R(r) + \left(\ln \frac{R'(r) - \pi_\gamma(r)}{R'(r) + \pi_\gamma(r)}\right)'. \end{aligned} \tag{14}$$

The transformation (14) introduces the new canonical variables T, Π_T, R, Π_R instead of old canonical variables γ, π_γ, R, π_R. We can check that it is indeed a canonical

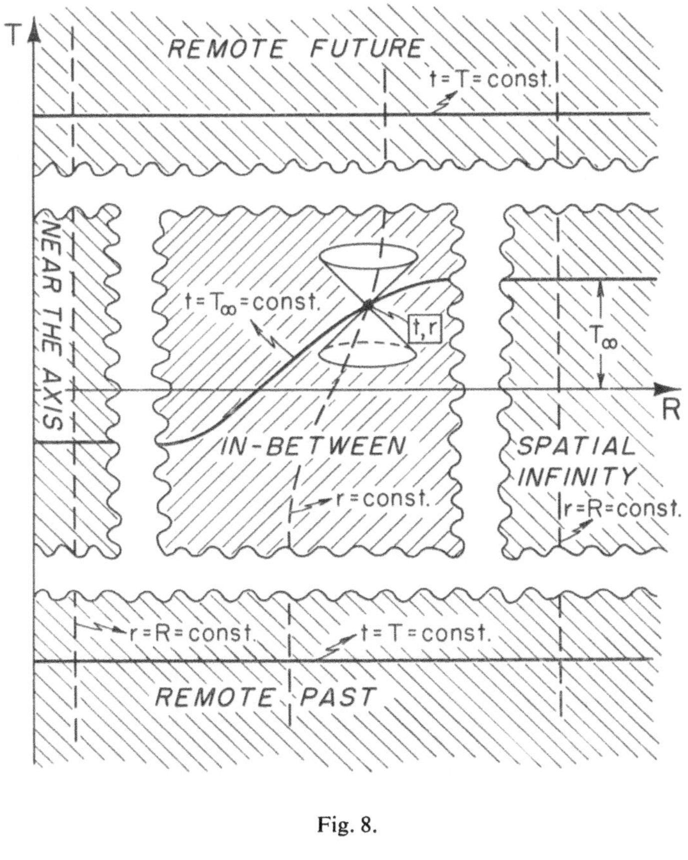

Fig. 8.

transformation, because

$$\int dt \int dr (\pi_\gamma \dot{\gamma} + \pi_R \dot{R}) = \int dt \int dr (\Pi_T \dot{T} + \Pi_R \dot{R}),$$

if we omit all boundary integrals arising from integrations by parts. This is permissible, if one uses the boundary conditions which I have mentioned, but did not write down. In terms of the new coordinates, they are conveniently summarized in Figure 8.

The notation which I have adopted suggests that I want to identify $T(r)$ with a privileged time and $R(r)$ with a privileged radial coordinate. Indeed, if we integrate the Hamilton equations corresponding to the action functional (12) written in the new canonical variables and reconstruct the spacetime, we find that T and R coincide with the invariantly defined privileged coordinates used by Einstein and Rosen.

Two features of the canonical transformation (14) deserve special attention. Firstly, it is a non-local transformation. To get the Einstein-Rosen time $T(r)$ at the point with the label r, one must know the old canonical variables not only at that point but all along the slice up to infinity. Secondly, the transformation (14) mixes the configuration space with the phase space, so that the new configuration space T, R, ϕ no longer corresponds to superspace. For example, the Einstein-Rosen time T is reconstructed from the momentum π_γ, not from the canonical coordinate γ.

To get an intuitive feeling for what is going on in such a construction of time, consider a simple analogue (Figure 9). Observe a mountain climber ascending the slope of a hill. How could he keep track of his elevation? After each step he could take his ice axe and dig a well down to sea level. Then he could sound a plummet and measure his elevation. This is obviouslytoo laborious, especially if the rock is solid and

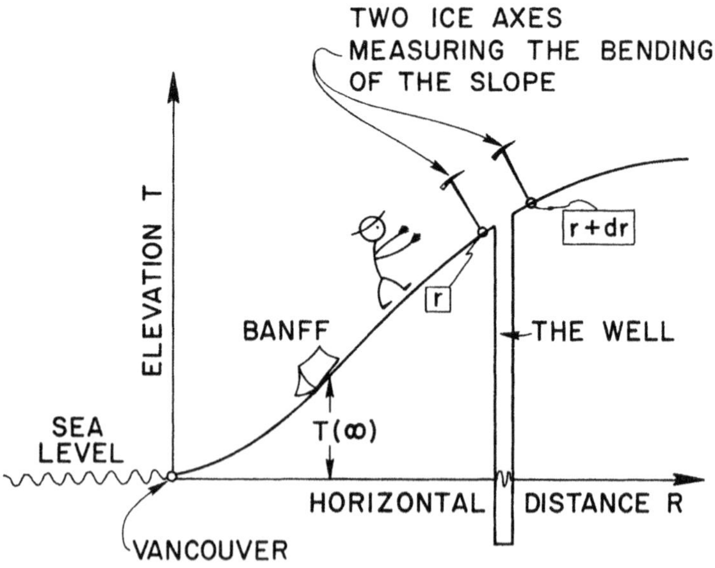

Fig. 9.

the elevation comparable to that of Banff. Moreover, digging wells is not in the spirit of mountaineering, just as leaving the spacelike hypersurfaces far into the embedding spacetime is not in the spirit of geometrodynamics. What our mountain climber could do is to take two ice axes instead of one and stick them in at each step perpendicular to the slope, measuring the distance between their bases and the distance between their tips. This gives him the bending of the slope in the surrounding space. Then, on a piece of paper, he could draw the shape of the mountain and, knowing the elevation of his starting point, he could deduce the elevation he had reached. This is exactly what the formula (14) does in spacetime, reconstructing $T(r)$ step by step from the extrinsic curvature contained in π_{γ}, and from the knowledge of the starting 'time elevation' $T(\infty)$ at infinity.

The proof of the pudding is in the eating, and the proof of the canonical transformation (14) is in its capacity to simplify the constraints. When we introduce the new canonical variables into the super-Hamiltonian and supermomentum (13), we get

$$\mathscr{H} = e^{\frac{1}{2}(\phi-\gamma)}(R'\Pi_T + T'\Pi_R + \tfrac{1}{2}(R^{-1}\pi_\phi{}^2 + R_\phi{}'^2)), \tag{15}$$

$$\mathscr{H}_1 = T'\Pi_T + R'\Pi_R + \phi'\pi_\phi . \tag{16}$$

The old variable γ in the exponent of Equation (15) should be replaced by the new canonical variables, using the inverted canonical transformation (14). The exponential factor in $\mathscr{H}$ does not matter anyhow, because we may divide the super-Hamiltonian constraint $\mathscr{H}=0$ through by it. Then, disregarding this factor, we find to our surprise that we have formally recovered the constraints of a parametrized scalar field theory in Minkowskian spacetime with one spatial dimension R. The supermomentum (16) is identical with the supermomentum (III.6). This indicates that the variables T, R, ϕ transform as scalars under the radial relabelings (8) of the slice, and the conjugate momenta π_T, π_R, π_ϕ transform as scalar densities. Disregarding the positive factors $e^{\frac{1}{2}(\phi-\gamma)}$ and $(X'^2-T'^2)^{-\frac{1}{2}}$ which may be divided out, we see that the super-Hamiltonian (15) is almost identical to the super-Hamiltonian (III.5), the only difference being the additional factors R^{-1} and R in its dynamical part. These factors are due simply to the fact that R is a cylindrical radial coordinate in a three-dimensional space, whereas X is a cartesian coordinate in a one-dimensional space. If I had parametrized a cylindrically symmetrical scalar field propagating in a three-dimensional space instead of a scalar field propagating in a one-dimensional space, I would have obtained exactly these additional factors R^{-1} and R.

The canonical transformation (14), casting the constraints into the form of a parametrized scalar field theory, enables me to transfer to my midi-superspace all results obtained in my last lecture, and many more which I had no time to discuss. There may be an important message conveyed to us by this procedure: to separate kinematical and dynamical variables in geometrodynamics, it may be necessary to mix the configuration space with the phase space, passing from superspace to a new dynamical arena. Time may be hidden in the extrinsic curvature of a slice, rather than in its intrinsic geometry. I have pursued the idea of an extrinsic time in my earlier papers, and I do not want to elaborate here further on this point.

We may now turn the new canonical momenta Π_T, Π_R, and Π_ϕ into variational derivative operators, and impose the constraints (15), (16) as restrictions on the state functional $\Psi[T, R, \phi]$. The supermomentum constraint carries the information that the state functional Ψ remains unchanged under the radial relabelling (8) of the hypersurface $T=T(r)$, $R=R(r)$. One may use it to eliminate the variational derivative $\delta\Psi/\delta R(r)$ from the Wheeler-DeWitt equation $\mathscr{H}\Psi=0$, and then use the special parametrization $R=r$. This yields an analogue of Equation (III.20),

$$i\frac{\delta\Psi}{\delta T(R)}=\frac{1}{2}\left(1-\left(\frac{\partial T}{\partial R}\right)^2\right)^{-1}\left(-iR^{-\frac{1}{2}}\frac{\delta}{\delta\phi(R)}-R^{\frac{1}{2}}\frac{\partial T}{\partial R}\frac{\partial\phi}{\partial R}\right)^2\Psi+ \\ +\frac{1}{2}R\left(\frac{\partial\phi}{\partial R}\right)^2\Psi. \tag{17}$$

Equation (17) is a many-fingered time counterpart of an ordinary Schrödinger equation. It's no surprise that we can derive from it the continuity equation for probability flux by the very same method used in particle dynamics. Take Equation (17), multiply it by Ψ^*, and subtract from the resulting equation its complex conjugate. By the usual rearrangements, we get the equation

$$\frac{\delta(\Psi\Psi^*)}{\delta T(R)}+\frac{\delta\Sigma_R}{\delta\phi(R)}=0, \tag{18}$$

where Σ_R is a functional of $T(R)$ and $\phi(R)$ depending on R as a label,

$$\Sigma_R=\frac{1}{2}i\left(1-\left(\frac{\partial T}{\partial R}\right)^2\right)^{-1}\left\{R^{-1}\left(\Psi\frac{\delta\Psi^*}{\delta\phi(R)}-\Psi^*\frac{\delta\Psi}{\delta\phi(R)}\right)+ \\ +2i\frac{\partial T}{\partial R}\frac{\partial\phi}{\partial R}\Psi\Psi^*\right\}. \tag{19}$$

Its structure is strikingly similar to the structure of the probability flux for a particle moving in an external electromagnetic field. The discrete index k labelling the Cartesian coordinates x^k of the particle is replaced by the continuous label R, the coordinates x^k are replaced by the field variables $\phi(R)$, the momentum operator $-i\partial/\partial x^k$ is replaced by the field momentum operator $-i\delta/\delta\phi(R)$, and the vector potential term $em^{-1}A^k$ is analogous to the factor $(\partial T/\partial R)(\partial\phi/\partial R)$. The functional Σ_R depends on the continuous label R; similarly, the probability flux for a particle depends on the discrete index k labeling the components of the flux vector. However, Equation (18) does not have the form of the equation of continuity. There is no integration in it over the label R so that the expression $\delta\Sigma_R/\delta\phi(R)$ is not analogous to the divergence of the probability flux. In fact, Equation (18) is not a single equation, as the equation of continuity should be, but represents an infinite system of equations, one equation for each value of the parameter R.

To get a true equation of continuity, we must pick out a one-parameter family of hypersurfaces $T=T(R, t)$ and ask how $\Psi\Psi^*$ changes when we pass from one hypersurface of the family to the next one. We get

$$\frac{\partial(\Psi\Psi^*)}{\partial t} = \int_0^\infty \mathrm{d}R \, \frac{\delta(\Psi\Psi^*)}{\delta T(R)} \frac{\partial T(R,t)}{\partial t} =$$

$$= - \int_0^\infty \mathrm{d}R \, \frac{\delta}{\delta\phi(R)} \left(\Sigma_R \Big|_{T(R)=T(R,t)} \frac{\partial T(R,t)}{\partial t} \right). \tag{20}$$

The last integral represents a divergence of the 'flux vector' $\Sigma_R \, \partial T(R,t)/\partial t$ in the infinitely dimensional space of functions $\phi(R)$. We can now integrate both sides of Equation (20) with respect to the function variable $\phi(R)$. The functional integral $\int D\phi$ of the divergence on the right-hand side of Equation (20) can be transformed by an infinitely dimensional version of the Gauss theorem into a functional surface integral over the boundary of the space of functions ϕ. Under the appropriate boundary conditions for the state functional Ψ, this functional surface integral vanishes. Therefore, the functional integral $\int D\phi(\Psi\Psi^*)$ does not depend on the label time. Because the slicing $T = T(R,t)$ is completely arbitrary, we may conclude that

$$\frac{\delta}{\delta T(R)} \int D\phi \, \Psi\Psi^* = 0. \tag{21}$$

More generally, if we have any two functionals $\Psi_1[T(R), \phi(R)]$ and $\Psi_2[T(R), \phi(R)]$ which satisfy the reduced Wheeler-DeWitt equation (17), we can define the inner product of these functionals by the formula

$$\langle \Psi_1, \Psi_2 \rangle_{T(R)} = \int D\phi \, \Psi_1^*[T(R), \phi(R)] \, \Psi_2[T(R), \phi(R)]. \tag{22}$$

By virtue of the evolution equation (17), this inner product is unaffected by an arbitrary deformation $\delta T(R)$ of the hypersurface $T = T(R)$. The inner product (22) turns the space of the state functionals into a Hilbert space. The operators $\phi(R)$ and $\pi_\phi(R) = -i\delta/\delta\phi(R)$ are Hermitian with respect to this product, as is the Hamiltonian density operator which stands on the right-hand side of the evolution equation (17).

Equation (21) leads naturally to the probabilistic interpretation of the state functional $\Psi[T(R), \phi(R)]$. If we normalize this state functional to unity,

$$\langle \Psi, \Psi \rangle_{T(R)} = 1, \tag{23}$$

we can identify $D\phi \, \Psi^*\Psi$ as the probability that the field ϕ falls within the range of values $D\phi$ lying around the definite distribution $\phi(R)$ on the hypersurface $T = T(R)$.

What does this all mean in terms of an idealized theory of measurement? We worked in the 'extrinsic time representation', in which the operators $T(r)$, $R(r)$, and $\phi(r)$ were diagonal. This representation is physically realized by a complete set of measuring apparatuses, one subset of apparatuses designed to measure $T(r)$, another subset designed to measure $R(r)$, and still another subset designed to measure $\phi(r)$.

Because these operators commute, the apparatuses can be constructed in such a way that the measurement of any one of these quantities does not disturb in an unpredictable way the measurement of any other of the quantities. What is the nature of our apparatuses? Things are easy for a scalar field in a Minkowskian spacetime. There we know exactly how to measure $T(r)$ and $R(r)$. The spacetime is not quantized and the measurement of time intervals and distances does not disturb the measurement of other quantities like the scalar field variable $\phi(r)$. We have a rigid inertial structure of reference points, each point carrying a label indicating its coordinates r, φ, z and being equipped with a 'label clock' showing the label time t. Moreover, at each reference point there is a standard clock showing the standard time T, different standard clocks being synchronized by light signals, and a net of standard measuring rods, connecting the reference point with all neighbouring points. The hypersurface $t=$ $=$const. is determined by the readings of the label clocks. The clocks installed at the reference points measure the time T and the rods stretching radially from the axis of symmetry of the $\phi(r)$ field to the point with the label r measure the R.

Measurements of the quantized Riemannian geometry are much more sophisticated. They again presuppose a structure of reference points defining our system of coordinates and the hypersurface of a constant label time. But the net of standard measuring rods may be set up only in the azimuthal and axial directions. If the rods were placed in the radial direction, they would measure the g_{11} component of the metric and with it the coefficient γ. But $-\gamma'$ is canonically conjugate to T, so that γ and T cannot be measured simultaneously. Laying the rods in all possible directions would be equivalent to fixing a representation quite different from the extrinsic time representation T, R, ϕ – namely, the metric representation γ, R, ϕ. This is why we can lay the rods only in the azimuthal and axial directions. They are the apparatuses measuring $g_{22}=R^2e^{-\phi}$ and $g_{33}=e^{\phi}$, and therefore $R(r)$ and $\phi(r)$.

Also our 'T-clocks' are very queer clocks indeed. What we require is a device measuring the peculiar combination $\pi_\gamma=\pi^1_1=g^{\frac{1}{2}}(K^2_2+K^3_3)$ of the components of the extrinsic curvature, without ever attempting to measure these components or the determinant g individually, plus a standard clock at spatial infinity, where the spacetime is asymptotically Minkowskian. This set of instruments permits us to measure T as defined by Equation (14).

I have thus fulfilled my promise to illustrate by means of a simple model the construction of a Hilbert space and to give a probabilistic interpretation of quantum geometrodynamics. Throughout these lectures, I was true to my program of illustrating difficult questions by means of simple models. This is surely not enough if we want to get exact answers to specific questions. But I hope it is just enough to let us smell a distant scent of quantum geometrodynamics and be attracted by it. This is the end of my four quartets on quantum geometrodynamics. It very well may be that quantum geometrodynamics is a dead end in itself, and that in order to marry the quantum theory to the general theory of relativity, one must go to other churches. What we call the beginning is often the end. George Leigh Mallory after making his memorable statement about Mt. Everest, disappeared forever on its slopes. We even do not know if he reached the summit, but he probably did not. I cannot be sure that I have chosen the correct route, and I am painfully aware that nothing is forgotten more easily than a few lectures delivered in haste. However, let me hope, quite irrationally, that it is not entirely so. Many years after Mallory got lost on Mt. Everest, his ice axe was found there. So, maybe, someone will find the simple tools I have used in these lectures useful enough to cut the ice on lesser summits, or try them again on the still unconquered relativistic Mt. Everest. If so, to make an end is to make a beginning. The end is where we start from.

Acknowledgements

These lectures were prepared in the American and delivered in the Canadian Rockies. I gratefully acknowledge the hospitality of the Aspen Center for Physics where I discussed the lectures *in statu nascendi* with Claudio Teitelboim and Percy Deift. I would also like to thank Percy Deift for his efforts to make my English easier to read in Princeton than it was to hear in Banff. The work on these notes was supported by National Science Foundation Grant GP30799X to Princeton University.

humbly inscribed and dedicated to

professor john a. wheeler, now of

joseph henry laboratories, princeton university;

the rightfull acknowledg'd authority on

quantumgeometrodynamics;

the source of the author's inspiration

Bibliography

To keep the lectures in the form in which they were delivered, no references were given in the text. The bibliography is therefore run as a commentary.

Page I

238 On the proposal to leave the gravitational field unquantized, see:

[1] Møller, C.: 1962, 'The Energy-Momentum Complex in General Relativity and Related Problems', in: *Les théories relativistes de la Gravitation,* (ed. by A. Lichnerowicz and M. A. Tonnelat), Centre Nationale de la Recherche Scientifique, Paris.

[2] Rosenfeld, L.: 1963, *Nucl. Phys.* **40**, 353.

238 On different methods of quantization for the gravitational field, see:

covariant quantization:

[3] DeWitt, B. S.: 1964, 'Dynamical Theory of Groups and Fields', in: *Relativity, Groups* and *Topology* (ed. by B. S. DeWitt and C. M. DeWitt), Gordon and Breach, New York-London.

[4] DeWitt, B. S.: 1967, *Phys. Rev.* **162**, 1195; **162**, 1239.

constraint quantization:

[5] Dirac, P. A. M.: 1958, *Proc. Roy. Soc. (London)* **A246**, 326; **A246**, 333.

[6] Dirac, P. A. M.: 1959, *Phys. Rev.* **114**, 924.

[7] DeWitt, B. S.: 1967, *Phys. Rev.* **160**, 1113.

[8] Wheeler, J. A.: 1968, 'Superspace and the Nature of Quantum Geometrodynamics', in: *Batelle Rencontres* (ed. by C. DeWitt and J. A. Wheeler), Benjamin, New York.

ADM method:

[9] Arnowitt, R. and Deser, S.: 1959, *Phys. Rev.* **113**, 745.

[10] Arnowitt, R., Deser, S., and Misner, C. W.: 1959, *Phys. Rev.* **116**, 1322.

[11] Arnowitt, R., Deser, S., and Misner, C. W.: 1960, *Phys. Rev.* **117**, 1595.

[12] Arnowitt, R., Deser, S., and Misner, C. W.: 1960, *Nuovo Cim.* **15**, 487.

[13] Arnowitt, R., Deser, S., and Misner, C. W.: 1960, *Phys Rev. Letters* **4**, 375.

[14] Arnowitt, R., Deser, S., and Misner, C. W.: 1960, *Phys. Rev.* **118**, 1100.

[15] Arnowitt, R., Deser, S., and Misner, C. W.: 1960, *J. Math. Phys.* **1**, 434.

[16] Arnowitt, R., Deser, S., and Misner, C. W.: 1960, *Phys. Rev.* **120**, 313.

[17] Arnowitt, R., Deser, S., and Misner, C. W.: 1960, *Phys. Rev.* **120**, 321.

[18] Arnowitt, R., Deser, S., and Misner, C. W.: 1960, *Ann. Phys.* **11**, 116.

[19] Arnowitt, R., Deser, S., and Misner, C. W.: 1961, *Nuovo Cim.* **19**, 668.

[20] Arnowitt, R., Deser, S., and Misner, C. W.: 1961, *Phys. Rev.* **121**, 1556.

[21] Arnowitt, R., Deser, S., and Misner, C. W.: 1961, *Phys. Rev.* **122**, 997.

the summary of the previous work is contained in

[22] Arnowitt, R., Deser, S., and Misner, C. W.: 1962, 'The Dynamics of General Relativity', in: *Gravitation: An Introduction to Current Research* (ed. by L. Witten), Wiley, New York.

gauge-independent quantization:

[23] Mandelstam, S.: 1962, *Ann. Phys.* **19**, 25.

[24] Mandelstam, S.: 1962, 'Quantization of the Gravitational Field', in: *Relativistic Theories of Gravitation* (ed. by L. Infeld), Pergamon, Gauthier-Villars, P. W. N., Warszawa.

[25] Mandelstam, S.: 1968, *Phys. Rev.* **175**, 1580.

source-theory formalism:

[26] Schwinger, J.: 1966, *Phys. Rev.* **152**, 1219.

[27] Schwinger, J.: 1967, *Phys. Rev.* **158**, 1391.

[28] Schwinger, J.: 1968, *Phys. Rev.* **173**, 1264.

Page

summing-over-histories

[29] Misner, C. W.: 1957, *Rev. Mod. Phys.* **29**, 497.
[30] Klauder, J. R.: 1962, *Nuovo Cimento* **25**, 542.
[31] Fadeev, L. D. and Popov, V. N.: 1967, *Phys. Letters* **25B**, 29.
[32] Fadeev, L. D.: 1971, *Actes, Congrès Intern. Math.*, Vol. **3**, p. 35, Gauthier-Villars, Paris.

on the equivalence of the summing-over-histories quantization and the canonical quantization:

[33] Leutwyler, H.: 1964, *Phys. Rev.* **134B**, 1155.

quantization on light cones:

[34] Komar, A.: 1969, 'On the Quantization of Gravitation', in: *Proceedings of the Eighth Annual Eastern Theoretical Physics Conference* (ed. by F. Rohrlich), Syracuse University.

twistor quantization:

[35] Penrose, R. and MacCallum, M. A. H.: 1970, *Twistor Theory: an Approach to the Quantization of Fields and Space-Time*, mimeographed lecture notes, Institute of Theoretical Astronomy, Cambridge.

For a recent review of different methods of quantization, see

[36] Brill, D. R. and Gowdy, R. H.: 1970, *Progress Theor. Phys.* **33**, 413.

On the early attempts to quantize gravity, see, e.g., the introduction to DeWitt's paper [7].

238 On the different versions of the canonical quantization of gravity, see references [5]–[22] and

[37] Bergmann, P. G.: 1956, *Helv. Phys. Acta Suppl.* **4**, 79.
[38] Schwinger, J.: 1963, *Phys. Rev.* **130**, 1253; 1963, *Phys. Rev.* **132**, 1317.
[39] Peres, A.: 1968, *Phys. Rev.* **171**, 1335.

as quoted in DeWitt's short review article on quantum theories of gravity,

[40] DeWitt, B. S.: 1970, *Gen. Rel. Grav.* **1**, 181.

239 For the conjecture that quantization of gravity can prevent singularities, see especially

[41] Wheeler, J. A.: 1964, 'Geometrodynamics and the Issue of the Final State', in: *Relativity, Groups and Topology* (ed. by B. S. DeWitt and C. M. DeWitt), Gordon and Breach, NewYork-London, pp. 328, 504.

and

[42] Harrison, B. K., Thorne, K. S., Wakano, M., and Wheeler, J. A.: 1965, *Gravitation Theory and Gravitational Collapse*, The University of Chicago Press, Chicago-London, pp. 141–143.

The analysis of simple cosmological models did not seem to indicate that quantization removes singularities: see

[43] Misner, C. W.: 1969, *Phys. Rev.* **186**, 1319; 1969, *Phys. Rev.* **186**, 1328.
[44] Ryan, M.: 1972, *Hamiltonian Cosmology*, Springer, Berlin–Heidelberg–New York.

Cf., however, with an opposite conclusion reached for simplified models in ref. [7] and in

[45] Liang, E. P. T.: 1972, *Phys. Rev.* **D5**, 2458.

239 On the Yang-Mills fields, see ref. [3].

242 On attempts to separate kinematical and dynamical variables in the general theory of relativity, see the articles by ADM [11], Sec. 2, [22], and

[46] Deser, S.: 1967, *Ann. Inst. Henri Poincaré* **7**, 149.
[47] Kuchař, K.: 1970, *J. Math. Phys.* **11**, 3322.
[48] York, J. W.: 1971, *Phys. Rev. Letters* **26**, 1556.
[49] York, J. W.: 1972, *Phys. Rev. Letters* **28**, 1082.
[50] York, J. W.: 1972, *J. Math. Phys.* **13**, 125.
[51] York, J. W.: 1972, *J. Math. Phys.*, in print.

Page

Natural separations have been achieved for different minisuperspace models. For cosmological models, see references [43], [44]. For the cylindrical gravitational waves, see

[52] Kuchař, K.: 1971, *Phys. Rev.* **D4**, 955.

and the final lecture IV. For the Schwarzschild field, see

[53] Berger, B. K., Chitre, D. M., Monchrief, V. E., and Nutku, Y.: 1972, *Phys. Rev.* **D5**, 2467.

On the general many-fingered time canonical formalism following from such a separation, see

[54] Kuchař, K.: 1972, *J. Math. Phys.* **13**, 768.

242 On the standard gravitational Cauchy problem and the assertion that ${}^4g_{\iota\kappa}$ and ${}^4g_{\iota\kappa,0}$ neither can be initially freely specified not determined afterwards by the equations of motion, see

[55] Lichnerowicz, A.: 1955, *Théories relativistes de la gravitation et de l'électromagtisme*, Masson, Paris.

[56] Adler, R., Bazin, M., and Schiffer, M.: 1965, *Introduction to General Relativity*, McGraw-Hill, New York, Chapter 7.

242–245 An excellent account of the parametrized canonical formalism in particle dynamics can be found in

[57] Lanczos, C.: 1949, *The Variational Principles of Mechanics*, University of Toronto Press, Toronto (4th edition 1970), chapters VI.10, IX. 6–IX.10.

See also ref. [22], chapters 7–2.1 – 7–2.3.

246–249 Parametrized canonical formalism for a field propagating on a Minkowskian background is discussed in

[58] Dirac, P. A. M.: 1964, *Lectures on Quantum Mechanics*, Yeshiva University, New York.

The scalar field is parametrized in ref. [10]; the constraints, however, are not projected into directions parallel and perpendicular to the hypersurface. The super-Hamiltonian and supermomentum of the scalar field propagating in one spatial dimension (the cylindrical case) is derived in ref. [52].

248 The lapse and shift functions were introduced by ADM in ref. [10]. A nice discussion of their geometrical significance can be found in ref. [41], pp. 345–348.

249 That constraints are always associated with generally covariant theories is discussed in

[59] Bergmann, P. G.: 1949, *Phys. Rev.* **75**, 680.

and in ref. [3].

II

250 This lecture is based on unpublished papers by C. Teitelboim, S. Hojman and K. Kuchař. My thanks are due to C. Teitelboim and S. Hojman for permission to use their results.

250 The argument that the Poisson brackets between the constraints may be calculated paying attention merely to their kinematical parts is Dirac's (see ref. [58], Lecture 3). The actual calculations may be found in

[60] Dirac, P. A. M.: 1951, *Canadian J. Math.* **3**, 1.

Equation (II.5) differs slightly from that found in the literature (compare references [7], [58]); the symmetrical form we display on the right hand side of the equation is achieved through well-known properties of the δ-function.

250 The commutation relations (II.5)–(II.7) drastically reduce the number of independent constraints; see

[61] Moncrief, V.: 1972, *Phys. Rev.* **D5**, 277.

[62] Moncrief, V. and Teitelboim, C.: 1972, *Phys. Rev.* **D6**, 966.

Page

They also imply the path-independence of geometrodynamical evolution; see ref. [54].

251 C. Teitelboim derived the Poisson brackets relations (II.5)–(II.7) in an arbitrary curved spacetime by a purely geometrical argument based on the composition of surface deformations. He also showed that the signature of spacetime can be read off from the commutator of two super-Hamiltonians. See

[63] Teitelboim, C.: 1972, Princeton University Preprint, submitted to *Ann. Phys. N.Y.*

251–258 The derivation of the commutation relations (II.24)–(II.26) using the language of infinitely dimensional continuous groups (see ref. [3], chapter 11) is part of an unpublished paper by the author. The set of deformations of hypersurfaces in a Riemannian spacetime is discussed under a different name and for slightly different purposes in

[64] Fischer, A. E. and Marsden, J. E.: 1972, *J. Math. Phys.* **13**, 546.

258–259 The concept of superspace was introduced by Wheeler in ref. [41] and extensively discussed by him in ref. [8]. For a history of the subject, see

[65] Misner, C. W.: 1972, 'Minisuperspace', in *Magic Without Magic* – J. A. Wheeler 60th Anniversary Volume (ed. by J. Klauder), Freeman, San Francisco.

The formalization of the concept of superspace is to be found in

[66] Fischer, A. E.: 1970, 'The Theory of Superspace', in: *Relativity* (ed. by L. Witten), Plenum Press, New York–London.

259 On separation of dynamical and kinematical variables, see comments to p. 147.

259–261 The derivation of the structure of geometrodynamical constraints from their commutation relations which removes the ansatz (II.34) is to be found in

[67] Hojman, S., Kuchař, K., and Teitelboim, C.: 1973, paper in preparation.

260 That $-2\int d^3x\, \pi^k{}_{i|k}\delta N^i$ generates transformations of intrinsic labels on a spatial hypersurface is well known. It is discussed, e.g., in ref. [7]. Here it is also noted that the right-hand side of Equation (II.7) yields the structure constants of the group of general transformations of spatial coordinates.

260 That the commutation relations (II.6) are equivalent to the statement that $\mathscr{H}(x)$ is a scalar density of weight 1 is discussed by Dirac in ref. [58].

261 On the standard derivation of ADM action functional (II.38) from the Hilbert action functional, see either the original ADM papers [9]–[22], or Wheeler's account in ref. [41], or the survey by Brill and Gowdy [36].

III

263 The general theory of quantization of a parametrized theory by imposing the constraints as restrictions on the state function was developed by Dirac and is explained by him, e.g., in ref. [58].

264 On the 'frozen formalism', see

[68] Bergmann, P. G.: 1956, *Nuovo Cimento* **3**, 1177.
[69] Komar, A. B.: 1958, *Phys. Rev.* **111**, 1182.
[70] Komar, A. B.: 1967, *Phys. Rev.* **164**, 1595.

and ref. [29]. See also the comments on this approach by DeWitt (ref. [7]) and Brill and Gowdy (ref. [36]).

265–167 On the 'field representation', in which the state is described by a functional of the canonical field coordinates $\phi(x)$, rather than being given in the more usual 'occupation number representation', see e.g.

Page

[71] Schweber, S. S.: 1961, *An Introduction to Relativistic Quantum Field Theory*, Row, Peterson, Evanston-Elmsford, Chapter 7, §5.

[72] Novozhilov, Yu. V. and Tulub, A. V.: 1961, *The Methods of Functionals in the Quantum Theory of Fields*, Gordon and Breach, New York.

265–267 The constraint quantization on curved and flat hypersurfaces is discussed in ref. [58]. On the constraint quantization of the scalar field and its relation to the half-parametrized and the ADM deparametrized formalism, see ref. [52].

268–269 The meaning of the geometrodynamical supermomentum constraint (that it ensures the labelling-independence of the state functional) was discovered by Misner (ref. [29]) and by Higgs:

[73] Higgs, P. W.: 1959, *Phys. Rev. Letters* **1**, 373.

268–269 For the Wheeler-DeWitt equation, see references [7] and [8]. For the geometrodynamical Hamilton-Jacobi equation which preceded it, see

[74] Peres, A.: 1962, *Nuovo Cimento* **26**, 53.

269–271 That factors in $\mathscr{H}$ and $\mathscr{H}_i$ should be ordered so as to close the commutators, see Dirac in ref. [58]. On an attempt along these lines, see Schwinger in ref. [38]. On difficulties in Schwinger's approach, see

[75] Dirac, P. A. M.: 1969, 'The Quantization of the Gravitational Field', in: *Contemporary Physics: Trieste Symposium 1968* (Vienna: IAEA), p. 539.

On a proposal to let any two field operators taken at the same point commute, see DeWitt in ref. [7].

270 On the idea of a covariant version of the Wheeler-DeWitt equation in mini-superspace, see ref. [65].

IV

271 This lecture is based, partly verbatim, on my paper [52] about the quantization of cylindrical gravitational waves.

271–270 For the definition of a Klein-Gordon inner product of state functionals which satisfy the Wheeler-DeWitt equation, see ref. [7].

274 On the first application of minisuperspace techniques to the Friedmann universe, see ref. [7]. Different homogeneous cosmologies were treated by minisuperspace methods by Misner and his collaborators; see, e.g., references [43], [65],

[76] Misner, C.: 1970, 'Classical and Quantum Dynamics of a Closed Universe', in: *Relativity* (ed. by M. Carmeli, S. Fickler, and L. Witten), Plenum Press, New York.

[77] Hughston, L. and Jacobs, K.: 1970, *Astrophys. J.* **160**, 147.

A comprehensive summary of this work is to be found in Ryan's book [44].

275 On Einstein-Rosen waves treated by traditional non-canonical methods, see

[78] Einstein, A. and Rosen, N.: 1937, *J. Franklin Inst.* **223**, 43.

[79] Rosen, N.: 1954, *Bull. Res. Council Israel* **3**, 328.

[80] Weber, J. and Wheeler, J. A.: 1957, *Rev. Mod. Phys.* **29**, 509.

[81] Marder, L.: 1958, *Proc. Roy. Soc. (London)* **A244**, 524; **A246**, 133.

[82] Ehlers, J. and Kundt, W.: 1962, in: *Gravitation: An Introduction to Current Research* (ed. by L. Witten), Wiley, New York–London, Chapter 2.

[83] Thorne, K. S.: 1965, *Phys. Rev.* **B138**, 251.

279 On the idea of an 'extrinsic time', see my papers [47], [52], [54], and references [22], [39], [48]–[51], [53].

281–282 On the theory of measurement of quantized gravitational fields by apparatuses constructed from clocks and elastic bodies, see

[84] DeWitt, B. S.: 1962, 'The Quantization of Geometry', in: *Gravitation: an Introduction to Current Research* (ed. by L. Witten), Wiley, New York-London.

DIFFERENTIAL GEOMETRY FROM A MODERN STANDPOINT

B. SCHMIDT

Institut für Theoretische Physik, Universität Hamburg, Jungiusstrasse 9, Hamburg

1. Manifolds

One is led to the concept of a differentiable manifold, if one tries to define spaces M more general than R^n* on which it is still possible to do calculus in some sense. For this purpose we demand first that any point of M – which is so far only a set – is contained in a subset U, which can be mapped onto an open set in R^n by a 1–1 map h.

Definition 1.1. An n-chart at a point p of M, where M is a set, is a pair (U, h) where $p \in U \subset M$ and h is a 1–1 map of U onto an open set in R^n.

An atlas on M is a collection of charts (U_α, h_α) such that $\bigcup U_\alpha = M$.

Suppose now we have an atlas on M and consider a real valued function $f: M \to R$ on M. If we select a chart (U_α, h_α) then $f \circ h_\alpha^{-1}$ is a realvalued function on the open subset $h_\alpha(U_\alpha)$ in R^n and we could try to call f differentiable if $f \circ h_\alpha^{-1}$ is differentiable. This is, however, not a proper definition, because it is possible that there are two charts (U_α, h_α), (U_β, h_β) such that $U_\alpha \cap U_\beta \neq \emptyset$ and $f \circ h_\alpha^{-1}$ is differentiable but not $f \circ h_\beta^{-1}$; this is impossible if $h_\alpha \circ h_\beta^{-1}$ is differentiable as well as $h_\beta \circ h_\alpha^{-1}$; these are the compatibility conditions for a differentiable structure.

Definition 1.2. Two charts (U_α, h_α), (U_β, h_β) are $C^k (C^0, C^\infty, C^\omega)$ related if $h_\alpha \circ h_\beta^{-1}$ and its inverse are k-times continuously differentiable maps $R^n \to R^n$ where they are defined.

Definition 1.3. An atlas is a C^katlas if any two charts are C^k related.

Note that every chart in the atlas is then an n-chart.

Definition 1.4. A C^k atlas is maximal if it contains any chart which is C^k related to its charts. The set M together with a maximal C^k atlas is a C^k-n-dimensional manifold.

A maximal C^k atlas is also called a C^k differentiable structure on M. On a C^k manifold we can now define differentiability up to the order k in a consistent way. From now on we will only use C^∞ manifolds.

Definition 1.5. If M is a C^∞ manifold and $f: M \to R$ a real-valued function, then f is differentiable if $f \circ h_\alpha^{-1}$ has derivatives of any order when (U_α, h_α) runs through an atlas of M.

A differentiable manifold has a natural topology which is uniquely determined by the condition that the coordinate maps h are continuous in both directions.

* $R^n = \{(x^1, \ldots, x^n); x^i \in R\}$ together with its usual topology and vector space structure.

W. Israel (ed.), Relativity, Astrophysics and Cosmology, 289–322.

Definition 1.6. The natural topology on a differentiable manifold has as a basis of its open sets the inverse images of open sets in R^n under the coordinate maps.

As is usually done in differential geometry we will assume that this topology is Hausdorff and has a countable basis, which is not implied by the definitions.

Examples of manifolds are R^n, the n-sphere S^n, the n-torus T^n. Obviously any open subset of a manifold is a manifold which shows that $Gl(n, R)$, the general linear group, is a manifold because it is an open subset of R^{n^2}. If M, M' are two manifolds one can define a manifold structure on the Cartesian product $M \times M'$, by taking $(U \times U', (h, h'))$ as charts in an obvious way.

2. The Tangent Space

On a manifold M we can not only define differentiable functions but also tangent vectors to a curve. This gives the possibility of attaching to any point $p \in M$ a vector space which is in some sense an 'infinitesimal version' of the vector space structure of R^n on which we have modelled our manifolds.

Let us consider differentiable curves through a point $p \in M$.

Definition 2.1. A differentiable curve σ is a map $(a, b) \to M$ where (a, b) is an open interval in R such that if (U, h) is any chart the map $h \circ \sigma$ is differentiable when it is defined.

According to this definition a curve has always a fixed parametrization. If h is a coordinate map and $\pi^i : R^n \to R$ the coordinate projection $(x^1, \ldots, x^n) \to x^i$, then $x^i \equiv \pi^i \circ h$ is a real-valued function on U which is differentiable. If σ is a curve in U, then $h \circ \sigma$ is a curve in R^n, and the real number

$$\left[\frac{\mathrm{d}}{\mathrm{d}t}(x^i \circ \sigma)\right]_{t=t_0} \tag{2.1}$$

determine the tangent to the curve $h \circ \sigma$ in R^n. One could now define a tangent vector at a point $p \in M$ as a set of curves through p and a coordinate map, such that the images of these curves have the same tangent in R^n. One would then have to show that this set is independent of the particular chart chosen. Instead of this approach, we can characterise tangent vectors independent of any particular chart in the following way.

Definition 2.2. Let $\sigma(t)$, $-\varepsilon < t < \varepsilon$ be a differentiable curve with $\sigma(0) = p \in M$. Then the tangent vector to $\sigma(t)$ at p is a map from the set of all differentiable functions defined around p into R defined by

$$X_p : f \to \left[\frac{\mathrm{d}}{\mathrm{d}t}(f \circ \sigma)\right]_{t=0}. \tag{2.2}$$

The tangent space to M at p is $T_p(M) = \{X_p;\ X_p \text{ tangent vector of a curve through } p\}$. The map (2.2) is the 'derivative of f in the direction of X_p'.

Let us calculate X_p in a particular chart (U, h):

$$\frac{\mathrm{d}}{\mathrm{d}t}(f \circ \sigma) = \frac{\mathrm{d}}{\mathrm{d}t}(f \circ h^{-1} \circ h \circ \sigma). \tag{2.3}$$

Now $f \circ h^{-1}$ is a map

$$(x^i) \in R^n \to (f \circ h^{-1})(x^i) \in R \tag{2.4}$$

and $h \circ \sigma$ is a map

$$t \in (-\varepsilon, \varepsilon) \to h \circ \sigma = (x^i \circ \sigma) \in R^n. \tag{2.5}$$

Therefore (2.3) implies

$$\left[\frac{\mathrm{d}}{\mathrm{d}t}(f \circ \sigma)\right]_{t=0} = \left[\left\{\frac{\partial}{\partial x^i}(f \circ h^{-1})\right\}_{h \circ \sigma} \cdot \frac{\mathrm{d}}{\mathrm{d}t}(x^i \circ \sigma)\right]_{t=0}. \tag{2.6}$$

From (2.6) we see that two curves σ, τ define the same tangent vector at p if

$$\left[\frac{\mathrm{d}}{\mathrm{d}t}(x^i \circ \sigma)\right]_p = \left[\frac{\mathrm{d}}{\mathrm{d}t}(x^i \circ \tau)\right]_p \tag{2.7}$$

which shows that the consistency of the definition with the considerations at the beginning of this section.

THEOREM 2.1. (1) If X_p, Y_p are tangent vectors at p, then the maps

$$f \to (X_p f) + (Y_p f) \tag{2.8}$$
$$f \to a(X_p f),\ a \in R \tag{2.9}$$

are again tangent vectors.

(2) A vector space structure is defined on $T_p(M)$ by

$$(X_p + Y_p) f \equiv (X_p f) + (Y_p f) \tag{2.10}$$
$$(aX_p) f \equiv a(X_p f). \tag{2.11}$$

(3) The dimension of $T_p(M)$ is n, if M is n-dimensional.

Proof. (1) We must show that there is a curve $\lambda(t)$, $-\varepsilon < t < \varepsilon$ such that

$$\left[\frac{\mathrm{d}}{\mathrm{d}t}(f \circ \lambda)\right]_{t=0} = (X_p f) + (Y_p f) \tag{2.12}$$

holds for any f. Choose curves $\sigma(t)$, $\tau(t)$ defining X_p, Y_p. Then $\tilde{\lambda}: t \to (h \circ \sigma + h \circ \tau)$ is a curve in R^n and $\lambda = h^{-1} \circ \tilde{\lambda}$ is a curve with the desired properties, if we have chosen the chart such that $x^i(p) = 0$. The same procedure works for (2.9).

(2) Is now trivial because of (1).

(3) Choose a local chart (U, h), then there are curves $\sigma_k(t)$, $(k = 1, \ldots, n)$ such that

$$(x^i \circ \sigma_k(t)) = \begin{cases} t + x^k(p) & i = k \\ x^i(p) & i \neq k \end{cases} \tag{2.13}$$

These curves are the 'coordinate lines'. If X_k is the tangent vector of σ_k at p, we find from (2.6)

$$X_k f = \frac{\partial}{\partial x^k}(f \circ h^{-1}). \tag{2.14}$$

Suppose now $a^k X_k = 0$. This implies $a^k(X_k f) = 0$ for any function f. If we choose in particular $f = x^i$, then

$$0 = a^k(X_k x^i) = a^k \frac{\partial}{\partial x^k}(x^i \circ h^{-1}) = a^k \frac{\partial}{\partial x^k}(\pi^i \circ h \circ h^{-1}) = a^i. \tag{2.15}$$

Hence (X_k) form a basis for $k = 1, \ldots, n$ in $T_p(M)$.

The basis in $T_p(M)$ which we just defined using a chart (U, h) is called the coordinate basis given by this chart. Any $X_p \in T_p(M)$ can be written as

$$X_p = \xi^i \frac{\partial}{\partial x^i} \tag{2.16}$$

because of (2.14).

Here I should make a remark about notation. (2.16) really means that $\xi^i(\partial/\partial x^i)$ acts on $(f \circ h^{-1})(x^1 \ldots x^n)$. What is usually done is that the function is written as $f(x^1 \ldots x^n)$, which is in principle meaningless because f is a map $M \to R$ and not defined on R^n. This becomes quite clear if one uses two different charts (U, h), (U', h') around a point. Then one gets $f \circ h^{-1}$ and $f \circ h'^{-1}$ which are different maps $R^n \to R$, whereas the $f(x^1 \ldots x^n)$ and $f(x^{1'} \ldots x^{n'})$ from the mathematical point of view are the same map. Nevertheless the precise notation is rather cumbersome and therefore I will use both.

For later use we need the following property of tangent vectors.

LEMMA 2.1. Any tangent vector satisfies

$$X_p(fg) = (X_p f)g(p) + f(p)(X_p g). \tag{2.17}$$

Proof. Choose a curve σ such that $\sigma(0) = p$ then

$$X_p(fg) = \frac{d}{dt}((fg) \circ \sigma) = \frac{d}{dt}((f \circ \sigma)(g \circ \sigma)) =$$
$$= \left(\frac{d}{dt}(f \circ \sigma)\right) g \circ \sigma + (f \circ \sigma) \frac{d}{dt}(g \circ \sigma) \tag{2.18}$$

which implies (2.17)

It is important to realize that it is precisely the differential structure which implies the existence of a tangent space at each point.

3. Dual Space of the Tangent Space Tensors

After we have attached a vector space to any point of M we can use the whole machinery of linear algebra, which I will do rather briefly:

DEFINITION 3.1. The dual of the tangent space (cotangent vector space) is defined as

$$T_p^*(M) = \{\omega; \omega: T_p(M) \to R, \omega \text{ linear}\}. \tag{3.1}$$

I will write $\omega(X)$ or $\langle\omega, X\rangle$ to denote the image of X under the map ω.

Of particular importance are the following elements of $T_p^*(M)$: Any function f defined around p defines a map

$$(\mathrm{d}f)_p : X_p \to X_{p.}f. \tag{3.2}$$

Clearly this is linear because of the definition of the vector space structure on T_p.

If (U, h) is a chart then $x^i = \pi^i \circ h$ are real valued functions and determine elements $\mathrm{d}x^i \in T_p^*$. The chart defines a coordinate basis $(\partial/\partial x^i)$ of T_p and

$$\mathrm{d}x^k\left(\frac{\partial}{\partial x^i}\right) = \frac{\partial}{\partial x^i}(x^k \circ h^{-1}) = \delta_i^k \tag{3.3}$$

holds. Hence $(dx^i)_{i=1,\dots,n}$ is the basis of T_p^* dual to the basis $(\partial/\partial x^i)_{i=1,\dots,n}$ of T_p. Therefore if f is any function there exists $a_i \in R$ and

$$(\mathrm{d}f)_p = a_i\, \mathrm{d}x^i. \tag{3.4}$$

Then

$$(\mathrm{d}f)_p\left(\frac{\partial}{\partial x^j}\right) = a_j \tag{3.5}$$

and by definition of $(\mathrm{d}f)_p$:

$$a_j = (\mathrm{d}f)_p\left(\frac{\partial}{\partial x^j}\right) = \frac{\partial}{\partial x^j}(f \circ h^{-1}). \tag{3.6}$$

Thus $(\mathrm{d}f)_p$ is the gradient of f.

THEOREM 3.1. If (U, h), (U, h') are two charts at $p \in M$, the coordinate bases in T_p, T_p^* are related by:

$$\mathrm{d}x^{i'} = \frac{\partial}{\partial x^j}(x^{i'} \circ h^{-1})\, \mathrm{d}x^j; \qquad \frac{\partial}{\partial x^{i'}} = \frac{\partial}{\partial x^{i'}}(x^j \circ h'^{-1}) \frac{\partial}{\partial x^j}. \tag{3.7}$$

Proof. We use simply (3.4) and (3.6) for the function $x^{i'}$ to get the first relation. To find the second we use the first relation with the roles of h, h' exchanged:

$$\mathrm{d}x^i = \frac{\partial}{\partial x^{j'}}(x^i \circ h'^{-1})\, \mathrm{d}x^{j'} \tag{3.8}$$

Suppose $(\partial/\partial x^{i'}) = A_{i'}^k(\partial/\partial x^k)$ and calculate

$$\mathrm{d}x^i\left(A_{i'}^k \frac{\partial}{\partial x^k}\right) = \mathrm{d}x^i\left(\frac{\partial}{\partial x^{i'}}\right) = \frac{\partial}{\partial x^{j'}}(x^i \circ h'^{-1})\, \mathrm{d}x^{j'}\left(\frac{\partial}{\partial x^{i'}}\right), \tag{3.9}$$

so

$$A_{i'}^i = \frac{\partial}{\partial x^{i'}}(x^i \circ h'^{-1}). \tag{3.10}$$

If one knows the transformation between two bases in a vector space one immediately gets the transformation of the components of a vector:

$$X = \xi^i \frac{\partial}{\partial x^i} = \xi^{i'} \frac{\partial}{\partial x^{i'}} \Rightarrow \xi^{i'} = \frac{\partial}{\partial x^i}(x^{i'} \circ h^{-1})\,\xi^i \tag{3.11}$$

$$\omega = a_i \, \mathrm{d}x^i = a_{i'} \, \mathrm{d}x^{i'} \Rightarrow a_{i'} = \frac{\partial}{\partial x^{i'}}(x^i \circ h'^{-1})\, a_i . \tag{3.12}$$

This is the transformation laws of co- and contra- variant vectors.

With any vector space one can associate the tensor algebra over this vector space. I assume that the reader is familiar with this procedure and only make some remarks to fix the notation. A tensor S over $T_p(M)$ can be defined as a map

$$S: \underbrace{T_p \times \cdots \times T_p}_{r \text{ copies}} \times \underbrace{T_p^* \times \cdots \times T_p^*}_{s \text{ copies}} \to R \tag{3.13}$$

which is linear in each argument. If there are r copies of T_p and s copies of T_p^* then S is an element of $\mathscr{T}_r^s$ the vector space of all tensors of covariant degree r and contravariant degree s. Because of the linearity S is uniquely determined by its values for a basis. For example, if e_i is a basis of $T_p(M)$

$$S(X, Y) = S(a^i e_i, b^k e_k) = a^i b^k S(e_i, e_k) = a^i b^k S_{ik} . \tag{3.14}$$

The numbers S_{ik} are the components of S in the basis e_i. The tensor product is a map

$$\mathscr{T}_r^s \times \mathscr{T}_{r'}^{s'} \to \mathscr{T}_{r+r'}^{s+s'} \tag{3.15}$$

defined by

$$(S \otimes R)(X_1, \ldots, X_r, Y_1, \ldots, Y_{r'}, \omega^1, \ldots, \omega^s, \eta^1, \ldots, \eta^{s'}) = \\ = S(X_1, \ldots, \omega^s) \cdot R(Y_1, \ldots, \eta^{s'}) \tag{3.16}$$

A basis (e_i) of T_p and the dual basis (e^i) of T_p^* define a basis

$$e^{k_1} \otimes \cdots \otimes e^{k_r} \otimes e_{i_1} \otimes \cdots \otimes e_{i_s} \tag{3.17}$$

of $\mathscr{T}_r^s$. Hence any $S \in \mathscr{T}_r^s$ can be written as

$$S = S_{k_1, \ldots, k_r}{}^{i_1, \ldots, i_s} e^{k_1} \otimes \cdots \otimes e_{i_s} \tag{3.18}$$

when

$$S_{k_1, \ldots, k_r}{}^{i_1, \ldots, i_s} = S(e_{k_1}, \ldots, e_{k_r}, e^{i_1}, \ldots, e^{i_s}) \tag{3.19}$$

are the components of S in the basis.

A tensor $S \in \mathscr{T}_r^0$ is symmetric if for any permutation π

$$S(X_1, \ldots, X_r) = S(X_{\pi(1)}, \ldots, X_{\pi(r)}) \tag{3.20}$$

holds. It is skew if

$$S(X_1, \ldots, X_r) = (\operatorname{sgn} \pi)\, S(X_{\pi(1)}, \ldots, X_{\pi(r)}) \tag{3.21}$$

holds.

A scalar product on T_p is defined by a symmetric tensor $g(X, Y)$. It is called positive definite if $g(X, X) > 0$ for any $X \neq 0$. It is nondegenerate if $g(X, Y) = 0$ for every X implies $Y = 0$. If $g(X, Y)$ is nondegenerate, there exists a basis (e_i) of T_p such that

$$g = \sum_{i=1}^{n} \varepsilon_i (e^i \otimes e^i), \quad \varepsilon^i = \pm 1 \tag{3.22}$$

or equivalently: $g(e_i, e_i) = \varepsilon_i$, $g(e_i, e_k) = 0$ for $i \neq k$. A non degenerate metric g defines uniquely an isomorphism

$$A: T_p \to T_p^* \tag{3.23}$$

by

$$A: X \to (\omega: Y \to g(Y, X)). \tag{3.24}$$

This is the invariant description of 'moving indices' up and down. If (e_i), (e^k) are dual bases in T_p, T_p^*, and if $g_{ik} = g(e_i, e_k)$ then the map (3.24) is given by

$$A: \xi^i e_i \to g_{ki} \xi^i e^k. \tag{3.25}$$

Using A one defines a scalar product on T_p^* by

$$g(A(X), A(Y)) = g(X, Y) \tag{3.26}$$

which is the 'contravariant metric' $g^{ik} = g(e^i, e^k)$. Clearly $g^{ik} g_{kl} = \delta^i_l$ holds because of the definition of A. More generally, A induces maps

$$\mathcal{T}^s_r \to \mathcal{T}^0_{r+s}, \quad \mathcal{T}^0_r \to \mathcal{T}^r_0 \tag{3.27}$$

in an obvious way:

$$T(X_1, \ldots, X_r, \omega^1, \ldots, \omega^s) \to T(X_1, \ldots, X_r, A^{-1}(\omega^1), \ldots, A^{-1}(\omega^s)) \tag{3.28}$$

or in components

$$T_{i_1, \ldots, i_r}{}^{j_1, \ldots, j_s} = T_{i_1, \ldots, i_r}{}^{j_1, \ldots, j_s} g_{j_1 k_1} \cdots g_{j_s k_s} \tag{3.29}$$

So far we have been working in one tangent space. Choosing tensors of the same type at each point of M, we get tensor fields on M.

DEFINITION 3.2. A tensor field on M is a map

$$S: M \to \bigcup_{p \in M} \mathcal{T}^s_{pr}. \tag{3.30}$$

S is differentiable if its components $S((\partial/\partial x^{i_1}), \ldots, \mathrm{d}x^{k_s})$ with respect to a chart are differentiable.

For example a vector field X is written in terms of a chart (U, h) as

$$X = \xi^i (x^k) \frac{\partial}{\partial x^i}. \tag{3.31}$$

If X is a vector field we denote its value at p by X_p and similarly for any tensor field.

4. Commutators and Exterior Derivative of Forms

So far we have only defined manifolds, the tangent space at any point and, using this, tensor fields. Now for the first time Differential Geometry comes into the game.

Consider two vector fields X, Y and a real valued function f on M. Then

$$p \in M \to X_p f \tag{4.1}$$

is again a differentiable function, which we denote by Xf.

LEMMA 4.1. The map

$$f \to [X, Y]_p f \equiv X_p(Yf) - Y_p(Xf) \tag{4.2}$$

defines an element of $T_p(M)$, called the commutator of X and Y. $[X, Y]$ is linear over R in both arguments and satisfies

$$[X, Y] = -[Y, X]; [[X, Y], Z] + [[Z, X], Y] + [[Y, Z], X] = 0. \tag{4.3}$$

Proof. (4.3) is obvious from (4.2). To show that the map (4.2) is really a tangent vector we calculate it in a particular chart (U, h) in which

$$X = \xi^i \frac{\partial}{\partial x^i}, \quad Y = \eta^k \frac{\partial}{\partial x^k}:$$

$$X_p(Yf) = \left(\xi^i \frac{\partial}{\partial x^i}\right)_p \left(\eta^k \frac{\partial}{\partial x^k}(f \circ h^{-1})\right)$$

$$= \xi^i \eta^k{}_{,i} \frac{\partial}{\partial x^k}(f \circ h^{-1}) + \xi^i \eta^k \frac{\partial^2}{\partial x^i \partial x^k}(f \circ h^{-1}) \tag{4.4}$$

therefore

$$[X, Y] = (\xi^i \eta^k{}_{,i} - \eta^i \xi^k{}_{,i}) \frac{\partial}{\partial x^k} \tag{4.5}$$

is a vector field.

A geometrical interpretation of $[X, Y]$ will be given in Section 8. Because of (4.3) vector fields on a manifold form an infinite dimensional Lie algebra over R.

As the elements in T_p^* are in some sense dual to vector fields one expects also to find an operator which forms new fields from covariant vector fields. The corresponding operation is exterior differentiation, which acts on totally skew covariant tensor fields, or forms, as these fields are also called.

An element $\omega \in \mathscr{T}_r^0$ is totally skew if

$$\omega(X_1, \ldots, X_r) = \frac{1}{r!} \sum_\pi (\operatorname{sgn} \pi) \omega(X_{\pi(1)}, \ldots, X_{\pi(r)});$$

$$\pi = \begin{pmatrix} 1, \ldots, r \\ \pi(1), \ldots, \pi(r) \end{pmatrix}. \tag{4.6}$$

If $T \in \mathscr{T}_r^0$ we define its skew part as

$$(AT)(X_1, \dots, X_r) = \frac{1}{r!} \sum_{\pi} (\operatorname{sgn} \pi)\, T(X_{\pi(1)}, \dots, X_{\pi(r)}). \tag{4.7}$$

This is an r-form.

For any two forms ω, η the exterior product $\wedge$ is defined by

$$\omega \wedge \eta = A(\omega \otimes \eta) \tag{4.8}$$

and this implies that any r-form can be written as

$$\omega = a_{i_1, \dots, i_r} \, dx^{i_1} \wedge \cdots \wedge dx^{i_r} \tag{4.9}$$

where $a_{i_1, \dots, i_r}$ are the components of the skew tensor ω. This shows that for example

$$(dx^{i_1} \wedge dx^{i_2})\left(\frac{\partial}{\partial x^i}, \frac{\partial}{\partial x^k}\right) = \frac{1}{2!}(dx^{i_1} \otimes dx^{i_2} - dx^{i_1} \otimes dx^{i_1})\left(\frac{\partial}{\partial x^i}, \frac{\partial}{\partial x^k}\right) \tag{4.10}$$

$$= \tfrac{1}{2}(\delta_i^{i_1}\delta_k^{i_2} - \delta_i^{i_2}\delta_k^{i_1}) = \begin{cases} \tfrac{1}{2} & \text{if } i_1 = i, i_2 = k \\ -\tfrac{1}{2} & \text{if } i_2 = i, i_1 = k \\ 0 & \text{in all other cases.} \end{cases}$$

THEOREM 4.1. There is a unique operation, the exterior differentiation d, which maps an r-form into an $r+1$ form such that

$$\begin{aligned} &(1) \quad (df)\,X = Xf \quad \text{if } f \text{ is a 0-form, i.e. a function} \\ &(2) \quad d(d\omega) = 0 \quad \text{for any } r\text{-form} \\ &(3) \quad d(\omega \wedge \eta) = d\omega \wedge \eta + (-1)^r \omega \wedge d\eta \quad \text{if } \omega \text{ is an } r\text{-form} \\ &(4) \quad d(\omega + \eta) = d\omega + d\eta \end{aligned} \tag{4.11}$$

Proof. We will first show that $d\omega$ is uniquely determined if we use a chart (U, h): suppose

$$\omega = a_{i_1, \dots, i_r} \, dx^{i_1} \wedge \cdots \wedge dx^{i_r} \tag{4.12}$$

then

$$d\omega = (da_{i_1, \dots, i_r}) \wedge (dx^{i_1} \wedge \cdots \wedge dx^{i_r}) + a_{i_1, \dots, i_r} \, d(dx^{i_1} \wedge \cdots \wedge dx^{i_r}) \tag{4.13}$$

by use of (3). Condition (2) shows that the second term vanishes and (1) gives*

$$d\omega = a_{i_1, \dots, i_r \mid k} \, dx^k \wedge dx^{i_1} \wedge \cdots \wedge dx^{i_r}. \tag{4.14}$$

Now we have to show that this form $d\omega$ is independent of the particular chart we used: If (U, h') is another chart, then, using the transformation law of the components;

$$a_{i'_1, \dots, i'_r \mid k'} = a_{i_1, \dots, i_r \mid k} \frac{\partial x^k}{\partial x^{k'}} \frac{\partial x^{i_1}}{\partial x^{i'_1}} \cdots \frac{\partial x^{i_r}}{\partial x^{i'_r}} + a_{i_1, \dots, i_r} \frac{\partial}{\partial x^{k'}}\left(\frac{\partial x^{i_1}}{\partial x^{i'_1}} \cdots \frac{\partial x^{i_r}}{\partial x^{i'_r}}\right). \tag{4.15}$$

Hence if we form skew parts, the second term vanishes and we find

$$\begin{aligned} d\omega &= a_{i_1, \dots, i_r \mid k} \, dx^k \wedge dx^{i_1} \wedge \cdots \wedge dx^{i_r} \\ &= a_{i'_1, \dots, i'_r \mid k'} \, dx^{k'} \wedge \cdots \wedge dx^{i'_r}. \end{aligned} \tag{4.16}$$

Therefore if we define 'd' by (4.14) it satisfies (4.11).

In which sense is 'd' now dual to the commutator?

* The stroke denotes partial derivatives.

LEMMA 4.2. If $\omega(X)$ is a 1-form, then

$$(\mathrm{d}\omega)(X, Y) = \tfrac{1}{2}\{X\omega(Y) - Y\omega(X) - \omega([X, Y])\}. \tag{4.17}$$

Proof. We first show that the right hand side is a skew tensor. Obviously the map $(X_p, Y_p) \to \{\ \ \}$ is skew. Hence we only have to show:

$$\{X\omega(fY) - (fY)\,\omega(X) - \omega([X, fY])\}_p = f(p)\,\{X\omega(Y) - \cdots\}. \tag{4.18}$$

This follows from

$$X(\omega(fY)) = X(f\omega(Y)) = (Xf)\,\omega(Y) + f(X\omega(Y)) \tag{4.19}$$

$$[X, fY] = f[X, Y] + (Xf)\,Y \tag{4.20}$$

which are immediate consequences of the definition.

Knowing that (4.17) is a tensor one has just to calculate its components in some chart using the coordinate basis to complete the proof.

Similar equations relate $\omega(X_1, \ldots, X_r)$ to $\mathrm{d}\omega(X_1, \ldots, X_r, Z)$; they can be found in Kobayashi and Nomizu (1963) and Hicks (1965). For example

$$(\mathrm{d}\omega)(X, Y, Z) = \tfrac{1}{3}\{X\omega(Y, Z) + \text{cyclic} - \omega([X, Y], Z) - \text{cyclic}\}.$$

5. Maps between Manifolds

Let us now consider maps between manifolds:

DEFINITION 5.1. A map $\phi: M^n \to \tilde{M}^k$ is called differentiable if the map $\tilde{h} \circ \phi \circ h^{-1}$: $R^n \to R^k$ is differentiable when defined, where (U, h) is a chart at $p \in M^n$ and $(\tilde{U}, \tilde{h})$ a chart at $\phi(p)$. If ϕ is a homeomorphism and ϕ as well as ϕ^{-1} are differentiable, ϕ is called a diffeomorphism.

THEOREM 5.1. Any differentiable map $\phi: M^n \to \tilde{M}^k$ induces a linear map

$$(\phi_*)_p : T_p(M^n) \to T_{\phi(p)}(\tilde{M}^k). \tag{5.1}$$

Proof. If $\sigma(t)$ is a curve in M^n and $\sigma(0) = p$, then $\phi \circ \sigma$ is a curve in $\tilde{M}^k$ and we define (ϕ_*) as the map which sends the tangent vector of σ at $\sigma(0)$ into the tangent vector of $\phi \circ \sigma$ at $(\phi \circ \sigma)(0)$, i.e.:

$$((\phi_*)_p\, \dot{\sigma}(0))\, f = \left[\frac{\mathrm{d}}{\mathrm{d}t}(f \circ \phi \circ \sigma)\right]_{t=0}. \tag{5.2}$$

To check linearity, we just calculate $(\phi_*)_p$ in local coordinates:

$$(\phi_*)_p\left(\xi^i \frac{\partial}{\partial x^i}\right) = \xi^i \frac{\partial}{\partial x^i}(\pi^A \circ \phi \circ h^{-1}) \frac{\partial}{\partial y^A} \tag{5.3}$$

where (U, h) is a chart around p in M^n with coordinates $x^i = \pi^i \circ h$, $i = 1, \ldots, n$ and

$(\tilde{U}, \tilde{h})$ is a chart around $\phi(p) \in \tilde{M}^k$ with coordinates $y^A = \pi^A \circ \tilde{h}$, $A = 1, \ldots, k$. This linear map $(\phi_*)_p$ is very important later on and we need the following properties:

THEOREM 5.2. Let $\phi: M^n \to \tilde{M}^k$ be a differentiable map. Then

(1) $((\phi_*)_p X) f = X_p(f \circ \phi)$; $f: \tilde{M}^k \to R$. (5.4)

(2) If $\tilde{\omega}$ is any r-form on $\tilde{M}^k$, $\phi^*\tilde{\omega}$ is an r-form on M^n defined by

$$(\phi^*\tilde{\omega})_p (X_1, \ldots, X_r) = \tilde{\omega}_{\phi(p)} ((\phi_*)_p X_1, \ldots, (\phi_*)_p X_r) \tag{5.5}$$

and

$$\phi^* (\tilde{\omega} \wedge \tilde{\eta}) = \phi^*\tilde{\omega} \wedge \phi^*\tilde{\eta} \tag{5.6}$$

$$\mathrm{d}(\phi^*\tilde{\omega}) = \phi^* (\mathrm{d}\tilde{\omega}). \tag{5.7}$$

(3) If X_1, X_2 are vector fields on M^n and if there exists vector fields $\tilde{X}_1$, $\tilde{X}_2$ on $\tilde{M}^k$ such that $(\phi_*)_p X_i = \tilde{X}_{i\,\phi(p)}$ $(i = 1, 2)$ for any p, then

$$(\phi_*)_p [X_1, X_2] = [(\phi_*)_p X_1, (\phi_*)_p X_2] = [\tilde{X}_1, \tilde{X}_2]_{\phi(p)}. \tag{5.8}$$

Proof. (1) Choose a curve σ in M^n such that $\dot{\sigma} = X_p$. Then $f \circ \phi: M^n \to R$ and

$$X_p(f \circ \phi) = \frac{\mathrm{d}}{\mathrm{d}t} (f \circ \phi \circ \sigma) = ((\phi_*)_p X_p) f \tag{5.9}$$

holds, because $\phi \circ \sigma$ is a curve in M^k with tangent vector $(\phi_*)_p X_p$.

(2) The linearity of $(\phi_*)_p$ implies that $\phi^*\tilde{\omega}$ is a skew covariant tensor on M^n. To prove (5.6), we just apply the definitions

$$\begin{aligned}(\phi^* (\tilde{\omega} \wedge \tilde{\eta})) (X_1, \ldots, X_{r+s}) &= (\tilde{\omega} \wedge \tilde{\eta}) (\phi_* (X_1) \ldots) = \\ &= (A (\tilde{\omega} \otimes \tilde{\eta})) (\phi_* (X_1) \ldots) \\ &= A (\phi^*\tilde{\omega} \otimes \phi^*\tilde{\eta}) (X_1, \ldots, X_{r+s}) \\ &= (\phi^*\tilde{\omega} \wedge \phi^*\tilde{\eta}) (X_1, \ldots, X_{r+s}).\end{aligned} \tag{5.10}$$

The second part of (2), relation (5.7), is proved by induction. Using (5.3), one checks easily that (5.7) is true for a 1-form. Suppose it is true for forms of degree smaller than $r+1$. Let ω be an $r+1$-form; then it can be written as a sum of products $\eta \wedge \chi$, where η, χ have degree 1, r respectively. Using the induction and (5.6)

$$\begin{aligned}\phi^* (\mathrm{d}(\eta \wedge \chi)) &= \phi^* (\mathrm{d}\eta \wedge \chi - \eta \wedge \mathrm{d}\chi) = \mathrm{d}(\phi^*\eta) \wedge \phi^*\chi - \phi^*\eta \wedge \mathrm{d}\phi^*\chi \\ &= \mathrm{d}(\phi^*\eta \wedge \phi^*\chi) = \mathrm{d}(\phi^* (\eta \wedge \chi)).\end{aligned} \tag{5.11}$$

Before we prove (3), let us note: If ϕ is a map $M^n \to \tilde{M}^k$ we can 'pull back' any form on $\tilde{M}^k$ to define a form on M^n. (In fact the definition (5.5) is meaningful for any covariant vector field on M^k). For vector fields (or more generally, contravariant tensor fields on M^n) X on M^n $(\phi_*)_p X_p$ is defined for any p, but in general this will not define a vector field on $\phi(M^k) \subset \tilde{M}^k$, where ϕ is not 1–1 on its image. If however there is a vector field $\tilde{X}$ on $\tilde{M}^k$ such that $(\phi_*)_p X_p = \tilde{X}_{\phi(p)}$ for any $p \in M^n$, X, $\tilde{X}$ are called ϕ-related. Then (3) states that if X_i, $\tilde{X}_i$ are ϕ-related for $i = 1, 2$, then $[\tilde{X}_1, \tilde{X}_2]$ is ϕ-related to $[X_1, X_2]$.

Proof of (3). (Use (1)!).

$$\begin{aligned}\{[\tilde{X}_1, \tilde{X}_2] f\}_{\phi(p)} &= (\tilde{X}_1)_{\phi(p)} (\tilde{X}_2 f) - (\tilde{X}_2)_{\phi(p)} (\tilde{X}_1 f) \\ &= ((\phi_*)_p X_1)(\tilde{X}_2 f) - ((\phi_*)_p X_2)(\tilde{X}_1 f) \\ &= (X_1)_p (\tilde{X}_2 f \circ \phi) - (X_2)_p (\tilde{X}_1 f \circ \phi) \\ &= (X_1)_p (X_2 (f \circ \phi)) - (X_2)_p (X_1 (f \circ \phi)) \\ &= [X_1, X_2]_p (f \circ \phi) = (\phi_*)_p \{[X_1, X_2] f\}_p .\end{aligned} \tag{5.12}$$

Hence

$$[(\phi_*)_p X_1, (\phi_*)_p X_2] = (\phi_*)_p [X_1, X_2]. \tag{5.13}$$

An immediate consequence of Theorem (5.2) is:

LEMMA 5.1. If the maps $\phi: M^n \to \tilde{M}^k$, $\psi: \tilde{M}^k \to N^l$ are differentiable, then $(\phi \circ \psi)_* = \phi_* \circ \psi_*$.

DEFINITION AND LEMMA 5.2. Let ϕ be a diffeomorphism $M \to M$, X a vector field. Then $\phi_* X$ is a vector field defined by $(\phi_* X)_{\phi(p)} = (\phi_*)_p X_p$ and we have $\phi_* [X, Y] = [\phi_* X, \phi_* Y]$.

6. Lie Groups and Lie Transformation Groups

We will now give the first useful application of the formalism developed so far. We will consider Lie groups, first because we will need them later, and secondly to demonstrate how one works with the tools provided so far.

DEFINITION 6.1. A Lie group G is a group which has a differentiable structure such that the maps

$$\begin{aligned} G \times G \to G; \quad (a, b) \to a \cdot b \\ G \to G; \quad a \to a^{-1} \end{aligned} \tag{6.1}$$

are differentiable.

We have already remarked that $Gl(n, R)$ is a manifold, and obviously the group operation is differentiable. The differentiability of the group operation implies that the map

$$L_a : x \to ax \tag{6.2}$$

is a diffeomorphism for any $a \in G$.

DEFINITION 6.2. A vector field X on G such that $(L_a)_* X = X$, i.e.,

$$(L_a)_* X_x = X_{ax} \quad \text{for any} \quad x \in G \tag{6.3}$$

is called left invariant.

As $(L_a)_*$ is linear and $(L_a)_* [X, Y] = [(L_a)_* X \ldots]$ holds (Theorem 5.2) it follows immediately that **G**, the set of all left invariant vector fields, forms a Lie algebra over R.

THEOREM 6.1. The set of all left invariant vector fields **G** forms a Lie algebra and $\dim \mathbf{G} = \dim G$.

The important point is that **G** is a finite dimensional subalgebra of the algebra of all vector fields, because this implies that if the vector fields $X_1, \ldots, X_n$ form a basis of **G**, then

$$[X_i, X_k] = C^l_{ik} X_l, \quad C^l_{ik} = \text{const.} \tag{6.4}$$

holds. Hence we obtain a remarkably simple 'finite structure' out of such an 'infinite group'.

Proof. We just construct the elements of **G**. Choose any vector $X_e \in T_e(G)$. Define a vector field on G by

$$X_a \equiv (L_a)_* X_e. \tag{6.5}$$

Clearly this is a property any element of **G** necessarily satisfies. The left invariance of the vector field defined by (6.5) follows because

$$(L_b)_* X_a = (L_b)_* (L_a)_* X_e = (L_b L_a)_* X_e = (L_{ba})_* X_e = X_{ba} \tag{6.6}$$

holds for any $b \in G$. Hence this vector field is contained in **G**.

As any element in **G** satisfies (6.5), we have constructed a vector space isomorphism $T_e(G) \to \mathbf{G}$, hence $\dim \mathbf{G} = \dim G$.

What is the geometrical meaning of the left invariant vector fields? To obtain this we need the interpretation of vector fields as generators of local 1-parameter groups of transformations.

DEFINITION 6.3. A curve $\sigma(t)$ is an integral curve of a vector field X if

$$\dot{\sigma}(t) = X_{\sigma(t)} \tag{6.7}$$

holds for any t for which σ is defined.

LEMMA 6.1. Let X be a vector field on M and $p \in M$. Then there exists a uniquely determined maximal integral curve $\sigma(t)$ with $\sigma(0) = p$ such that any other integral curve $\lambda(t)$ with $\lambda(0) = p$ is a restriction of σ.

Proof. Choose a chart (U, h) at p. Then (6.7) implies for

$$i = 1, \ldots, n, \qquad \frac{\mathrm{d}}{\mathrm{d}t}(x^i \circ \sigma) = \xi^i (x^k \circ \sigma), \qquad (x^i \circ \sigma)(0) = x^i(p) \tag{6.8}$$

which is a system of ordinary differential equations on $h(U)$ for the maps $t \to x^i \circ \sigma$. Because $\xi^i(x^k)$ is differentiable there is a unique solution $x^i \circ \sigma$, $i = 1, \ldots, n$ satisfying (6.8).

To get the maximal solution we choose a covering (U_α, h_α) of M and match the local solutions on the intersections of the charts.

Remark. An integral curve of X has by definition a fixed parametrization. Integral curves are not necessarily closed subsets of M. It is possible that the closure is the whole space, if the curve winds round sufficiently nastily. (One can easily construct such an example on a torus.)

LEMMA 6.2. Let X be a vector field on M, and $\sigma_p(t)$ the maximal integral curve with $\sigma_p(0)=p$. Let p_0 be any point in M. Then there exists a neigborhood U of p_0 and an $\varepsilon>0$ such that the maps

$$\psi_t : q \in U \to \sigma_q(t), \quad t \in (-\varepsilon, \varepsilon) \tag{6.9}$$

are diffeomorphisms $U \to \psi_t(U)$ which satisfy

$$\psi_t \circ \psi_{t'} = \psi_{t+t'} \tag{6.10}$$

where this expression is defined, i.e., for $|t+t'|<\varepsilon$, $|t|<\varepsilon$, $|t'|<\varepsilon$. ψ_t is called the local 1-parameter group of local transformations generated by X.

Proof. The first part will not be worked out explicitly. One chooses a chart and uses the fact that the solutions of (6.8) are differentiable functions of the initial values. As ψ_t is the identity for $t=0$, it follows that there is an ε such that ψ_t is invertible.

To prove (6.10) we write $f^k(t, x^i)$ for the general solution of (6.8) with initial values $f^k(0, x^i)=x^k$. This implies

$$(x^i \cdot \psi_t \circ h^{-1})(x^k) = f^i(t, x^k). \tag{6.11}$$

Now one finds immediately that $g^i(t)=f^i(t+t', x^k)$ is also a solution of (6.8) satisfying $g^i(0)=f^i(t', x^k)$. Hence because of the uniqueness

$$g^i(t) = f^i(t, f^k(t', x^l)) = f^i(t + t', x^l) \tag{6.12}$$

holds, and this relation is (6.10) given in terms of the coordinates.

LEMMA 6.3. If X generates ψ_t and $\Psi: M \to M$ is a diffeomorphism then $\Psi_* X$ generates $\Psi \circ \psi_t \circ \Psi^{-1}$.

Proof. The definition of $\Psi_* X$ implies that $\Psi \circ \sigma$ is an integral curve of $\Psi_* X$ if σ is an integral curve of X. Hence $t \to (\Psi \circ \psi_t)(p)$ is an integral curve of $\Psi_* X$. Because of $(\Psi \circ \psi_0)(p) = \Psi(p)$ we get the expression $\Psi \circ \psi_t \circ \Psi^{-1}$ for the 1-parameter group generated by $\Psi_* X$.

Let us now go back to the left invariant vector fields of a Lie group. The following theorem gives the geometrical interpretation.

THEOREM 6.2. Suppose $g(t)$ is a curve in G whose points form a subgroup of G. Then $X_x = \dot{\sigma}_x(0)$ with $\sigma_x(t)=x \cdot g(t)$ is a left invariant vector field.

Conversely, any $X \in \mathbf{G}$ determines uniquely a curve whose points form a subgroup of G.

Proof. We have to show $(L_a)_* X_x = X_{ax}$. This is immediate:

$$\begin{aligned}(L_a)_* X_x = (L_a)_* \dot{\sigma}_x(0) = (L_a \circ \sigma_x(t))'_{t=0} = (axg(t))'_{t=0} = \\ = \dot{\sigma}_{ax}(0) = X_{ax}.\end{aligned} \tag{6.13}$$

The proof of the second part is a consequence of Lemma 6.3. If $X \in \mathbf{G}$ generates ψ_t, then $\psi_t \circ L_a = L_a \circ \psi_t$ and this implies that any ψ_t is a right translation $\psi_t = R_{a(t)}$. In particular, $\psi_t(e) = a(t)$ is the integral curve of X through e and $\psi_{t+t'} = \psi_t \circ \psi_{t'}$ implies

$$R_{a(t+t')} = R_{a(t)} \cdot R_{a(t')} = R_{a(t') \cdot a(t)},$$

and hence $a(t+t') = a(t) \cdot a(t')$.

So far this has been local, but it is obvious that $a(t)$ is defined for any t, hence $t \to a(t)$ is a differentiable homomorphism $R \to G$.

The importance of the Lie algebra **G** of a Lie group lies in the fact that **G** determines uniquely the local structure of the Lie group; moreover, it determines uniquely a Lie group which is simply connected. Without proof we mention:

THEOREM 6.3. Suppose G, G' are two Lie groups with isomorphic Lie algebras. Let $A: T_e(G) \to T_e(G')$ be a Lie algebra isomorphism. Then there exists neighborhoods U of e in G and U' of e in G' and a diffeomorphism $\phi: U \to U'$ such that

$$(\phi_*)_e = A\,; \quad \phi(ab) = \phi(a) \cdot \phi(b) \tag{6.14}$$

where defined.
If G, G' are simply connected $U = G$, $U' = G'$ holds, i.e., the groups are isomorphic.

The proof can be found in Cohn (1961) or Sternberg (1964).

Let us consider an example which we will need later. The General Linear Group $Gl(n, R)$ is a Lie group of dimension n^2:

$$Gl(n, R) = \{(a^i_k)\,;\quad \det(a^i_k) \neq 0\}. \tag{6.15}$$

We have a chart on Gl, the restriction of the chart on R^{n^2}.

$$h: \begin{matrix} Gl(n, R) \to R^{n^2} \\ a \to a^i_k \end{matrix}. \tag{6.16}$$

Unfortunately, we have to use a labeling of the coordinates with 'one index up and one down' so that we must be careful to distinguish between the components of vectors and co-vectors.

The product is defined by

$$(ab)^i_k = (a^i_r b^r_k). \tag{6.17}$$

Let us now calculate a basis of the Lie algebra **Gl**. The vector fields

$$\frac{\partial}{\partial a^i_k}, \quad i, k = 1, \ldots, n \tag{6.18}$$

form a basis of the tangent space at any point. The field $X = c^i_k(a^r_s)\,(\partial/\partial a^i_k)$ is an element of **Gl** provided

$$(L_a)_* X_e = X_a. \tag{6.19}$$

It is easy to calculate $(L_a)_* X_e$. The curve $\delta^i_k + \mathrm{d}^i_k t$ has the tangent $\mathrm{d}^i_k(\partial/\partial a^i_k)$ at $t=0$;

$a_r^i(\delta_k^r + d_k^r t)$ is the image of this curve under L_a. Hence

$$(L_a)_* X_e = a_r^i d_k^r \frac{\partial}{\partial a_k^i} \tag{6.20}$$

is a left invariant vector field and

$$E_r^k = a_r^i \frac{\partial}{\partial a_k^i}, \quad k, r = 1, \ldots, n \tag{6.21}$$

is a basis of **Gl**.

This gives the commutation relations

$$\begin{aligned}
[E_r^k, E_{r'}^{k'}] &= a_r^i \frac{\partial}{\partial a_k^i}\left(a_{r'}^j \frac{\partial}{\partial a_{k'}^j}\right) - a_{r'}^j \frac{\partial}{\partial a_{k'}^j}\left(a_r^i \frac{\partial}{\partial a_k^i}\right) = \\
&= a_r^i \delta_i^j \delta_{r'}^k \frac{\partial}{\partial a_{k'}^j} - a_{r'}^j \delta_j^i \delta_r^{k'} \frac{\partial}{\partial a_k^i} = \\
&= a_r^j \delta_{r'}^k \frac{\partial}{\partial a_{k'}^j} - a_{r'}^i \delta_r^{k'} \frac{\partial}{\partial a_k^i} = \\
&= \delta_{r'}^k E_r^{k'} - \delta_r^{k'} E_{r'}^k
\end{aligned} \tag{6.21*}$$

which we need later.

One can study the geometry of Lie groups itself, but more important in Differential Geometry are Lie groups acting as transformation groups on manifolds.

DEFINITION 6.4. A Lie group G acts as a Lie transformation group on the right on a manifold M, if there is a differentiable map $\phi: M \times G \to M$ such that $R_a: x \to \phi(x, a)$ is a differeomorphism for any $a \in G$ and $R_a \circ R_b = R_{ba}$ holds for any $a, b \in G$. We also write $R_a(x) = xa$ if no confusion is possible about which point is in M and which is in G.

Left action is characterized by $L_a \circ L_b = L_{ab}$.

DEFINITION 6.5. The action of G is called

effective if $R_a x = x$ for all x implies $a = e$.
free if $R_a x = x$ for some x implies $a = e$.
transitive if $M = \{R_a(x) \mid a \in G\}$.

The set $O(x) = \{R_a(x) \mid a \in G\}$ is called the orbit of G through x.

In the following we assume always that G acts effectively on M.

A Lie transformation group defines a finite-dimensional Lie algebra of vector fields on M in the following way:

LEMMA 6.4. The map $\sigma_x: G \to M$ defined by $\sigma_x(a) = \phi(x, a)$ is differentiable. For any $A \in \mathbf{G}$

$$(\sigma_x)_* A_a \equiv A_{xa}^* \tag{6.22}$$

is a vector field on M not identically zero and

$$[A, B]^* = [A^*, B^*]. \tag{6.23}$$

The group acts freely if and only if $A^* \neq 0$ everywhere whenever $A \neq 0$.

Proof. The differentiability of σ_x follows from the differentiability of ϕ. Suppose (6.22) defines a vector field. Then A^* is σ_x related to A for any x and this implies (6.23) if we use Theorem 5.2. To show that (6.22) defines a vector field we have to use the left invariance of A:

$$(\sigma_x)_* A_a = (\sigma_x)_* (L_a)_* A_e = (\sigma_x \circ L_a)_* A_e = (\sigma_{xa})_* A_e. \tag{6.24}$$

Hence $xa = yb$ implies $(\sigma_x)_* A_a = (\sigma_y)_* A_b$ and A^* is well defined. By the very definition of A^*, $R_{a(t)}$ is the local 1-parameter group generated by A^* if $a(t)$ is generated by A. (This follows from $(\sigma_{xa})_* A_e = (\sigma_x)_* A_a$). Hence $A^*_x = 0$ is equivalent to $R_{a(t)}(x) = x$. Therefore, $A^* \neq 0$ if and only if G acts freely.

As an application we consider isometries of pseudo-Riemannian spaces. A manifold M together with a non-degenerate metric $g(X, Y)$ is a pseudo-Riemannian space (M, g).

DEFINITION 6.6. A diffeomorphism $\phi: M \to M$ is an isometry if $\phi^* g = g$ i.e., if $g_x(X, Y) = g_{\phi(x)}((\phi_*) X, (\phi_*) Y)$ for all $x \in M$.

The following theorem will not be proved. (Compare Kobayashi and Nomizu, 1963).

THEOREM 6.4. The group of isometries of (M, g) is a Lie transformation group. The vector fields A^*, which are defined on M by the action of the isometry group are called Killing fields or infinitesimal isometries.

In Section 12 the Killing equation will be derived, which characterizes Killing fields.

7. The Frame Bundle of a Manifold

We have defined differentiable tensor fields on manifolds, but we are so far not able to differentiate a tensor field. The main point of this and the following section is to show that an additional structure is needed on a manifold to get an operation of differentiation, which maps tensor fields into tensor fields. The reason for this is the following: The basic idea of a derivative of a map is to approximate it by a linear map

$$f(x + \Delta x) = f(x) + f'(x)\,\Delta x + \theta(\Delta x^2). \tag{7.1}$$

This is only possible if we can compare the value of the map at different points. Consider as an example a vector field X on a manifold. How could we compare X_p and X_q? Obviously $X_p - X_q$ is meaningless. If however there were a preferred basis in each tangent space, or more generally a preferred isomorphism $\psi: T_p \to T_q$, then one could form $\psi(X_p) - X_q$. This is precisely what is done in calculus on R^n. The natural coordinate system provides a preferred basis, and in this basis one forms the difference quotients of tensor fields.

I shall use the frame bundle to define Lie derivatives as well as covariant derivatives because I believe this approach gives the clearest geometrical motivation for the structures one usually introduces.

DEFINITION 7.1. A frame $u=(p, X_i)$ is an ordered basis $(X_i)_{i=1,\ldots,n}$ of $T_p(M)$. The set of all frames of M is denoted by $L(M)$.

The general linear group $Gl(n, R)$ acts in a natural way freely on the right on $L(M)$: If $a=(a^i_k)\in Gl$, then

$$R_a:(p, X_i)\to(p, X_k a^k_i) \tag{7.2}$$

is a map $L(M)\to L(M)$. There is a projection π defined by

$$\pi: L(M)\to M: \quad \pi(p, X_i)=p. \tag{7.3}$$

There exists on $L(M)$ a differentiable structure such that π is differentiable and $Gl(n, R)$ is a Lie transformation group of $L(M)$.

LEMMA 7.1. If (U_α, h_α) is a C^∞-atlas on M, then $(\pi^{-1}(U_\alpha), \Psi_\alpha)$ is a C^∞-atlas on $L(M)$, where

$$\Psi_\alpha: \pi^{-1}(U_\alpha)\to U_\alpha\times Gl(n, R)\subset R^n\times R^{n^2}$$

is defined by

$$\Psi_\alpha(p, X_i)=(x^i_\alpha(p), X^k_i), \quad \text{where} \quad X_i=X^k_i\frac{\partial}{\partial x^k}.$$

Proof. It is obvious that $(\pi(U_\alpha), \Psi_\alpha)$ are charts which cover $L(M)$. The compatibility on the intersection follows from

$$X_i=X^k_i\frac{\partial}{\partial x^k}=X^k_i\frac{\partial x^{k'}}{\partial x^k}\frac{\partial}{\partial x^{k'}}=X^{k'}_i\frac{\partial}{\partial x^{k'}} \tag{7.4}$$

and the differentiability of $\partial x^{k'}/\partial x^k$. Clearly the topology is Hausdorff and has a countable basis.

DEFINITION 7.2. $L(M)$ together with the manifold structure defined in the Lemma above is called the frame bundle.

The following is now obvious.

LEMMA 7.2. The projection $\pi: L(M)\to M$ is differentiable and $Gl(n, R)$ is a Lie transformation group on $L(M)$.

Having now made $L(M)$ into a manifold, we can apply the whole machinery developed so far to $L(M)$ as well as to M. The important point now is how geometrical objects on M can be described on $L(M)$.

Consider, for example, a vector field X on M. Then the components ξ^i of X in an arbitrary frame are defined by

$$X_p = \tilde{\xi}^i(u)\, X_i\,, \quad \text{where} \quad u = (p, X_i)\,. \tag{7.5}$$

Using the chart $(\pi^{-1}(u), \Psi)$ defined above we can calculate $\tilde{\xi}^i(x^k, X^s_r)$

$$X_p = \xi^i(x)\frac{\partial}{\partial x^i} = \xi^i Y^r_i X^k_r \frac{\partial}{\partial x^k} \quad \text{where} \quad (Y^k_r) = (X^k_r)^{-1}\,. \tag{7.6}$$

Therefore,

$$\tilde{\xi}^r(x^i, X^t_s) = \xi^i(x^k)\, Y^r_i(X^t_s)\,. \tag{7.7}$$

Obviously the function $\tilde{\xi}^i$ are determined on the fibre $\pi^{-1}(p)$ if their values are known at one point u. This is expressed by the transformation property

$$\tilde{\xi}^k(ua) = (a^{-1})^k_r\, \tilde{\xi}^r(u)\,. \tag{7.8}$$

It is clear how this can be generalized to arbitrary tensor fields on M.

LEMMA 7.3. Any tensor field S on M is determined by the components of S in an arbitrary frame on M and so corresponds to a collection of real valued functions on $L(M)$. These functions transform under the action of $Gl(n, R)$ on $L(M)$ according to a representation of Gl determined by the type of S.

With respect to a chart $(\pi^{-1}(U), \Psi)$ these functions are

$$\tilde{S}_{j_1\ldots j_r}{}^{l_1\ldots l_s} = S_{i_1\ldots i_r}{}^{k_1\ldots k_s}(x^l)\, X^{i_1}_{j_1}\ldots X^{i_r}_{j_r} Y^{l_1}_{k_1}\ldots Y^{l_s}_{k_s}\,. \tag{7.9}$$

The action of the group is given by

$$\tilde{S}_{j_1\ldots j_r}{}^{l_1\ldots l_s}(ua) = (a^{-1})^{l_1}_{k_1}\ldots(a^{-1})^{l_s}_{k_s}\, \tilde{S}_{i_1\ldots i_r}{}^{k_1\ldots k_s}(u)\cdot a^{i_1}_{j_1}\ldots a^{i_r}_{j_r}\,. \tag{7.10}$$

We will write $\tilde{S}$ for (7.9) and

$$\tilde{S}(ua) = \varrho(a^{-1})\, \tilde{S}(u) \tag{7.11}$$

for (7.10), without really formalising the notation. (This is quite easy to do if one defines functions with values in a vector space).

What is the advantage of this cumbersome description of tensor fields on M? It lies in the fact that the tensor field is now given as a set of functions and we have already defined the directional derivative of a function.

THEOREM 7.1. Let $\hat{X}$ be a vector field on $L(M)$ invariant under the action of $Gl(n, R)$, i.e., $(R_a)_*\hat{X} = \hat{X}$ for any $a \in Gl(n, R)$. If S is any tensor field on M, there exists a unique tensor field T of the same type such that $\tilde{T} = \hat{X}\tilde{S}$.

Proof. The functions $\hat{X}\tilde{S}$ are uniquely determined. We have only to check that they have the right transformation property. Choose any $a \in Gl$. Then $(R_a)_*\hat{X} = \hat{X}$ implies

$$(\hat{X}f)_{ua} = (\hat{X}f(ua))_u \tag{7.12}$$

for any a.

Applying this to the functions given by $\tilde{S}$ we get

$$\tilde{T}(ua) = (\hat{X}\tilde{S})_{ua} = (\hat{X}\tilde{S}(ua))_u . \tag{7.13}$$

Using now

$$\tilde{S}(ua) = \varrho(a^{-1})\,\tilde{S}(u) \tag{7.14}$$

it follows that

$$\tilde{T}(ua) = \hat{X}(\varrho(a^{-1})\,\tilde{S}(u))_u = \varrho(a^{-1})\,(\hat{X}\tilde{S}(u))_u = \varrho(a^{-1})\,\tilde{T}(u). \tag{7.15}$$

Hence T has the correct transformation property to define a tensor of the same type as S.

What is the geometrical significance of T? Because $\hat{X}$ is invariant under R_a, $\pi_*(\hat{X}) = X$ is a vector field on M. If $\hat{X}$ generates $\hat{\psi}_t$, $\hat{\psi}_t$ commutes with any R_a, hence it defines a vector space isomorphism between the tangent spaces along the integral curves of X in M. The definition of T uses this isomorphism to define a directional derivative of S with respect to X.

8. The Lie Derivative

Now we will use the results of the last section to define Lie derivatives. Let X be a vector field on M, ψ_t the corresponding 1-parameter subgroup around $p \in M$. The differential $(\psi_t)_{*q}$ is a linear transformation $T_q \to T_{\psi_t(q)}$. Hence $(\psi_t, (\psi_t)_*)$ defines a local 1-parameter group in the frame bundle, which in turn defines a vector field $\hat{X}$, the natural lift of X into $L(M)$.

Lemma 8.1. Let X be a vector field on M, then there is a unique vector field $\hat{X}$ on $L(M)$ invariant under $Gl(n, R)$ such that $\hat{X}$ is π-related to X.

Proof. The construction of $\hat{X}$ was described above. We calculate $\hat{X}$ in local coordinates (x^i, X^k_l) on $L(M)$. If $X = \xi^i(\partial/\partial x^i)$, then ψ_t satisfies

$$\psi_t : x^i \to x^i + \xi^i(x^k)\,t + O^i(t^2) \tag{8.1}$$

$$(\psi_t)_* : \frac{\partial}{\partial x^i} \to \frac{\partial}{\partial x^i} + \xi^k{}_{|i}(x^l)\,\frac{\partial}{\partial x^k}\,t + O^i(t^2)\,\frac{\partial}{\partial x^i}. \tag{8.2}$$

Hence the 1-parameter group $\hat{\psi}_t$ in the frame bundle satisfies

$$\hat{\psi}_t : (x^i, X^r_s) \to (x^i + \xi^i t + O^i(t^2),\, X^l_s(\delta^r_l + \xi^r{}_{|l}t + O^r_l(t^2))). \tag{8.3}$$

Therefore, we find

$$\hat{X} = \xi^i(x^k)\,\frac{\partial}{\partial x^i} + X^l_s \xi^r{}_{,l}(x^j)\,\frac{\partial}{\partial X^r_s}. \tag{8.4}$$

Using theorem (7.1) we can define the Lie derivative of tensor S with respect to a vector field X.

DEFINITION 8.1. $\mathscr{L}_X S$ is the uniquely determined tensor of the same type as S satisfying

$$\overline{\mathscr{L}_X S} = \hat{X}\tilde{S}\,. \tag{8.5}$$

LEMMA 8.2. The Lie derivative obeys the relations

$$\begin{aligned}
&(1) \qquad \mathscr{L}_X (S \otimes R) = (\mathscr{L}_X S) \otimes R + S \otimes \mathscr{L}_X R\\
&(2) \qquad \mathscr{L}_X f = X f\\
&(3) \qquad \mathscr{L}_X Y = [X, Y]\,.
\end{aligned} \tag{8.6}$$

Proof. The definition of the tensor product implies

$$S \otimes R = \tilde{S} \cdot \tilde{R} \tag{8.7}$$

in our symbolic notation. Then $\hat{X}(fg) = (\hat{X}f)g + f(\hat{X}g)$ implies (1). (2) is obvious as a function considered as a tensor has one and the same component in any frame. (3) Here we use the expression for $\hat{X}$ we have just derived

$$\hat{X}\tilde{Y}^k = \left(\xi^i \frac{\partial}{\partial x^i} + \xi^l{}_{,i}\, X^i_r \frac{\partial}{\partial X^l_r}\right)(\eta^s Y^k_s)\,. \tag{8.8}$$

Using $(\partial/\partial X^l_r) Y^b_d = - Y^b_l\, Y^r_d$ it follows that

$$\hat{X}\tilde{Y}^k = (\xi^i \eta^s{}_{,i} - \eta^i \xi^s{}_{,i})\, Y^k_s = \overline{[X, Y]}^k\,. \tag{8.9}$$

This is one of the geometrical interpretations of $[X, Y]$. From the definition of $\hat{X}$ it is clear that definition of $\mathscr{L}_X$ is the same as the one usually given in terms of ψ_*:

$$(\mathscr{L}_X Y)_p = \lim_{t \to 0} \frac{1}{t} [((\psi_t)_* Y)_p - Y_p]\,. \tag{8.10}$$

The definition given above formalises the linearization $(\psi_t)_*$ as well as the limit $t \to 0$.

As an application we consider Killing fields.

THEOREM 8.1. Let X be a vector field on a peudo-Riemannian space (M, g). Then $\mathscr{L}_X g = 0$ holds if and only if the 1-parameter group ψ_t generated by X is an isometry. Such a vector field is called a Killing field and the set of all Killing vector fields forms a Lie algebra over R.

Proof. Let $\hat{X}$ be the natural lift of X and $\tilde{g}$ the functions on $L(M)$ defined by g. Then $L_X g = 0$ implies $\hat{X}\tilde{g} = 0$, hence $\tilde{g}$ is constant along the integral curves of $\hat{X}$. By the definition of $\hat{X}$, these integral curves are, if ψ_t is generated by X,

$$u(t) = (x(t), X_i(t)) = (\psi_t(p), (\psi_t)_* X_i) \tag{8.11}$$

and, therefore, $\tilde{g}(u(t)) = \text{const.}$ implies $g_p(X, Y) = g_{\psi_t(p)}((\psi_t)_* X, (\psi_t)_* X)$. It is obvious that this sequence of conclusions can be reversed. Finally we must show

$$L_X g = 0\,, \quad L_Y g = 0 \Rightarrow L_{[X, Y]} g = 0\,. \tag{8.12}$$

We remarked above that X, $\hat{X}$ are π-related, hence $\widehat{[X, Y]} = [\hat{X}, \hat{Y}]$. Hence for any tensor field

$$\begin{aligned}\widetilde{L_{[X,Y]}S} &= \widehat{[X, Y]}\,\tilde{S} = [\hat{X}, \hat{Y}]\,\tilde{S} = \hat{X}\hat{Y}\tilde{S} - \hat{Y}\hat{X}\tilde{S} \\ &= \hat{X}\widetilde{L_Y S} - \hat{Y}\widetilde{L_X S} = \widetilde{L_X L_Y S} - \widetilde{L_Y L_X S}. \end{aligned} \tag{8.13}$$

Therefore

$$L_{[X,Y]}S = L_X L_Y S - L_Y L_X S \tag{8.14}$$

holds quite generally and this implies (8.12).

We will later see that the Lie algebra of Killing fields is at most of dimension $n(n+1)/2$ if $\dim M = n$.

9. Covariant Derivative – Linear Connection

It is obvious that $(\mathscr{L}_X S)_p$ depends not only on X_p but on X in a neighbourhood of p. If we want a directional derivative depending only on X_p we must give a rule which produces a lift $\hat{X}$ such that $\hat{X}_{\pi^{-1}(p)}$ is uniquely determined by X_p. To get linearity with respect to the direction in which it is differentiated we must demand $\hat{X} + \hat{Y} = \widehat{X+Y}$. Thus one is led quite naturally to the definition of a linear connection on M.

Definition 9.1. A linear connection Γ on M is a map which assigns to any $u \in L(M)$ an n-dimensional subspace $H_u \subset T_u(L(M))$ such that

(1) $\pi_* H_u = T_{\pi(u)}(M)$

(2) $(R_a)_* H_u = H_{ua}$

(3) H_u is differentiable, which means that it can be spanned by differentiable vector fields.

Now the following Lemma is trivial.

Lemma 9.1. If X is a vector field on M and Γ a linear connection on M, then there exists a unique horizontal lift $\hat{X}$ of X invariant under $Gl(n, R)$ and such that $\hat{X} + \hat{Y} = \widehat{X+Y}$.

Proof. $\hat{X}_u$ is uniquely defined by the condition $\hat{X}_u \in H_u$ and

$$\pi_* \hat{X}_u = X_{\pi(u)} \tag{9.2}$$

because π_* is an isomorphism $H_u \to T_{\pi(u)}(M)$.

Using Theorem 7.1 one defines the covariant derivative of a tensor field S in the direction of X_p.

Definition 9.2. If S is a tensor field defined near p and $X_p \in T_p(M)$ then $\nabla_{X_p} S$ is a tensor at p defined by

$$\widetilde{\nabla_{X_p} S} = (\hat{X}_p \tilde{S})_{\pi^{-1}(p)}. \tag{9.3}$$

This definition implies:

THEOREM 9.1. The directional derivative satisfies

$$\begin{array}{lll}(1) & \nabla_{X_p}S + \nabla_{Y_p}S = \nabla_{X_p+Y_p}S\,, \quad \nabla_{\alpha X_p}S = \alpha\nabla_{X_p}S & \\ (2) & \nabla_{X_p}(fS) = f\nabla_{X_p}S + (Xf)_p S_p\,, \quad \nabla_{X_p}(S+R) = \nabla_{X_p}S + \nabla_{X_p}R & (9.4)\\ (3) & \nabla_{X_p}(S\otimes R) = (\nabla_{X_p}S)\otimes R_p + S_p\otimes\nabla_{X_p}R\,. & \end{array}$$

Because of (1) one can define the covariant derivative ∇S, which is contained in $\mathscr{T}^r_{s+1}$ if $S\in\mathscr{T}^s_r$.

DEFINITION 9.3.

$$(\nabla S)\,(X, X_1, \ldots, X_r, \omega^1, \ldots, \omega^s) = (\nabla_X S)\,(X_1, \ldots, X_r, \omega^1 \ldots \omega^s)\,. \tag{9.5}$$

In the next section we will see how one calculates these derivatives.

We have seen that we can lift vector fields from M to $L(M)$ if a connection is given on M. The horizontal subspace gives also the possibility of lifting curves.

THEOREM 9.2. Let $\sigma(t)$, $0\leqslant t\leqslant 1$ be a differential curve, (i.e., the restriction of a differentiable curve defined for $-\varepsilon<t<1+\varepsilon$) and $u_0\in\pi^{-1}(\sigma(0))$. Then there exists a unique curve $\hat\sigma(t)$ in $L(M)$ such that $\hat\sigma(0)=u_0$, $\pi(\hat\sigma(t))=\sigma(t)$, $\dot{\hat\sigma}(t)\in H_{\sigma(t)}$.

We will postpone the proof of this theorem to the next section. For a piece $0\leqslant t\leqslant t_1$ the statement is obvious because we can lift vector fields. The essential point is to show that one can lift the whole curve.

DEFINITION 9.3. A curve $u(t)$ in $L(M)$ is called horizontal if $\dot u(t)\in H_{u(t)}$. If $u(t)= =(x(t), X_i(t))$ is horizontal, the frame $X_i(t)$ is called parallel along the curve $x(t)$.

It is easy to find the relation between parallel propagation and ∇_X.

LEMMA 9.2. Let S be a tensor field and $x(t)$ a curve and $u(t)$ a horizontal lift of $x(t)$. Then

$$\widetilde{(\nabla_{\dot x(0)}S)}_{u(0)} = \left[\frac{\mathrm{d}}{\mathrm{d}t}(\tilde S(u(t)))\right]_{t=0}. \tag{9.6}$$

Proof. By definition of the horizontal lift $\dot u(0)$ is horizontal. Hence

$$\widetilde{(\nabla_{\dot x(0)}S)}_{u(0)} = (\dot u(0)\,\tilde S)_{u(0)} = \left[\frac{\mathrm{d}}{\mathrm{d}t}(\tilde S(u(t)))\right]_{t=0}. \tag{9.7}$$

Hence to get $\nabla_X S$ we can differentiate the components of S in a basis which is parallel along a curve which fits X. Consequently we say that a vector $X(t)$ is parallel along a curve $x(t)$ if

$$\nabla_{\dot x}X = 0\,.$$

10. Descriptions of a Linear Connection

In this section we will show how one can describe a linear connection and how to calculate ∇_X and parallel propagation. Let us start with the horizontal subspace.

DEFINITION 10.1. Let Γ be a connection on M. Let E^i_k be a basis of $\mathbf{G}l$ and E^{*i}_k the corresponding vector fields on $L(M)$. Then any $X \in T_u(L(M))$ can be uniquely decomposed into vertical and horizontal parts:

$$X = vX + hX, \; hX \in H_u, \quad vX \text{ tangent to the fibres}.$$

Writing vX in terms of its components in the basis E^{*i}_k, one can define n^2 1-forms on $L(M)$ by $vX = \omega^i_k(X) E^{*k}_i$.

The forms ω^i_k are called the connection forms of Γ on $L(M)$!

There are n further 1-forms on the frame bundle which are independent of Γ.

DEFINITION 10.2. The canonical 1-forms Θ^i on the frame bundle are defined by

$$\pi_*(X_u) = \Theta^i(X_u)\, X_i \quad \text{if} \quad u = (p, X_1, \ldots, X_n). \tag{10.2}$$

The last two definitions show that a connection on M gives a basis of $T^*_u(L(M))$ at any $u \in L(M)$. Naturally we ask now for the vector fields dual to the 1-forms. These are the vector fields E^{*i}_k and the standard horizontal vector fields, which we now define.

DEFINITION AND LEMMA 10.3. Let ξ^i be an n-tuple of real numbers; then there exists a unique *horizontal* vector field $B(\xi^i)$ on $L(M)$ such that

$$\pi_* B_u(\xi^i) = \xi^i X_i \quad \text{if} \quad u = (p, X_1, \ldots, X_n). \tag{10.3}$$

By B_k we denote the vector fields $k = 1, \ldots, n$ corresponding to the natural basis $(\xi^i) = (1,0,0,\ldots)$ and so on.

Proof. The existence is obvious because $\pi_* : H_u \to T_{\pi(u)}$ is a vector space isomorphism.

LEMMA 10.1. The fields E^{*i}_k, B_j and the 1-forms ω^i_k, Θ^i are dual and satisfy

$$\omega^i_k(E^{*r}_s) = \delta^i_s \delta^r_k\,; \qquad \omega^i_k(B_r) = 0$$
$$\Theta^i(B_k) = \delta^i_k\,; \qquad \Theta^i(E^{*r}_s) = 0\,. \tag{10.4}$$

This is an immediate consequence of the definitions. For example: the definition of Θ^i implies

$$\pi_*(B_k)_u = \Theta^i(B_k)\, X_i \quad \text{if} \quad u = (p, X_1, \ldots, X_n) \tag{10.5}$$

and the definition of B_k gives $\pi_*(B_k) = X_k$, hence

$$X_k = \Theta^i(B_k)\, X_i \Rightarrow \Theta^i(B_k) = \delta^i_k\,. \tag{10.6}$$

The invariance of H_u under $Gl(n, R)$ implies the following transformation properties of the fields and forms.

THEOREM 10.1. For any $a = (a^i_k) \in Gl$ one has

(1) $\quad (R_a)_* B_i = (a^{-1})^k_i B_k$

(2) $\quad (R_a)_* E^{*s}_r = a^s_l E^{*l}_j (a^{-1})^j_r$

(3) $(R_a)^* \Theta^i = (a^{-1})^i_k \Theta^k$

(4) $(R_a)^* \omega^s_r = a^s_l \omega^l_j (a^{-1})^j_r$, (10.7)

where E^i_k is the basis of $\mathbf{Gl(n, R)}$ defined in Section 6.

Proof. Let us first calculate $(R_a)_* B(\xi)$. As $B(\xi^i)_u$ is horizontal, so is $(R_a)_* B(\xi^i)_u = C^i(B_i)_{ua}$. Hence projection into M gives

$$\pi_*(R_a)_* B(\xi^i)_u = \pi_*(B(\xi^i)_u) = \pi_* C^i(B_i)_{ua}. \tag{10.8}$$

By the definition of B and $u = (\pi(u), X_i)$, $ua = (\pi(u), X_k a^i_k)$ so

$$\xi^k X_k = C^i X_k a^k_i \Rightarrow C^i = \xi^k (a^{-1})^i_k. \tag{10.9}$$

Hence

$$(R_a)_* B(\xi^i)_u = (a^{-1})^i_k \xi^k (B_i)_{ua} = B((a^{-1})^i_k \xi^k)_{ua} \tag{10.10}$$

and this implies (1). Note that $(R_a)_* B_i$ is a linear combination of B_i with coefficients depending only on a.

The transformation of E^{*i}_r is a property of $Gl(n, R)$, as $Gl(n, R)$ acts freely on $L(M)$. Let $u = (x^i, X^j_k)$ be any point in $L(M)$. Then

$$((R_a)_* E^{*r}_s)_u = (R_a)_* (E^{*r}_s)_{ua^{-1}}. \tag{10.11}$$

Because of

$$E^{*r}_s(X^j_k) = X^i_s \frac{\partial}{\partial X^i_r} \tag{10.12}$$

we find

$$((R_a)_* E^{*r}_s)_u = X^l_t (a^{-1})^t_s (R_a)_* \left(\frac{\partial}{\partial X^l_r}\right)_{ua^{-1}}. \tag{10.13}$$

One calculates easily

$$(R_a)_* \frac{\partial}{\partial X^l_j} = a^j_c \frac{\partial}{\partial X^l_c} \tag{10.14}$$

and that implies (2).

The transformation property of ω^i_k, Θ^i follows from (1) and (2) because these are dual to E^{*i}_j, B_k:

$$\begin{aligned}((R_a)^* \Theta^i)(B_j) &= \Theta^i((R_a)_* B_j) = \Theta^i((a^{-1})^k_j B_k) = (a^{-1})^k_j \Theta^i(B_k)\\ &= (R^*_a \Theta^i)(E^{*r}_s) = 0.\end{aligned} \tag{10.15}$$

Similarly one gets (4).

The last theorem shows that the connection is unique if we know ω^i_k or B_i at one point of each fibre. The following theorem expresses all fields in local coordinates.

THEOREM 10.2. Let (x^i) be a coordinate system in U and (x^i, X^i_k) the coordinates defined in $\pi^{-1}(U)$. Suppose a connection is given on M. Then there are functions Γ^l_{ik} on U such that:

(1) $$\frac{\widehat{\partial}}{\partial x^i} = \frac{\partial}{\partial x^i} - \Gamma^l_{ik} X^k_s \frac{\partial}{\partial X^l_s}$$

(2) $\omega^i_j = Y^i_k \, \mathrm{d}X^k_j + Y^i_k \Gamma^k_{lm} X^m_j \, \mathrm{d}x^l$

(3) $\Theta^i = Y^i_k \, \mathrm{d}x^k$

(4) $B_i = X^k_i \left(\frac{\partial}{\partial x^k} - \Gamma^l_{km} X^m_s \frac{\partial}{\partial X^l_s} \right)$

(5) $\nabla_{\partial/\partial x^i} \frac{\partial}{\partial x^k} = \Gamma^l_{ik} \frac{\partial}{\partial x^l} .$ (10.17)

Proof. Suppose $\widehat{\partial/\partial x^i}$ is horizontal at (x^i, δ^i_k), then there are functions such that

$$\frac{\widehat{\partial}}{\partial x^i} = \frac{\partial}{\partial x^i} - \Gamma^l_{ik} \frac{\partial}{\partial X^l_k} \tag{10.18}$$

and $(R_a)_*(\widehat{\partial/\partial x^i})$ implies (1) if we use Theorem 10.1. Now we calculate B_r, which is a linear combination of $\widehat{\partial/\partial x^i}$:

$$(B_r)_u = C^i_r \left(\frac{\partial}{\partial x^i} - \Gamma^l_{ik} X^k_s \frac{\partial}{\partial X^l_s} \right). \tag{10.19}$$

If $u = (x^i, X^i_j)$, then u is the frame $(x^i, (\partial/\partial x^t) X^t_j)$; hence

$$\pi_* (B_r)_u = \frac{\partial}{\partial x^t} X^t_r = C^t_r \frac{\partial}{\partial x^t} \Rightarrow C^t_r = X^t_r . \tag{10.20}$$

Therefore,

$$B_r = X^i_r \left(\frac{\partial}{\partial x^i} - \Gamma^l_{ik} X^k_s \frac{\partial}{\partial X^l_s} \right) \tag{10.21}$$

which is (4).

From $E^{*r}_s = X^i_s (\partial/\partial X^i_r)$ and the fact that ω^i_k, Θ^k are dual to E^{*r}_s, B_i one can calculate (2) and (3) from (4) and E^{*r}_s.

Let us finally check (5) which now relates the frame bundle approach to the usual one.

The components of $\partial/\partial x^k$ are $\overline{\partial/\partial x^k} = Y^s_k$. Hence

$$\frac{\widehat{\partial}}{\partial x^i} \left(\overline{\frac{\partial}{\partial x^k}} \right) = \left(\frac{\partial}{\partial x^i} - \Gamma^l_{ij} X^j_t \frac{\partial}{\partial X^l_t} \right) (Y^s_k) =$$

$$= \Gamma^l_{ij} X^j_t Y^t_k Y^s_l = \Gamma^l_{ik} Y^s_l = \overline{\Gamma^l_{ik} \frac{\partial}{\partial x^l}} . \tag{10.22}$$

Therefore (5) holds.

Proof of Theorem 9.2. Suppose first that $\sigma(t)$ is contained in a coordinate neighborhood with compact closure. Using Theorem 10.2 any horizontal lift is determined by the differential equation

$$\frac{\mathrm{d}X^k_j}{\mathrm{d}t} + \Gamma^k_{lm}(x^i(t)) X^m_j \dot{x}^l(t) = 0 . \tag{10.23}$$

Because $\Gamma^k_{lm}\dot{x}^l$ is bounded for $0 \leqslant t \leqslant 1$ one gets a solution $X^k_j(t)$ defined for $0 \leqslant t \leqslant 1$ for any set of initial values $X^k_j(0)$. For any arbitrary curve $\sigma(t)$; $0 \leqslant t \leqslant 1$ we can find t_i, $i = 1, \ldots, k$, $0 < t_1, \ldots, t_k < 1$ such that any piece of $\sigma(t)$, $t_i \leqslant t \leqslant t_{i+1}$ is contained in a coordinate neighborhood with compact closure, hence we get a horizontal lift of σ by joining lifts of the pieces with the right initial values.

The next Lemma gives the transformation property of the Γ^l_{ik}, which are not tensors. The essential point about using the frame bundle is that on $L(M)$ a connection corresponds to a tensorial object.

Lemma 10.2. Let (U, h), (U, h') be two charts around p. Then

$$\Gamma^{l'}_{i'k'} = \Gamma^l_{ik} \frac{\partial x^i}{\partial x^{i'}} \frac{\partial x^k}{\partial x^{k'}} \frac{\partial x^{l'}}{\partial x^l} + \frac{\partial^2 x^k}{\partial x^{i'} x^{k'}} \frac{\partial x^{l'}}{\partial x^k}. \tag{10.24}$$

From $(\partial/\partial x^{i'}) = (\partial x^i/\partial x^{i'})\,(\partial/\partial x^i)$ and the properties of ∇_X, it follows that

$$\begin{aligned} \nabla_{\partial/\partial x^{i'}} \frac{\partial}{\partial x^{k'}} &= \frac{\partial x^i}{\partial x^{i'}} \nabla_{\partial/\partial x^i}\left(\frac{\partial x^l}{\partial x^{k'}} \frac{\partial}{\partial x^l}\right) = \\ &= \frac{\partial x^i}{\partial x^{i'}} \frac{\partial x^l}{\partial x^{k'}} \Gamma^j_{il} \frac{\partial}{\partial x^j} + \frac{\partial x^i}{\partial x^{i'}} \frac{\partial^2 x^l}{\partial x^{j'} \partial x^{k'}} \frac{\partial x^{j'}}{\partial x^i} \frac{\partial}{\partial x^l} \\ &= \frac{\partial x^i}{\partial x^{i'}} \frac{\partial x^l}{\partial x^{k'}} \Gamma^j_{il} \frac{\partial x^{s'}}{\partial x^j} \frac{\partial}{\partial x^{s'}} + \frac{\partial^2 x^l}{\partial x^{i'} \partial x^{k'}} \frac{\partial x^{s'}}{\partial x^l} \frac{\partial}{\partial x^{s'}}, \end{aligned} \tag{10.25}$$

and this implies the Lemma.

Theorem 10.2 and the Lemma above show

Theorem 10.3. A connection on M can be defined by a map which assigns to any pair of vector fields X, Y a vector field $\nabla_X Y$ such that

$$\begin{aligned} \nabla_{X+\bar{X}} Y &= \nabla_X Y + \nabla_{\bar{X}} Y, \quad \nabla_X (Y + \bar{Y}) = \nabla_X Y + \nabla_X \bar{Y} \\ \nabla_{fX} Y &= f \nabla_X Y, \quad \nabla_X f Y = f \nabla_X Y + (Xf)\, Y \end{aligned} \tag{10.26}$$

holds.

11. Curvature Tensor and Structure Equations

We have seen that a connection Γ on M defines a parallelisation on $L(M)$ by the vector fields E^{*i}_k, B_i. All information about Γ is contained in the fields B_i. What can we now do with these fields? The only possible way to obtain further properties of Γ is from their commutators. Therefore, we define functions $\tilde{T}^l_{ik}$, $\tilde{R}^r_{sik}$ on $L(M)$ by

$$[B_i, B_k] = -\tilde{T}^l_{ik} B_l - \tilde{R}^r_{sik} E^{*s}_r. \tag{11.1}$$

The notation already indicates what we will now show:

Theorem 11.1. The functions $\tilde{T}^l_{ik}$, $\tilde{R}^r_{sik}$ defined by (1) are the components of two tensor fields $T(X, Y, \omega)$, $R(X, Y, Z, \omega)$, called the torsion and curvature tensors.

Proof. According to the considerations in Section 7 we have to show

$$\tilde{R}^r_{sik}(ua) = (a^{-1})^r_u \tilde{R}^u_{vwt}(u)\, a^v_s a^w_i a^t_k \tag{11.2}$$

$$\tilde{T}^l_{ik}(ua) = (a^{-1})^l_u \tilde{T}^u_{vw}(u)\, a^v_i\, a^v_k. \tag{11.3}$$

This follows by a simple calculation from the known transformation properties of B_i, and the following: apply (R_a) to (1) at a point u in $L(M)$:

$$(R_a)_* [B_i, B_k]_u = -\tilde{T}^l_{ik}(u)\,(R_a)_*(B_l)_u - \tilde{R}^r_{sik}(u)\,(R_a)_*(E^{*s}_r)_u \tag{11.4}$$

$$[(R_a)_* B_i, (R_a)_* B_k]_{ua} = -\tilde{T}^l_{ik}(u)\,(a^{-1})^j_l\,(B_j)_{ua} - \tilde{R}^r_{sik}(u)\,(a^{-1})^j_r\,(E^{*l}_j)_{ua}\, a^s_l \tag{11.5}$$

$$[(a^{-1})^l_j B_l,\, a_k^{-1n} B_n]_{ua} = \cdots \tag{11.6}$$

$$(a^{-1})^l_j [B_l, B_n]_{ua} = \cdots \tag{11.7}$$

$$(a^{-1})^l_j \{-\tilde{T}^r_{ln}(ua)\,(B_r)_{ua} - \tilde{R}^v_{wln}(ua)\,(E^{*w}_v)_{ua} = \cdots. \tag{11.8}$$

This implies the right transformation properties for $\tilde{R}$, $\tilde{T}$.

The torsion and curvature tensors are usually written as $T(X, Y)$, $R(X, Y)Z$ with the definitions:

$$T(X_p, Y_p) \in T_p(M) \quad \text{such that} \quad \omega_p(T(X_p, Y_p)) = T(X_p, Y_p, \omega_p) \tag{11.9}$$

$$R(X_p, Y_p)\, Z_p \in T_p(M) \quad \text{such that}$$
$$\omega_p(R(X_p, Y_p)\, Z_p) = R(X_p, Y_p, Z_p, \omega_p). \tag{11.10}$$

The notation $R(X, Y)$ indicates that $Z \to R(X, Y)Z$ is a linear transformation in $T_p(M)$, called the curvature transformation.

The commutators between the vectors E^{*i}_k, B_i give no further information besides (1) about Γ. We need them, however, to obtain the corresponding 1-form relations.

LEMMA 11.1. *If Γ is a connection on M and E^{*i}_k is the canonical basis in $Gl(n, R)$, then*

$$[E^{*r}_s, B_k] = \delta^r_k B_s \tag{11.11}$$

$$[E^{*r'}_{s'}, E^{*r}_s] = \delta^{r'}_s E^{*r}_{s'} - \delta^r_{s'} E^{*r'}_s. \tag{11.12}$$

Proof. Both relations can be calculated directly from the expression for B_i given in Theorem 10.2.

Let us now pass from the vector fields to the 1-forms.

THEOREM 11.2. *If Γ is a connection on M, then*

$$\mathrm{d}\omega^i_k = -\omega^i_j \wedge \omega^j_k + \tfrac{1}{2}\tilde{R}^i_{klm}\Theta^l \wedge \Theta^m \tag{11.13}$$

$$\mathrm{d}\Theta^i = -\omega^i_k \wedge \Theta^k + \tfrac{1}{2}\tilde{T}^i_{lm}\Theta^l \wedge \Theta^m \tag{11.14}$$

where ω^i_k is defined with respect to the natural basis in $Gl(n, R)$. These equations are called the structure equations on $L(M)$.

Proof. Suppose e_a is a parallelisation and ω^a are the dual 1-forms, i.e., $\omega^b(e_a)=\delta^b_a$. Then

$$[e_a, e_b] = \gamma^c_{ab} e_c \tag{11.15}$$

implies

$$d\omega^a = -\tfrac{1}{2}\gamma^a_{bc}\omega^b \wedge \omega^c \tag{11.16}$$

because

$$2(d\omega^a)(e_b, e_c) = e_b\omega^a(e_c) - e_c\omega^a(e_b) - \omega^a([e_b, e_c]) \tag{11.17}$$

as we showed in Section 4. Hence

$$2(d\omega^a)(e_b, e_c) = -\omega^a(\gamma^d_{bc}e_d) \tag{11.18}$$

i.e.

$$2d\omega^a(e_b, e_c) = -\gamma^a_{bc} \tag{11.19}$$

which is what we had to show.

Using these relations for ω^i_k, Θ^l, E^{*r}_s, B_i one obtains the structure equations. We can now pull back the structure equations to M.

DEFINITION 11.1. A local cross section σ is a differentiable map $\sigma: U \subset M \to L(M)$ such that $\sigma \circ \pi = id_u$.

A local cross section σ is just a frame at each point of U which varies differentiably. For example any coordinate system on U induces a cross-section $(x^i) \to (x^i, (\partial/\partial x^i))$.

We now can use Theorem 5.2 to pull the 1-forms from $L(M)$ back to M. $\sigma^*\omega^i_k$ are the connection 1-forms on M, $\sigma^*\tilde{R}^i_{kln} = R^i_{klm}$ are the components of the curvature tensor in the frame $\sigma(x)$. To prove this let us first show that the 2-forms $\frac{1}{2}\tilde{R}^i_{kln}\Theta^l \wedge \Theta^m$ are independent of the parallelisation.

THEOREM 11.3. $\Omega^i_k(X, Y) \equiv d\omega^i_k(hX, hY)$, where hX is the horizontal part of X, are 2-forms on $L(M)$, called curvature 2-forms.

Proof. $\Omega^i_k(X, Y) = d\omega^i_k(hX, hY) = \frac{1}{2}\tilde{R}^i_{klm}(\Theta^l \wedge \Theta^m)(hX, hY) =$

$$= \tfrac{1}{2}\tilde{R}^i_{klm}(\Theta^l \wedge \Theta^m)(X, Y)$$

because $\omega\ (hX, hY) = 0$ and $\Theta^i(hX) = \Theta^i(X)$.

The corresponding theorem holds for the torsion. Now we prove

THEOREM 11.4. If σ is a local cross-section then the structure equations on M are

$$\begin{aligned} d(\sigma^*\omega^i_k) &= -\sigma^*\omega^i_j \wedge \sigma^*\omega^j_k + \sigma^*\Omega^i_k\,; & \sigma^*\Omega^i_k &= R^i_{klm}\sigma^*\Theta^l \wedge \sigma^*\Theta^m \\ d(\sigma^*\Theta^i) &= -\sigma^*\omega^i_j \wedge \sigma^*\Theta^j + \sigma^*T^i\,; & T^i &= d\Theta^i(hX, hY)\,; \\ & & \sigma^*T^i &= T^i_{kl}\sigma^*\Theta^k \wedge \sigma^*\Theta^l \end{aligned} \tag{11.21}$$

where $R^i_{klm}(x)$ are the components of R in the frame $\sigma(x)$.

Proof. The structure equations on M are implied by those on $L(M)$ because of the properties of σ^*. (Theorem 5.2). R^i_{klm} is defined by $\tilde{R}^i_{klm}(\sigma(x))$, hence these are the components of $R(X, Y)$ in the frame $\sigma(x)$.

Finally let us give the forms on M in terms of a coordinate system using the natural basis $\partial(\partial x_i)$:

LEMMA 11.2. Let (x^i) be a coordinatesystem, σ the cross-section

$$(x^i) \to \left(x^i, \frac{\partial}{\partial x}i\right);$$

then

$$\sigma^* \theta^i = \mathrm{d}x^i, \tag{11.22}$$

and

$$\sigma^* \omega^i_j = \Gamma^i_{lj}\, \mathrm{d}x^l. \tag{11.23}$$

Proof. This follows immediately from Theorem 10.2.

Remark. The structure equation on M can be used to calculate the Riemann tensor. Just calculate $\mathrm{d}(\sigma^*\omega^i_k)$ and add $\sigma^*\omega^i_s \wedge \sigma^*\omega^s_k$.

We have seen that the a connection Γ can also be defined by $\nabla_X Y$, the directional derivative on M. The following theorem shows how the curvature and torsion tensor are expressed in terms of $\nabla_X Y$.

THEOREM 11.5. Let $R(X, Y)Z$, $T(X, Y)$ be the curvature and torsion tensors of a connection. Then

$$R(X, Y)Z = \nabla_X \nabla_Y Z - \nabla_Y \nabla_X Z - \nabla_{[X, Y]} Z, \tag{11.24}$$

$$T(X, Y) = \nabla_X Y - \nabla_Y X - [X, Y]. \tag{11.25}$$

Proof. A proof using coordinates can be given by a direct calculation of the right-hand side and comparing it with the expression $R(X, Y, Z)$ one gets from the structure equations on M (Theorem 11.4).

If one develops the theory of connections with $\nabla_X Y$, the relations above are the definitions of $\mathrm{R}(X, Y)Z$ and $T(X, Y)$. Theorem 11.5 contains also the analytic interpretation of $\mathrm{R}(X, Y)Z$: it describes the non-commutativity of $\nabla_X \nabla_Y$. A geometrical interpretation of $R(X, Y)Z$ can be found in Ehlers' lectures.

Finally a remark about the Bianchi identities, which we will not derive in detail. For a torsion free connection there is only one identity:

$$(\nabla_X R)(Y, Z) + (\nabla_Z R)(X, Y) + (\nabla_Y R)(Z, X) = 0 \tag{11.26}$$

for all vector fields X, Y, Z which is equivalent to the relation obtained by applying the exterior derivative to the structure equation, using $\mathrm{d}^2 = 0$.

12. The Connection Induced by a Metric

Any non-degenerate metric $g(X, Y)$ on M induces a connection on M.

DEFINITION 12.1. A connection Γ is called a metric connection if $\nabla g=0$. It is called a Levi Civita or a Riemannian connection if in addition the torsion of Γ vanishes.

The geometrical significance of $\nabla g=0$ gives the following:

LEMMA 12.1. $\nabla g=0$ holds if and only if parallel propagation preserves the scalar product.

Proof. $\nabla g=0$ is equivalent to $(\nabla_X g)(Y,Z)=0$ for any X, Y, Z. This is equivalent to $\hat{X}\tilde{g}=0$, the statement that the scalar product is preserved under parallel propagation.

THEOREM 12.1. Let g be a non-degenerate metric. Then there exists a unique torsion-free metric connection Γ.

Proof. From Section 10 it follows, if Γ is any connection, that

$$(\nabla_X g)(Y,Z)=Xg(Y,Z)-g(\nabla_X Y,Z)-g(Y,\nabla_X Z). \tag{12.1}$$

Hence a necessary and sufficient condition for $\nabla g=0$ is

$$Xg(Y,Z)-g(\nabla_X Y,Z)-g(Y,\nabla_X Z)=0 \tag{12.2}$$

for any vector fields X, Y, Z. Let us write this relation twice again with cyclic permutation of X, Y, Z.

$$Zg(X,Y)-g(\nabla_Z X,Y)-g(X,\nabla_Z Y)=0 \tag{12.3}$$

$$Yg(Z,X)-g(\nabla_Y Z,X)-g(Z,\nabla_Y X)=0. \tag{12.4}$$

Suppose now that Γ is torsion-free, then

$$\nabla_X Y-\nabla_Y X=[X,Y] \tag{12.5}$$

holds. Now add (12.2) and (12.4) and subtract (12.3):

$$\begin{aligned} &Xg(Y,Z)+Yg(Z,X)-Zg(X,Y)-g(\nabla_X Y,Z)-g(Y,\nabla_X Z)-\\ &-g(\nabla_Y Z,X)-g(Z,\nabla_Y X)+g(\nabla_Z X,Y)+g(X,\nabla_Z Y)=0. \end{aligned} \tag{12.6}$$

Using (12.5) one finds

$$\begin{aligned} 2g(\nabla_X Y,Z)=&\,Xg(Y,Z)+Yg(X,Z)-Zg(X,Y)+\\ &+g([X,Y],Z)+g([Z,X],Y)-g([Y,Z],X). \end{aligned} \tag{12.7}$$

Hence from the assumption that Γ is metric and torsion-free, we have derived an expression for Γ which determines Γ uniquely in terms of g. Therefore, we can define Γ by (12.7) and it is easy to check that $\nabla_X Y$ defined by (12.7) satisfies the conditions of Theorem 10.3 and therefore defines a Riemannian connection.

There are two particular cases in which the expression (12.7) simplifies considerably. If we use a coordinate basis $\partial/\partial x^i$ one finds because of $[\partial/\partial x^i, \partial/\partial x^k]=0$

$$2g\left(\nabla_{\partial/\partial x^i}\frac{\partial}{\partial x^k},\frac{\partial}{\partial x^l}\right)=\frac{\partial g_{kl}}{\partial x^i}+\frac{\partial g_{il}}{\partial x^k}-\frac{\partial g_{ik}}{\partial x^l}$$
$$2g\left(\Gamma^j_{ik}\frac{\partial}{\partial x^j},\frac{\partial}{\partial x^l}\right)=\cdots \tag{12.8}$$
$$2\Gamma^j_{ik}g_{jl}=\cdots.$$

If one uses a basis e_a of vector fields such that $g(e_a, e_b)=g_{ab}=\text{const.}$, for example an orthonormal basis, then $\nabla_{e_a}e_b=\Gamma^c_{ab}\,e_c$ defines the components of Γ in this basis and Γ is expressed in terms of γ^c_{ab} defined by $[e_a, e_b]=\gamma^c_{ab}e_c$, by relation (12.7).

Now we can derive Killing's equation for vector fields which generate local 1-parameter groups of isometries.

THEOREM 12.2. Let X be an infinitesimal isometry, i.e., $L_Xg=0$. Then X satisfies for any Y, Z

$$g(\nabla_Y X, Z)+g(Y,\nabla_Z X)=0 \tag{12.9}$$

which is in components

$$\xi_{i;k}+\xi_{k;i}=0. \tag{12.10}$$

Proof. For any vector field X one has $\nabla_X g=0$, hence $L_X g=0$ implies $(L_X-\nabla_X)g=0$. Now

$$((L_X-\nabla_X)\,g)(Y,Z)=(L_X-\nabla_X)\,g(Y,Z)- \\ -g((L_X-\nabla_X)Y,Z)-g(Y,(L_X-\nabla_X)\,Z) \tag{12.11}$$

holds and using $(L_X-\nabla_X)g(Y, Z)=0$ we get

$$g((L_X-\nabla_X)\,Y,Z)+g(Y,(L_X-\nabla_X)\,Z)=0. \tag{12.12}$$

As the torsion vanishes

$$L_XY-\nabla_XY=[X,Y]-\nabla_XY=-\nabla_YX. \tag{12.13}$$

Therefore, (12.12) we can be written as

$$g(-\nabla_YX,Z)=-g(Y,-\nabla_ZX) \tag{12.14}$$

and this is (12.9).

13. Geodesics

Let Γ be a connection on M; then one can define a preferred class of curves on M.

DEFINITION 13.1 A curve $\sigma(t)$; $a<t<b$ is called a geodesic if its tangent vector $\dot\sigma(t)$ is parallel along $\sigma(t)$, i.e., if $\nabla_{\dot\sigma}\dot\sigma=0$.

LEMMA 13.1. Let $\tau(\lambda)$ be a curve and $\lambda(t)$ a reparametrisation such that $\tau(t)$ is a geodesic. Then t is determined up to a linear transformation $t\to\alpha t+\beta$.

Proof. Suppose $t \to \tau(t)$, $s \to \tau(s)$ are two parametrisations, i.e., $s = s(t)$ then the tangent vectors are related by $\dot{\tau}(s) = \dot{\tau}(t)\,(\mathrm{d}t/\mathrm{d}s)$. If both curves are geodesics then

$$0 = \nabla_{\dot{\tau}(s)}\dot{\tau}(s) = \nabla_{\dot{\tau}(t)\mathrm{d}t/\mathrm{d}s}\left(\dot{\tau}(t)\frac{\mathrm{d}t}{\mathrm{d}s}\right) = \\ = \left(\frac{\mathrm{d}t}{\mathrm{d}s}\right)^2 \nabla_{\dot{\tau}(t)}\dot{\tau}(t) + \left(\frac{\mathrm{d}t}{\mathrm{d}s}\dot{\tau}(s)\right)\dot{\tau}(s)\left(\frac{\mathrm{d}t}{\mathrm{d}s}\right) \tag{13.1}$$

and therefore $\mathrm{d}t/\mathrm{d}s =$ constant.

The following theorem relates geodesics to curves in $L(M)$.

THEOREM 13.1. Let $u(t)$ be an integral curve of $B(\xi)$, a standard horizontal vector field on $L(M)$. Then $\pi(u(t))$ is a geodesic and any geodesic can be obtained in this way.

Proof. If $u(t)$ is an integral curve of $B(\xi)$ then

$$\dot{u}(t) = B(\xi)_{u(t)} \tag{13.2}$$

and

$$\dot{x}(t) = \pi_* \dot{u}(t) = \pi_* B(\xi)_{u(t)} = \xi^i X_i(t), \quad \text{if} \quad u(t) = (x(t), X_i(t)). \tag{13.3}$$

Now $X_i(t)$ is parallel along $x(t)$ and ξ^i is constant; therefore, $\dot{x}(t)$ is parallel along $x(t)$, i.e., $\nabla_{\dot{x}}\dot{x} = 0$.

Consider conversely a geodesic $x(t)$ and choose any horizontal lift $u(t) = (x(t), X_i(t))$. Then there are numbers ξ^i such that $\dot{x} = \xi^i X_i$ because $\dot{x}$ is parallel along $x(t)$. Suppose now

$$\dot{u}(t) = a^i(t)\, B_i \tag{13.4}$$

and we get

$$\xi^i X_i = \dot{x} = \pi_* \dot{u} = a^i(t)\, X_i\,. \tag{13.5}$$

Hence $a^i = \xi^i =$ constant and $\dot{u}(t) = \xi^i B_i = B(\xi)$.

From Theorem 13.1 one obtains the following:

THEOREM 13.2. Let x_0 be any point in M and $X_0 \in T_{x_0}(M)$. Then there exists a unique geodesic $\tau(t)$ such that $\tau(0) = x_0$, $\dot{\tau}(0) = X_0$.

DEFINITION 13.2. A connection is geodesically complete if any maximally extended geodesic is defined for parameter values $-\infty < t < +\infty$.

A geodesically complete connection Γ on M can not be the restriction of Γ to an open sub-manifold.

Finally some remarks on normal neighborhoods.

DEFINITION 13.3. Let $x(t)$ be the geodesic with $x(0) = x_0$, $\dot{x}(0) = X$. Then we define the exponential map by

$$\operatorname{ext} tX = x(t). \tag{13.6}$$

Clearly there exists a neighborhood N of the zero vector in T_{x_0} such that exp is defined. (The set of directions is compact) If the connection is complete, exp is defined on T_{x_0}.

THEOREM 13.3. For any $x_0 \in M$ there exists $N \subset T_{x_0}$, a neighborhood of the zero vector of T_{x_0}, such that exp is a diffeomorphism $N \to \exp(N)$.

Proof. This will not be worked out explicitly. In local coordinates x^i around x_0, the equation of a geodesic is

$$\nabla_{\dot{x}}\dot{x} = \frac{d^2 x^i}{dt^2} + \Gamma^i_{kl}\dot{x}^k\dot{x}^l = 0. \tag{13.7}$$

Let $f^i(t, x_0^i, \dot{x}_0^i)$ be the general solution satisfying

$$f^i(0, x_0^i, \dot{x}_0^i) = x_0^i, \quad \left[\frac{d}{dt} f^i(t, x_0^i, \dot{x}_0^i)\right]_{t=0} = \dot{x}_0^i \tag{13.7}$$

because Γ^l_{ik} are C^∞ functions, f^i are C^∞ functions of all arguments. If T_{x_0} is considered as a manifold $(\dot{x}_0^i)$ is a coordinate system on T_{x_0} and exp is given by

$$\dot{x}_0^i \to f^i(1, x_0^i, \dot{x}_0^i); \tag{13.8}$$

hence exp is differentiable. Furthermore $(\exp)_*|_{\dot{x}^i_0 = 0}$ is clearly invertible; hence there is $N \subset T_{x_0}$ such that exp is a diffeomorphism.

A neighborhood of x on which exp is a diffeomorphism is called a normal neighborhood. Stronger results on normal neighborhoods can be found for example in Kobayashi and Nomizu (1963).

References

Cohn, P.: 1961, *Lie Groups*, Cambridge University Press, Cambridge.

Hicks, N. J.: 1965, *Notes on Differential Geometry*, Van Nostrand Company, Inc., Princeton, New Jersey.

Kobayashi, S. and Nomizu, K.: 1963, *Foundations of Differential Geometry*, John Wiley, Sons, New York, Vol. 1.

Sternberg, S.: 1964, *Lectures on Differential Geometry*, Prentice-Hall, Inc., New Jersey.

INDEX OF SUBJECTS

ASTROPHYSICS AND SPACE SCIENCE LIBRARY

Edited by

J. E. Blamont, R. L. F. Boyd, L. Goldberg, C. de Jager, Z. Kopal, G. H. Ludwig, R. Lüst, B. M. McCormac, H. E. Newell, L. I. Sedov, Z. Švestka, and W. de Graaff

1. C. de Jager (ed.), *The Solar Spectrum. Proceedings of the Symposium held at the University of Utrecht, 26–31 August, 1963.* 1965, XIV + 417 pp.
2. J. Ortner and H. Maseland (eds.), *Introduction to Solar Terrestrial Relations. Proceedings of the Summer School in Space Physics held in Alpbach, Austria, July 15–August 10, 1963 and Organized by the European Preparatory Commission for Space Research.* 1965, IX + 506 pp.
3. C. C. Chang and S. S. Huang (eds.), *Proceedings of the Plasma Space Science Symposium, Held at the Catholic University of America, Washington, D.C., June 11–14, 1963.* 1965, IX + 377 pp.
4. Zdeněk Kopal, *An Introduction to the Study of the Moon.* 1966, XII + 464 pp.
5. Billy M. McCormac (ed.), *Radiation Trapped in the Earth's Magnetic Field. Proceedings of the Advanced Study Institute, Held at the Chr. Michelsen Institute, Bergen, Norway, August 16–September 3, 1965*, XII + 901 pp.
6. A. B. Underhill, *The Early Type Stars.* 1966. XIII + 282 pp.
7. Jean Kovalevsky, *Introduction to Celestial Mechanics*, 1967, VIII + 427 pp.
8. Zdeněk Kopal and Constantine L. Goudas (eds.), *Measure of the Moon. Proceedings of the Second International Conference on Selenodesy and Lunar Topography held in the University of Manchester, England, May 30–June 4, 1966.* 1967, XVIII + 479 pp.
9. J. G. Emming (ed.), *Electromagnetic Radiation in Space. Proceedings of the Third ESRO Summer School in Space Physics, held in Alpbach, Austria, from 19 July to 13 August, 1965.* 1968, VIII + 307 pp.
10. R. L. Carovillano, John F. McClay, and Henry R. Radoski (eds.), *Physics of the Magnetosphere. Based upon the Proceedings of the Conference held at Boston College, June 19–28, 1967.* 1968, X + 686 pp.
11. Syun-Ichi Akasofu, *Polar and Magnetospheric Substorms.* 1968, XVIII + 280 pp.
12. Peter M. Millman (ed.), *Meteorite Research. Proceedings of a Symposium on Meteorite Research held in Vienna, Austria, 7–13 August, 1968.* 1969, XV + 941 pp.
13. Margherita Hack (ed.), *Mass Loss from Stars. Proceedings of the Second Trieste Colloquium on Astrophysics, 12–17 September, 1968.* 1969, XII + 345 pp.
14. N. D'Angelo (ed.), *Low-Frequency Waves and Irregularities in the Ionosphere. Proceedings of the 2nd ESRIN-ESLAB Symposium, held in Frascati, Italy, 23–27 September, 1968.* 1969, VII + 218 pp.
15. G. A. Partel (ed.), *Space Engineering. Proceedings of the Second International Conference on Space Engineering, held at the Fondazione Giorgio Cini, Isola di San Giorgio, Venice, Italy, May 7–10, 1969.* 1970, XI + 728 pp.
16. S. Fred Singer (ed.), *Manned Laboratories in Space. Second International Orbital Laboratory Symposium.* 1969, XIII + 133 pp.
17. B. M. McCormac (ed.), *Particles and Fields in the Magnetosphere. Symposium Organized by the Summer Advanced Study Institute, held at the University of California, Santa Barbara, Calif. August 4–15, 1969.* 1970, XI + 450 pp.
18. Jean-Claude Pecker, *Experimental Astronomy.* 1970, X + 105 pp.
19. V. Manno and D. E. Page (eds.), *Intercorrelated Satellite Observations related to Solar Events. Proceedings of the Third ESLAB/ESRIN Symposium held in Noordwijk, The Netherlands, September 16–19, 1969.* 1970, XVI + 627 pp.
20. L. Mansinha, D. E. Smylie and A. E. Beck, *Earthquake Displacement Fields and the Rotation of the Earth. A NATO Advanced Study Institute Conference Organized by the Department of Geophysics, University of Western Ontario, London, Canada, June 22–28, 1969.* 1970, XI + 308 pp.
21. Jean-Claude Pecker, *Space Observatories.* 1970, XI + 120 pp.

22. L. N. Mavridis (ed.), *Structure and Evolution of the Galaxy, Proceedings of the Nato Advanced Study Institute, held in Athens, September 3–19, 1969.* 1971, VII + 312 pp.
23. A. Muller (ed.), *The Magellanic Clouds. A European Southern Observatory Presentation: Principal Prospects, Current Observational and Theoretical Approaches' and Prospects for Future Research. Based on the Symposium on the Magellanic Clouds, held in Santiago de Chile, March 1969, on the Occasion of the Dedication of the European Southern Observatory.* 1971, XII + 189 pp.
24. B. M. McCormac (ed.), *The Radiating Atmosphere. Proceedings of a Symposium Organized by the Summer Advanced Study Institute, held at Queen's University, Kingston, Ontario, August 3–14, 1970.* 1971, XI + 455 pp.
25. G. Fiocco (ed.), *Mesospheric Models and Related Experiments. Proceedings of the 4th ESRIN-ESLAB Symposium, held at Frascati, Italy, July 6–10, 1970.* 1971, VIII + 298 pp.
26. I. Atanasijević, *Selected Exercises in Galactic Astronomy.* 1971, XII + 144 pp.
27. C. J. Macris (ed.), *Physics of the Solar Corona. Proceedings of NATO Advanced Study Institute on Physics of the Solar Corona, held at Cavouri-Vouliagmeni, Athens, Greece, 6–17 September 1970.* 1971, XII + 345 pp.
28. F. Delobeau, *The Environment of the Earth.* 1971, IX + 113 pp.
29. E. R. Dyer (general ed.), *Solar-Terrestrial Physics 1970. Proceedings of the International Symposium on Solar-Terrestrial Physics, held in Leningrad, U.S.S.R., 12–19 May 1970.* 1972, VIII + 938 pp.
30. V. Manno and J. Ring (eds.), *Infrared Detection Techniques for Space Research, Proceedings of the Fifth ESLAB-ESRIN Symposium held in Noordwijk, The Netherlands, June 8–11, 1971.* 1972. XII + 344 pp.
31. M. Lecar (ed.), *Gravitational N-Body Problem, Proceedings of IAU Colloquium No. 10, held in Cambridge, England, August 12–15, 1970.* 1972, XI + 441 pp.
32. B. M. McCormac (ed.), *Earth's Magnetospheric Processes. Proceedings of a Symposium Organized by the Summer Advanced Study Institute and Ninth ESRO Summer School, held in Cortina, Italy, August 30–September 10, 1971.* 1972, VIII + 417 pp.
33. Antonin Rükl, *Maps of Lunar Hemispheres.* 1972, V + 24 pp.
34. V. Kourganoff, *Introduction to the Physics of Stellar Interiors.* 1973, XI + 115 pp.
35. B. M. McCormac (ed.), *Physics and Chemistry of Upper Atmospheres. Proceedings of a Symposium Organized by the Summer Advanced Study Institute, held at the University of Orléans, France, July31–August 11, 1972.* 1973, VIII + 389 pp.
36. J. D. Fernie (ed.), *Variable Stars in Globular Clusters and in Related Systems. Proceedings of the IAU Colloquium No. 21, held at the University of Toronto, Toronto, Canada, August 29–31, 1972.* 1973, IX + 234 pp.
37. R. J. L. Grard (ed.), *Photon and Particle Interactions with Surfaces in Space. Proceedings of the 6th ESLAB Symposium, held at Noordwijk, the Netherlands, 26–29 September, 1972.* 1973, XV + 577 pp.

SOLE DISTRIBUTORS FOR U.S.A. AND CANADA:

Vols 2–6, and 8: Gordon & Breach Inc., 150 Fifth Ave., New York, N.Y. 10011
Vols 7 and 9–28: Springer Verlag New York, Inc., 175 Fifth Ave., New York, N.Y. 10011

GPSR Compliance

The European Union's (EU) General Product Safety Regulation (GPSR) is a set of rules that requires consumer products to be safe and our obligations to ensure this.

If you have any concerns about our products, you can contact us on ProductSafety@springernature.com

In case Publisher is established outside the EU, the EU authorized representative is:

Springer Nature Customer Service Center GmbH
Europaplatz 3
69115 Heidelberg, Germany

Batch number: 10761697

Printed by Printforce, the Netherlands